Levels of Organic Life and the Human

forms of living
Stefanos Geroulanos and Todd Meyers, *series editors*

Levels of Organic Life and the Human

An Introduction to Philosophical Anthropology

Helmuth Plessner

Introduction by J. M. Bernstein
Translated by Millay Hyatt
Editorial assistance by Phillip Honenberger

FORDHAM UNIVERSITY PRESS
NEW YORK 2019

This book was first published in German in 1928 as *Die Stufen des Organischen und der Mensch: Einleitung in die philosophische Anthropologie*, by Helmuth Plessner © 1975 Walter de Gruyter, GmbH Berlin Boston. All rights reserved.

Fordham University Press gratefully acknowledges financial assistance and support provided for the publication of this book by the Helmuth Plessner Society.

The translation of this work was supported by a grant from the Goethe-Institut, which is funded by the German Ministry of Foreign Affairs.

Copyright © 2019 Fordham University Press

All rights reserved. No part of this publication may be reproduced, stored in a retrieval system, or transmitted in any form or by any means—electronic, mechanical, photocopy, recording, or any other—except for brief quotations in printed reviews, without the prior permission of the publisher.

Fordham University Press has no responsibility for the persistence or accuracy of URLs for external or third-party Internet websites referred to in this publication and does not guarantee that any content on such websites is, or will remain, accurate or appropriate.

Fordham University Press also publishes its books in a variety of electronic formats. Some content that appears in print may not be available in electronic books.

Visit us online at www.fordhampress.com.

Library of Congress Control Number: 2019937395

Printed in the United States of America

21 20 19   5 4 3 2 1

First edition

CONTENTS

Foreword from the Helmuth Plessner Society	vii
Translator's Preface and Acknowledgments	ix
Preface to the First Edition (1928)	xv
Preface to the Second Edition (1965)	xix
Introduction J. M. Bernstein	xxxvii

1. Aim and Scope of the Study ... 1
   The Development of Intuitionist *Lebensphilosophie* in Opposition to Experience, 3 • *Lebensphilosophie* and the Theory of the Humanities, 11 • Working Plan for the Foundation of a Philosophy of the Human, 22

2. The Cartesian Objection and the Nature of the Problem ... 34
   Extension vs. Interiority and the Problem of Appearance, 34 • Appearance as Originating in Interiority, 38 • The Prior Givenness of Interiority and the Forward Displacement of Myself: The Proposition of Immanence 41 • Extension as Outer World; Interiority as Inner World, 46 • The Proposition of Representation and the Element of Sensation, 51 • The Inaccessibility of Other I's according to the Principle of Sensualism, 55 • The Need for a Revision of the Cartesian Dichotomy in the Interest of a Science of Life, 58 • A Methodological Reformulation of the Opening Question, 64

3. The Thesis ... 75
   The Question, 75 • The Dual Aspect in the Appearance of Ordinary Perceptual Things, 76 • Against the Misinterpretation of This Analysis: A Closer Focus on the Subject Matter, 81 • The Dual Aspect of Living Perceptual Things: Köhler contra Driesch, 84 • How Is Dual Aspectivity Possible? The Nature of the Boundary, 93 • The Task of a Theory of the Essential Characteristics of the Organic, 99 • Definitions of Life, 104 • Nature and Object of a Theory of the Essential Characteristics of the Organic, 110

vi   Contents

4. The Modes of Being of Vitality   115
   Essential Characteristics Indicating Vitality, 115 • The Positionality of
   Living Being and Its Spacelikeness, 118 • Living Being as Process and
   Type; the Dynamic Character of the Living Form; the Individuality of the
   Living Thing, 123 • Living Process as Development, 129 • The Curve of
   Development: Aging and Death, 137 • The Individual Living Thing as a
   System, 144 • The Self-Regulation of the Individual Living Thing and the
   Harmonious Equipotentiality of Its Parts, 149 • Individual Living Things
   as Organized: The Dual Meaning of Organs, 154 • The Temporality of
   Living Being, 159 • The Positional Union of Space and Time and the
   Natural Place, 168

5. The Organizational Modes of Living Being:
   Plants and Animals   172
   The Circle of Life, 172 • Assimilation—Dissimilation, 182 •
   Adaptedness and Adaptation, 186 • Reproduction, Heredity,
   Selection, 196 • The Open Form of Organization of the Plant, 202 •
   The Closed Form of Organization of the Animal, 209

6. The Sphere of the Animal   219
   The Positionality of the Closed Form: Centrality and Frontality, 219 •
   The Coordination of Stimulus and Response in the Case of an Inoperative
   Subject (Decentralized Type of Organization), 227 • The Coordination of
   Stimulus and Response by a Subject (Centralized Type of Organization), 231
   • The Animal's Surrounding Field Organized into Complex Qualities and
   Things, 242 • Intelligence, 252 • Memory, 257 • Memory as the Unity of
   Residue and Anticipation, 262

7. The Sphere of the Human   267
   The Positionality of the Excentric Form: "I" and Personhood, 267 •
   Outer World, Inner World, Shared World, 272 • The Fundamental Laws
   of Anthropology: The Law of Natural Artificiality, 287 • The Law of
   Mediated Immediacy: Immanence and Expressivity, 298 • The Law of
   the Utopian Standpoint: Nullity and Transcendence, 316

Appendix   *323*
Glossary   *337*
Notes   *345*
Index   *359*

FOREWORD FROM THE HELMUTH PLESSNER SOCIETY

It is with great pleasure that we present the English translation of Plessner's *Levels* more than ninety years after the first publication of the German original. In our view, this book is one of the great philosophical texts of twentieth-century philosophy and has already shaped significant debates in Germany and in other European countries (for instance, in the Netherlands, Italy, and Poland), resulting in a significant body of literature, especially in the last two decades. But a more extensive international debate was hampered by the lack of an English translation.

The time for a renewed discussion of Plessner's philosophical anthropology is ripe: internationally we can observe a significant interest in philosophical anthropology and philosophical reflection on "life" and "nature." At the same time, the divide between so-called analytic and Continental traditions seems to have become less antagonistic and more open for fruitful transgressions of borders. In conjunction with these factors, we hope this English translation will enable a broader discussion of Plessner's approach to philosophical anthropology.

The Helmuth Plessner Society is grateful to a variety of persons and institutions that made this translation possible. First of all, we thank Millay Hyatt, who was responsible for the translation as a whole. She did a splendid job in translating this complex text into readable English and organized the entire process with a high sense of responsibility. Phillip Honenberger read the first and final drafts and made suggestions for improving accuracy and readability, including collaborating with Millay Hyatt on solving a number of difficult terminological problems throughout the manuscript.

Jan Beaufort, Ralf Becker, Jos de Mul, Thomas Ebke, Joachim Fischer, Austin Harrington, Hans-Peter Krüger, Gesa Lindemann, Olivia Mitscherlich, Gerard Raulet, Nils F. Schott, Volker Schürmann, Georg Toepfer, Jasper van Buuren, and Matthias Wunsch contributed in different ways to the solution of difficult questions of the translation. Former presidents of the Helmuth Plessner Society Joachim Fischer and Hans-Peter Krüger worked for several years to make this current translation possible; Gesa Lindemann and the current president, Volker Schürmann, helped with the organization. Marcus Düwell, vice president of the society, was responsible for the coordination of the project during the final, critical phase of the translation process.

We also thank Fordham University Press for enabling the publication and in particular Thomas Lay for his relentless support. We thank Norbert Richter for preparing the index of the book. We are particularly grateful to Jay Bernstein for writing a preface that offers an insightful introduction to the book and offers illuminating perspectives for future investigations and academic discussions.

Finally, we gratefully acknowledge that the translation would not have been possible without financial support from various institutions: The Helmuth Plessner Fonds Groningen, Plessner's stepdaughters Katharina Günther and Dorothea Krätzschmar-Hamann, the Goethe Institute, and the Helmuth Plessner Society supported this translation with substantial contributions.

We thank all of them and hope that this translation receives the attention it deserves.

The Executive Committee of the Helmuth Plessner Society

TRANSLATOR'S PREFACE

Rendering Plessner's magnum opus into English has been a daunting and rewarding task. As translators are wont to do, we've aimed to make the translation maximally readable in the target language without losing or changing the meaning of the original. As Plessner himself might put it, such endeavors inevitably fall short of their ideal (see the subsection entitled "The Law of Mediated Immediacy: Immanence and Expressivity," in Chapter 7).

Scholarly apparatus has been kept to a minimum. All footnotes are from the original German except where indicated. Citations to books and essays have been altered to refer to English editions where such exist; otherwise, they're left unchanged.

The original German is presented in square brackets wherever an important meaning in the original might otherwise be lost—for instance, in cases of parallel constructions that don't carry over to the English rendering; or where subtle distinctions between two German words that could only be translated as the same word in English might otherwise be missed. We've tried hard to keep such insertions to a minimum.

A glossary of important terms has been provided as an appendix (in both an English-to-German and German-to-English ordering).

While we believe the text can mostly be interpreted on its own, the reader may benefit from a few advance indications.

First, Plessner's use of "intuition" [*Anschauung*] and "intuitive/intuitional" [*anschaulich*] is a technical philosophical one that doesn't perfectly correspond to everyday English senses of these words. According to Kant's *Critique of Pure Reason* and discussions that take off from it, *Anschauung* is a cognitive faculty that supplies the "immediate" and "sensory" parts of experience and knowledge. Plessner inherits yet modifies this use, treating *Anschauung* primarily as a faculty of access to the qualitative features of things. These qualitative features may include sensory qualities like color and shape, but also properties of the boundaries of living and nonliving things (discussed from Chapter 3 onward). Note that Plessner's use of the term carries no dualistic implications of distance from the object (via the immediacy of intuition) or of enclosedness within consciousness (via a "merely" sensory or phenomenal status).

Plessner uses "*die Intuition*" and its variants (indicated in square brackets in the text) primarily in relation to Henri Bergson and Oswald Spengler, where he means something closer to our everyday understanding of the word; in all other cases "intuition" and "intuitive" should be read in the technical sense. "Intuitional" and "intuitive" are used interchangeably.

"Experience" renders *Erfahrung*, but usually means (especially in the earlier chapters) something closer to the clunky English noun-phrase "the empirical" than experience in general. This is the "experience" of the empiricist epistemological tradition from Locke to the logical positivists. Sometimes *Erfahrung* calls to mind everyday sensory experience; other times, it refers to the entirety of scientific observation, experiment, and empirically based theory. *Erlebnis* points to the wider phenomenon of "lived experience," which is how, grammar permitting, it has mostly been translated.

We rendered *Geisteswissenschaften* as "humanities" throughout, but the slightly wider scope of the term to cover such fields as psychology and sociology (at least sometimes and in some ways) should be borne in mind. The apparent ambiguity in the boundaries of the *Geisteswissenschaften* may be a mirror of the time: comparisons and contrasts between the *Geisteswissenschaften* and the natural sciences [*Naturwissenschaften*], as well as the boundaries, character, and epistemic status of each, were major themes of nineteenth- and twentieth-century German philosophy, for instance in Wilhelm Dilthey and neo-Kantian philosophers like Heinrich Rickert. Plessner's *Levels* is clearly intended as (among other things) a contribution to these debates.

Because *Wirklichkeit* and *Realität* are often used interchangeably to mean whatever is real in a broad and nonspecific sense, both terms have frequently been rendered as "reality." This does, however, mask the slightly different connotations of *Wirklichkeit* and *Realität*, which have been variously exploited in German philosophy. *Wirklichkeit*, sometimes translated as "actuality," carries connotations of "having an effect, being effectual." It also implies the kind of realness that has a direct effect on consciousness or action. *Realität*, on the other hand, carries the connotation of something existing independently of being an object of cognition or experience, or at least something that is existent but accessed otherwise than by direct involvement.

We've employed several devices to minimize misreadings here.

First, the adjectival forms *wirklich* and *real* have been differently rendered as "actual/actually" and "real/really," respectively. All cases of "real" or "really" are renderings of *real* (and variants), except where noted. All cases of "actual" and "actually" are renderings of *wirklich* and vice versa, except where noted.

Realität is always rendered "reality." *Wirklichkeit* is rendered as "actuality" wherever such a rendering wouldn't mislead, but is also rendered as "reality" in some cases, always noted in the text.

Where "reality," "real," or variants appear more than once in a sentence, if the German is given for the first instance, all subsequent instances (and variants) in the sentence are the same German root, unless otherwise indicated.

The adjective/adverb *reell* has been translated as "real," but it should be noted that Plessner's use of the word derives from the technical Husserlian usage as "immediately present or known to consciousness." To prevent confusion, German is included in parentheses for all instances of *reell*.

Plessner uses two main terms to refer to bodies: *Körper* and *Leib*. Though the meaning of this distinction is perhaps subject to some interpretive controversy, the basic difference may be summarized as that between the "physical body"—that is, the body (including the body of an organism) as a thing in physical time and space—and the "lived body," or the body as it is experienced or lived through by the living subject whose body it is. In our translation *Leib* is rendered as "lived body." *Körper* is rendered simply as "body," or as "physical body" in cases where disambiguation would be helpful. The adjectival forms *leiblich* and *körperlich* are rendered as "embodied" and "bodily," respectively. Occasionally the latter term is rendered as "physical" or "physically."

Finally, something must be said about Plessner's idiosyncratic—yet illuminating and innovative—use of spatial concepts and terminology in the *Levels*. The title itself gives evidence of this spatial theme. In the text, *räumlich* ("spatial") is distinguished from *raumhaft* ("spacelike"): the former describes structures in physical space, the space of the natural sciences; the latter describes structures that are similar to physical structures, but aren't contained or locked into the same strict set of relations characteristic of physical space. These have special significance for living being, as described in the text. *Zeitlich* ("temporal") is similarly distinguished from *zeithaft* ("timelike").

Plessner's key term "positionality" [*Positionalität*] in English links etymologically across a broader range of terms than does the German. Thus *Stellung* (position), *Gestelltheit* (positioning), *Entgegensetzung* (contraposition), *gegen* and nouns constructed with *Gegen-* (oppose, opposite, opposing), *setzen* (sometimes but not always rendered as posit), *gestellt* (positioned or placed), and *Lage* (in some cases position; this is always indicated in the text) all circle more closely around *Positionalität* than do the German expressions. We like to think that Plessner would be pleased.

Grenze is rendered as "boundary," *begrenzt* as "bounded," and *Begrenzung* as "boundedness." We have treated these terms as technical terms and translated them consistently throughout, even when constructions with "boundedness" may be slightly awkward in English.

Plessner employs a wide variety of prepositional phrases to describe relations between things (living and nonliving) and their bodies, cores, boundaries, and environments, as well as to describe reflexive relations between living things and themselves. These prepositional expressions include standard German prepositions with multiple shades of meaning, such as *mit* (with; alongside of), *über* (above; across/over; about), *vor* (before, either spatially or temporally), *nach* (after; according to; following), *zu* ("to" in the sense of "for the sake of"), *gegen* (against; contrary to; in opposition to), and *für* ("for" in the sense of "as appearing to" or "as of concern to"). They also include idiosyncratic expressions such as *ihm entgegen* (over against it), *über es hinaus* (out beyond it), and *in es hinein* (into it). What makes these expressions unusual is their avoidance of a reflexive construction (e.g., *in es hinein* rather than *in sich hinein* [into itself]). Plessner is very careful to save grammatical reflexivity in these contexts for the level of the organic capable of self-reflection: excentricity. In cases where the referent of "it" is ambiguous in English, we have followed Plessner's great reader and translator Marjorie

Grene and repeated the object—for instance, "the body is into the body that it is"—where Plessner simply has *"der Körper ist in ihm hinein."*

Prepositional expressions such as those in the last category are best read as technical terms. They are sometimes crucial to the argument and theory presented in the book, particularly from Chapter 3 forward. Where such prepositional expressions are particularly important for Plessner's argument, we give the original German the first time they appear in the text, but not in subsequent uses. Readers are encouraged to use the glossary and index to check the meanings of any such words or phrases they find confusing.

A dialogue between translator, the editors at Fordham, and representatives of the Helmuth Plessner Foundation was instrumental in making decisions about the title. The full German title is *Die Stufen des Organischen und der Mensch: Einleitung in die philosophische Anthropologie*. For *Stufen*, "Levels" was preferred to "Stages" since the former fits the important sense of a hierarchy of complexity without the temporal or genealogical implications of the latter. *Organischen* was rendered as "organic life" rather than the arguably more literal "organisms" or "the organic" since these options were either too narrow or too broad to capture Plessner's meaning. We selected "human" rather than "man" for *Mensch* as more accurately reflecting the intended scope of Plessner's inquiry.

We've employed the male pronoun throughout as a reflection of the original German and the time in which the book was written. As in the title, however, *Mensch* has been translated as "human."

Millay Hyatt and Phillip Honenberger

Acknowledgments

This translation would not have been possible without the generous and masterful assistance of many scholars.

Millay Hyatt was responsible for the translation as a whole. Phillip Honenberger read the first and final drafts and made suggestions for improving accuracy and readability. Expert assistance with specific passages and chapters was provided by Jan Beaufort, Ralf Becker, Jasper van Buuren, Thomas Ebke, Austin Harrington, Hans-Peter Krüger, Gesa Lindemann,

Olivia Mitscherlich, Gerard Raulet, Volker Schürmann, Georg Toepfer, and Matthias Wunsch. Marcus Düwell, Joachim Fischer, Gesa Lindemann, and Volker Schürmann helped with coordination among translator, assistants, and publisher. Hans-Peter Krüger's help with crucial interpretive questions throughout the later stages of translation was extensive and invaluable. Thanks also to Nils F. Schott for early input on terminology issues. Any inaccuracies remaining in the text are entirely the responsibility of Millay Hyatt, whose debt to all of these consultants is great.

PREFACE TO THE FIRST EDITION (1928)

This book was most significantly inspired by the profound tensions between natural science and philosophy I encountered as a zoology student in Heidelberg under Otto Bütschli, Curt Herbst, Wilhelm Windelband, Ernst Troeltsch, Hans Driesch, and Emil Lask. As someone who did not want to sacrifice the one to the other, these tensions impelled me to reflect on new possibilities of understanding nature from a philosophical perspective in a way that would be as able to withstand the severe criticism brought by the philosophy of the time as it would be receptive to impulses coming in particular from the new biology of Driesch and Jakob von Uexküll. I believe I took the first step in this direction with my *Die Einheit der Sinne* (The Unity of the Senses).[1] It was while I was writing this book that I conceived of the plan for the current volume, which I originally intended to publish in the form of a short pamphlet, as a sort of addendum to the final chapter of *Die Einheit der Sinne*. Once I began writing, however, it became clear that this plan was impossible. The work required a broader scope and had to have its own foundation and methodology. In 1924 I announced it in the foreword to *The Limits of Community* under the title "Plants, Animals, Man—Elements of a Cosmology of Living Form." External circumstances along with the great difficulty of having to work on terrain neglected by the recent philosophical tradition delayed completion of the manuscript until the fall of 1926.

The questions of philosophical biology and anthropology treated in this book emerged from the systematic continuation of my reflections on a philosophy of the senses. The theory of sensory modality as a specific form of relation between a lived body [*Leib*] and a psyche, or between materiality and meaning-giving [*Sinngebung*], demanded—in particular as regards the so-called lower senses and the problem of the objectivity of sense qualities—a more fundamental formulation of the relationship between the lived body and its environment [*Umwelt*]. This in turn led to new insights into certain laws of the correlation between the form of the lived body and that of the environment [iv],[2] which evidently constitute the organizational laws of life. From there it was only one step to a theory of the levels of correlation between form of life and sphere of life that encompasses the plant, animal, and human types of life.—In addition to this development of questions deriving from the elementary theory of knowledge, my studies in social philosophy led me straight to the anthropological problem.

The preface is not the place to review the historical background of the new questions being asked. Wherever this becomes necessary, I do so in the work itself. Of course, this book would not have been possible without the recent revolutions in the fields of psychology, sociology, and biology, and, especially, philosophical methodology. It is too early, incidentally, to decide which forces have been most significant in the emergence of the new philosophical disciplines, whether it is psychoanalysis or *Lebensphilosophie*, cultural sociology or phenomenology, intellectual history [*Geistesgeschichte*] or the crises in medicine. Given this topicality of the new book, it is, in any case, important to keep in mind the originality of its conception. Nothing can harm the cause of philosophical biology and anthropology more than applying to them the standards of a synthesis derived from the most varied sciences in response to the concerns of our times.

The fact that the following investigations are grounded in the conviction that these disciplines are sciences with their own methodology and original intuitive foundations brings them into line with the views of that brilliant researcher who, as far as I can see, to this day works alone in this field: it is Max Scheler's indisputable achievement to have made a wealth of discoveries in his work on emotion, structural laws of the person, and the structural connections between person and world that belong to the thematic stock of philosophical biology and anthropology. Furthermore, his activities in recent years show that he is on the point of establishing a foundation for philosophical anthropology that takes into account earlier bio-philosophical analyses known in part to the older Munich and Göttingen phenomenology circle that, for instance, influenced Hedwig Conrad-Martius's *Metaphysische*

Gespräche [v]. While I hope for substantial agreement between the subject matter of my work and Scheler's research, I also do not want to brush over the crucial differences in our approaches to the problems being treated. Notwithstanding the metaphysical tendencies of his philosophy, Scheler, in all foundational questions, is a phenomenologist. His most important works, including his most recent publications, show his focus to be a primarily phenomenological one. Since I published my work on methodology in 1918, however, I have resisted using phenomenology as a foundation-securing research approach. I will not elaborate on this point here. Phenomenological work, in my view, requires for philosophy a certain methodological guidance that should have neither an empirical nor a metaphysical source.

Of the great thinkers of the recent past, none knew this more deeply than Wilhelm Dilthey. Dilthey's philosophy and historiography constitute, both in terms of methodology and subject matter, essential inspiration for philosophical anthropology's new presentation of the problem. It is thanks to Georg Misch that today we are able to recognize the revolution in philosophy brought about by Dilthey's ideas, whose principles Misch most recently set forth in a programmatic way in his essay "Die Idee der Lebensphilosophie in der Theorie der Geisteswissenschaften."[3] If I nevertheless have to distance myself to a certain degree from the work of Heidegger, which is heavily informed by Dilthey's ideas, then that is mainly because I cannot accept Heidegger's principle in *Being and Time* (which did not come to my attention until this volume was going to press) that the study of extra-human being must necessarily be preceded by an existential analysis of the human.[4] This idea shows him to still be caught in the spell of that old tradition (which has found expression in a great variety of forms of subjectivism) according to which the philosophical questioner is existentially closest to himself and therefore sees himself when looking toward the object of his questioning. By contrast, I defend the notion—which is meaning and legitimation of my natural-philosophical approach—that the human in his being is distinguished from all other being by *being neither closest to nor furthest from himself*. By virtue of this very excentricity of his form of life, he finds himself as an element in a sea of being and thus, despite the non-ontic [vi] character of his existence, to be of a piece with all the things in this world. It was the merit of Josef König's book *Der Begriff der Intuition* to identify, for the first time, this situation of excentricity (albeit not in these terms and not as a form of life) as the ground and medium of philosophy.[5] There are thus surprising connections between his systematic investigations and my own work—surprising, because neither in our respective approaches to the problem nor in our goals for our respective studies did we have an inkling of them. They remain to be elaborated upon in future publications.

PREFACE TO THE SECOND EDITION (1965)

[vii] If an author decides to publish a new, unrevised edition of a book that was published in 1928, thirty-six years ago, he owes the reader an explanation. Indolence and hubris will hardly suffice; adhering to the old version must be justified by its subject matter. This is not to declare as sacrosanct the *way* in which this subject matter is presented. The point here is logical coherence. If it holds, the text does not require correction, even if it is shaped by the state of research at the time it was written and the author might characterize the matter differently today on this or that point. Hardly any suggestions of this sort came his way, however. The *Levels* has not been subjected to serious criticism.

There are different reasons for this. In the same year, Scheler's *The Human Place in the Cosmos* was published, an outline sketch of the highly anticipated anthropology he had announced years prior. Originally conceived as a lecture at the School of Wisdom in Darmstadt, it quickly found a large readership, thanks to its brevity and skillful use of biological and psychological facts. The most obvious thing to do, then, was to take the unwieldy work of an unknown author to be the explication of Scheler's ideas, particularly as, superficially speaking, it seemed to follow the latter's model of levels. Theodor Litt, Theodor Haering, and especially Nicolai Hartmann very quickly and emphatically opposed such a flippant suspicion, but *semper aliquid haeret*, something always sticks—and so the *Levels* remained for the time being in the shadow of Scheler as the founder of philosophical anthropology.

One would think that the five years between the book's publication and the inhibition of any form of discussion in 1933 should have sufficed for a

xix

revision of this judgment, even if the book was difficult. But, insofar as we can even speak of an interest in philosophical anthropology at the time, these five years stood entirely under the influence of Heidegger and Jaspers. The discovery of the concept of existence seemed to be the key to solving the difficulties (and thus to deciding the fate of philosophical anthropology) that could not be mastered by the human sciences [*Wissenschaften vom Menschen*] suffering in particular from their separation into natural science methodology on the one hand and humanities [*geisteswissenschaftliche*] methodology on the other: psychology and psychopathology [viii] as well as the branches of internal medicine concerned with psychosomatic questions, ethnology, prehistory, and human phylogeny.

It seemed to be. Any doubts about the key role of the concept of existence that came up within this or that science were dismissed by the philosophy avant-garde as derivative. It was rather the removal of the historically evolved difficulties, it was argued, that was pressing and, in the strict sense, original. These difficulties were caused by ontologizing the essential nature of the human, a habit since antiquity that is aided and abetted by the human sciences. A properly understood philosophical anthropology, the argument continued, must pierce the scientific horizon of the established disciplines in order to develop a fundamental understanding of being. Whether this was done following Heidegger or Jaspers made little difference in this respect.

In 1927 and 1928, such notions hampered the interest in anthropology and had an inhibiting effect on the reception of the *Levels*. The greatest impediment was the book itself. Somebody had dared to treat biological subjects in a philosophical manner? In Germany, philosophers, trained philosophers, rarely have a relationship to natural science. If they do, they are theoretical physicists and are concerned with the epistemology of quantum physics. Botanists and zoologists, who tend to be simpler souls, people of intuition and not as sophisticated as physicists, are not drawn to the hairsplitting of conceptual analysis if only because they work with solid objects whose status as real does not pose them any problems. A philosophy of the organic? The times of Driesch were over: the question of vitalism had lost its currency just as the idea of producing living processes in a test tube had lost its terror. Biochemistry and theoretical chemistry had long been taken for granted as avenues for the study of genetics and viruses. Where there was still room for speculation, in phylogeny and in particular in anthropology, even the neo-Darwinists operated with great caution. The centennial celebrations in 1959 were marked by this atmosphere everywhere. Phenomena of regulation, control, and memory, once regarded as arcana of living matter, lost their special status in the light of cybernetics—perhaps too quickly, but electronic models do invite analogies. And these too are fertile.

Given these tendencies in biological research, a book with the title the *Levels of the Organic* seemed to betray anachronistic [ix] sympathies. Levels? Is the author hostile to evolution, perhaps even a proponent of idealistic morphology? Does "levels" not sound like a hierarchy of the forms "plant," "animal," "human" modeled on Aristotle?

Scheler's sketch, as indicated earlier, had breathed new life into Aristotelian thought. The development of this study was to form the culmination of his life's work—quite rightly, as anyone will admit who surveys his prolific output, beginning with his work on feelings of sympathy and his analyses of the material ethics of value, particularly the second part of that work,[1] and extending to his final opus, *Die Wissensformen und die Gesellschaft*. All of these works have as their subject and frame of reference the human as action center—that is, they avoid Husserl's move of reducing their subject, for reasons of phenomenological methodology, to consciousness, of making consciousness into the horizon of the transcendental constitution of every possible phenomenon, thus also of the human, and thereby reverting to the transcendental-idealistic starting position. Scheler, one of the original phenomenologists together with all of Theodor Lipps's students, did not go along with Husserl's later turn to idealism, maintaining rather that this turn violated Husserl's postulation of the primacy of consciousness and of the pure I. It is from this perspective that Scheler's actual achievement—his discovery of the cognitive import of emotional acts, his emphasis on the specific apriority of the emotional in a renewed reference to Pascal's "logic of the heart"—acquires its significance for the anthropological idea. It is this work that determines the radius and form of this idea's concretion. It is here that embodiment [*Leiblichkeit*] and environmentality, love and hate, remorse and rebirth are thematically located. Heidegger was right to dedicate, if not *Being and Time*, at least his book about Kant to Scheler, as the latter broke the spell of the cognitive and opened for Heidegger perhaps not the route to his fundamental ontology nor the entry point to that route, but certainly the region where this entry point could be sought. We cannot blame Heidegger for making provisions against the mistaken interpretation of his undertaking as a philosophical anthropology—after all, the analysis of the human mode of being was for him only a means to an end.

Being and Time was published in 1927. The power and density of the ideas, the originality and dark tone of the language, the unwavering progression through its thematic structure all immediately attracted attention. Husserl's methodological rigor and meticulous descriptions were seen as having been, with great virtuosity, put in the service of a fundamental problem that the master had not been able to help solve, even as he approached the question of the constitution [x] of transcendental philosophy: the problem of historicity. In a single push, it seemed at the time, Heidegger had broken through the old front lines of the neo-Kantian and phenomenological idealism of consciousness, of a Platonizing study of essences à la Scheler and of a historical relativism à la Dilthey. The turn to the object and the renewal of ontology, compelled by the older phenomenological school and by Nicolai Hartmann against the old idealistic tradition, was affirmed and at the same time superseded by the discovery of the dimension of existence or Dasein (the human).

This, in any case, must have been the first impression made on the expert. It was significant and disconcerting for a thinker like Scheler, who had an audience extending beyond professional circles, that Heidegger's work was beginning to elicit an even greater response. The methodological atheism of this destruction, which spared no conceptual certainties in the history of Western thought, appealed more immediately to the generation shattered by the war than Scheler's theism and the "refracted color" that guided his thinking.[2] Here there might be a world—but there was the affliction of existence. Here there were transcendental supports—there the individual was alone. Here norms and values—there pure decision in the face of death, finitude, and self-choice.

Heidegger's analysis of the mode of being specific to the human was not an end in itself; otherwise one could have taken it to be a form of philosophical anthropology, all reservations notwithstanding. It rather purported to be a procedure for finding the meaning of being, as a method aiming at fundamental ontology. The meaning of being [*Sinn von Sein*] can also be understood—and not necessarily incorrectly—as the meaning of Being [*Sinn des Seins*], thereby acquiring the enormous claim of an interpretation of the world as in the style of Schopenhauer. Insight into the ontological difference between being [*Sein*] and that which is [*Seiendem*] blocked this from debate, however. At stake was rather that which is meant by the predicate "is" in its original meaning. Heidegger's undertaking attempts to secure an interpretation of the temporal structure of existence by explicating its finitude through the modes of its temporalization.

Existential analysis, however, should not be seen as a mere procedure in the service of fundamental ontology. According to Heidegger, being human, the essence or nature of the human, is only determined by his (historically mutable) relationship to being. This is in accord with Scheler's early essay "On the Idea of Man" from 1911, in which he, following Nietzsche, describes the human as a transition [xi] and, in so many words, as a figure of transcendence.³ Biological differences mean nothing. The difference between Thomas Edison, as the ideal type of *homo faber*, and a chimpanzee is only gradual. By nature there is no such thing as the human, who comes to be only by his relationship to God. The theomorphism of the human in Scheler corresponds to the ontomorphism in Heidegger.

In his sketch of 1928, Scheler tempers this claim, which no longer conformed to his religious convictions. He did not, however, abandon it. With his characterization of forms of life in terms of drive structures, from the ecstatic impulsion of plants to spirit, which needs the vital forces but can only channel them, Scheler here seems to posit the difference between animal and human without recourse to God. Spirituality, on which the human has a monopoly, is seen as disengaged from drives and urges, producing the capacity for understanding and hence for world-openness; it manifests as drive displacement and the ability to say no. The specific anatomical structure of hominids—upright gait, freed hands, cerebralization—may play a supporting role here, but it is not the deciding factor. Why, from this perspective, should a bird's body not also be a site of drive displacement and world-openness—if the spirit enters it?

Scheler was not able to carry out the plan set forth in this sketch, and this made the academic gossip that the *Levels* was his legacy all the more believable. Did its author not also live in Cologne, and was he not Scheler's student? He was not, as great as the affinity was. He had undertaken something that was contrary to Scheler's manner, something that Scheler abominated: to understand the levels of the organic world from a single point of view—with the intention, *nota bene*, of finding and testing an approach that would make it possible to characterize the specific modes in which animated bodies appear, while avoiding, precisely, historically encumbered designations such as feelings, urges, drives, and spirit. Such a characterization must abstain from using the conceptual instruments of natural science or psychology that Scheler trotted out in the old panpsychistic manner (and with a fascination for Freud).⁴

[xii] The philosophically unschooled do not notice these shortcomings. They take the will for the deed. Many at the time believed in Scheler's synthetic outline without recognizing that if it were to be representative of the undertaking of philosophical anthropology, it would be an all-too-easy target for philosophy. Karl Löwith, who passed through Heidegger's school and is an uncontroversial witness, writes on this point (with not only Scheler's sketch in mind but especially my *Levels*): "Heidegger's ontological analysis of Dasein overtook the attempts that had been made to develop a philosophical anthropology. The dictum that existing Dasein differs qualitatively from mere being-at-hand [*Vorhandensein*] and being ready-to-hand [*Zuhandensein*] and that life's mode of being is only accessible privatively by way of existing Dasein, made it seem as if human birth, life, and death could be reduced to 'thrownness,' 'existing,' and 'being-towards-the-end.' In the same way, the world became an 'existential.' The living world, rediscovered by Nietzsche at great sacrifice . . . was lost again in existentialism along with the embodied human." What the human can be and the way he is situated in the world ought not to be understood in such a way as to relieve him of his natural ties to procreation, birth, and death. "The disembodied and genderless 'Dasein in man' cannot be primordial."⁵

Heidegger, of course, was free to restrict his existential analysis to its methodological meaning. Disregarding the physical [xiii] conditions of "existence" was reasonable if his aim was to use existence to show what is meant by "being." This disregard only becomes ill-fated—and here indeed is the catch—if it justifies itself with and becomes linked to the claim that the mode of being of life, of body-bound life, is only accessible privatively, by way of existing Dasein. It is with this claim that the inward orientation—philosophy's dearly loved habit since the days of German idealism—

regained the upper hand. In the "in-each-case-mineness" [*Jemeinigkeit*] of the constitution of existence, Augustine's admonition is heeded once again: do not go outward; in the inward man dwells truth. Kant's Copernican revolution, which established consciousness as the horizon under which objects constitute themselves and which was renewed by Husserl for the entire area of possible intentionality, was reasserted by the hypothesis of the methodological primacy of existence.

Existence expresses a human potential: that of taking oneself seriously. Morally speaking, this is the point at which a human being collects himself and becomes a self, himself. According to Heidegger, this potential corresponds to finitude, the human's finitude. The fact that he must become aware of it is strange enough, and one wonders how he is to do so. Heidegger describes this process as one in which mood, care, and fear are stages of awareness. But does this description merely have a methodological purpose, or is it meant to show the association of existence with something other, something from which it sets itself apart but on which it remains dependent? What is implied by the ability, for instance, to be in a mood or to be afraid? Surely a living being. The analysis of existence, however, only takes notice of this living being insofar as certain modes of its vitality—which itself remains in the shadows—become significant by laying open existence.

For the psychologist or psychiatrist this is not a problem, for he expects to encounter persons with temperaments, dispositions, and physical characteristics and is not troubled that he has to do here with empirical facts. This, however, is simply to disregard what is the crucial question: whether "existence" can not only be set apart from "life" but also be separated from it, and to what extent existence is based on life. Löwith mentions in the context cited earlier his essay "Phaenomenologische Ontologie und protestantische Theologie," in which he claims to be the first to have questioned the separation of existence from life.[6]

[xiv] Such an alliance could have promoted an understanding of the *Levels*, at which the Heideggerians did not deign to so much as cast a glance, even if the dictum that life's mode of being is only accessible privatively by way of existing Dasein was held onto as unshakeable. The method, after all, ought not to triumph over the subject matter. Once one has become convinced of the impossibility of a free-floating dimension of existence, it

becomes necessary to find a way to ground it. What might this foundation look like and what power does it have? How strong is its bond with the lived body? This is a justified question, as only embodied being can be in a mood or be afraid. Angels do not know fear. Even animals are subject to moods and to fear.

The analysis of a free-floating existence, however, encounters no biological facts, and Löwith's question of the separability or inseparability of existence from life does not need to disturb it as it goes about its business. It is for this reason that there is no path from Heidegger to philosophical anthropology, either before the turn or after it.

Conversely, anthropological research—somatic anthropology, human paleontology, proto- and prehistory—finds itself confronted by questions of how to delimit what is human. The answers to these questions remain incomplete insofar as they, at best, correlate biological and cultural findings but are unable to relate them to a common ground. The overarching dimension is missing. Scientists can permit themselves to ignore this dimension and to, for instance, eliminate or leave aside the psychophysical problem—but the problem remains, and who is to attend to it if not philosophers? The facts of the evolutionary history of life on earth force us to assume an evolutionary history of intelligence and of consciousness for which human intelligence and human consciousness cannot furnish the measure (naïve evolutionism, of course, unhesitatingly held on to such a measure, not only considering evolution to end with the human, but to be fulfilled by him, as if he were its goal and its purpose). It was the achievement of Uexküll and modern ethology to have done away with anthropomorphic analogies. The discovery of animal behavior and its being-in-the-world form the basis for understanding human behavior. With remarkable skill, Arnold Gehlen developed a biological model of human behavior (for the first time in 1940), stressing Herder's concept of the human as deficient being [xv] and citing numerous sources of inspiration, such as the anatomist Louis Bolk, the biologists Adolf Portmann and Konrad Lorenz, Sigmund Freud, and especially Max Scheler. This model's usefulness, however, is limited. Philosophical anthropology should not avoid critically examining it.

It seems reasonable to apply the biological principle of behavior to the human, all the more so as, "despite centuries of thinking about the question of how body and soul, or body, soul, and spirit ultimately and metaphysically relate to each other . . . there have been few answers. Thus we could try to suspend every form of question and conceptualization that led toward a dualism of this kind. Could we not . . . find a kind of key issue that would not introduce the body-soul-problem in the first place? It would have to be one treatable by the empirical sciences if we wanted to take advantage of the opportunity of excluding, along with this dualism, all metaphysical, i.e., unanswerable, questions in general. A good candidate for such an approach was action, that is, an understanding of the human as a primarily active being, where 'acting' is, in an initial approximation, defined as an operation aimed at changing nature for human purposes."[7]

This suggestion is not new. American pragmatism has been working on this "key issue" since William James and F. C. S. Schiller. John Dewey, in his work *Human Nature and Conduct* (1922), again placed great emphasis on it. In German sociology, incidentally, it was Max Weber who elevated the category of social action to the guiding concept in the analysis of social reality. An important motivation of this approach was that it allows us to understand the solidification of this reality in institutions in terms of actor motives (the subjectively intended meaning), making it possible to explain social occurrences.

The concept of action also avoids the fateful cleavage of human being into a bodily and a nonbodily region. Whether it is merely evaded and banished from view, as it were, is another question. If, like Gehlen, one wants to be an empiricist, one has the right to do just that. His theories are well-known and can all be grouped around the notion of compensation, for which Herder provided the label "deficient being." Gehlen's skillful combination of Hermann Klaatsch's notions of the characteristic ancientness and relative lack of specialization of the build of the human body with [xvi] Bolk's ideas about retardation and fetalization, Portmann's about the extrauterine spring, and Scheler's about weak instincts, surplus drives, and world-openness add up to a creature to whom Herder's "invalid of its higher powers" seems less fitting than my characterization of a combatant of his lower ones. Gehlen conceives of the *homo* species exclusively in terms of its potential to act. The differentiation of static foot and prehensile hand, reduced body hair, defenselessness of the newborn, delay of sexual maturity, lack of specified instincts, language, and institutionalization together characterize the vital demeanor of a distinctive, action-based organism—whether freed to act or forced to act is a matter of interpretation. How evolution managed to generate him has no bearing on our insight into the correspondence system of his characteristics.

An organism is always—that is, by definition—a system, an ensemble of reciprocal functions, and it is useful to apply the notion of a functional system, familiar to the physiologist, to the human as a living being with the capacity for specific accomplishments; to the human who, furthermore, is clearly endowed with this capacity in part by virtue of his body and its development. Gehlen was not the first to attempt this; consider, for instance, Paul Alsberg's book *In Quest of Man*, in which the notion of organ elimination—Gehlen's "relief"—becomes the guiding idea of the entire work (the developmental principle of "*extra-bodily* adaptation").[8] The invention of tools and of language approximate each other here to a significant degree, although the instrumental effect of language is distinguished by the fact that actively addressing things does not change them. The representational function of words brings about an intermediate world of what I would call an institutional nature, a norm-guided, objective system of "meanings," whose relieving function—and here we have the limitation of this approach—becomes a new burden on another level.

This is not an objection. It is only that conceiving of language as action does not get us very far. Every relief effected by sparing physical effort corresponds to an increased burden arising from the growing indirectness of language-guided behavior. What then is providing relief for whom? Anyone who can converse with another by means of linguistic communication enjoys the advantage of the reciprocity of perspectives that holds for both parties. As separate as they may be, each is part of the other. This, however, is only granted through the travails [xvii] of precise articulation, which as expression—at constant risk of misinterpretation—in turn creates a distance from the speaker, from the agent of language. Expression [*Äußerung*] conjures internalization [*Innerung*] and is only made possible by internalization, by a deepening and closure of the acting subject "into himself." The advantage of indirect communication with words does not compensate for its evident disadvantage.

Gehlen argues that verbal communication is only possible—a good and true thought—for organisms whose motor skills have a high degree of plasticity and do not stay within largely inherited channels corresponding to particular instincts. Instinct relief as a result of extensive instinct reduction, the replacement, as the zoologist Otto Storch put it, of inherited motor skills with acquired motor skills, and the ability to speak thus belong to the overall design of a being that is "world-open." Instinct reduction and the freeing of motor skills, however, have their limits. There are instinct residues, vestiges of human phylogeny, which reappear on certain occasions: in facial expressions, in the forms of the other sex, and in certain primeval releasing devices of the trenchant, the symmetrical, and the garish that stand out from the usual appearance of things. All releasing devices (Konrad Lorenz introduced the concept) are strangely captivating: we immediately understand elementary mimetic gestures; we are fascinated by erotic and conspicuous appearance qualities.

This conceptual arsenal allows Gehlen to characterize human behavior as something that is observable. The turn "inward"—or, to be more precise, the opening up of an inside—has a second starting point in addition to language: the freeing of motor skills uncoupled them from human drives, which translate the answer into a burst of feeling. This burst of feeling "creates the hiatus, the gap between the present excitation and deferred action into which consciousness springs."[9] If this state of affairs is projected onto the dedifferentiation of drives, by which the same thing is meant as instinct reduction and which corresponds to a surplus of drives, a unifying inner life emerges of its own accord, established not only in the circuitry of the brain, but equally in the drive structure all the way down into the vegetative darkness of the unconscious. This vindicates Freud, as it posits paneroticism as a characteristic of human drives, acquired in the process of the dedifferentiation of the originally organ-bound drives. What is more, spiritual possibilities [xviii], even in the nonlinguistic area, come into view, as the uncoupling of drives from preformed motor skills means an increase of biological ambiguity in behavior, which ultimately leads to a total emancipation from the behavior's utility, to the dissolving and formalization of, for instance, the releasing schemata in perception, and to the release of pure qualities of appearance.[10]

In other words, thanks to his open drive structure, thanks to his language, which, in turn, is compatible with this structure, the human became emancipated from biologically unambiguous behavior as it is exhibited by all animals and acquired biological ambiguity. The pragmatic suit with its behaviorist cut does not fit him. Human behavior cannot be made to conform to a schema—not to the schema of chain reflexes or to the schema of goal-oriented action. It is this emancipation of human behavior from biologically unambiguous action, which Gehlen identified by himself holding fast to the pragmatic perspective, that enables anthropology to abandon the very perspective Gehlen recommended. This is no tragedy. That is the point, after all, of any experimental introduction of a model or "key issue." It does not mean that Gehlen contradicted himself, only that he carried a hypothesis to its limit.

A negative result is still a good result for an empiricist, even if it was attained by meandering on a path made crooked by ad hoc hypotheses and concealed information. Human behavior in the vast range of its possibilities cannot be grasped by looking at just one aspect. Frederik J. J. Buytendijk's *Algemene theorie der menselijke houding en beweging* made this evident.[11] Specific phenomena such as laughing and crying, which I have explored, manifest as behavioral boundary reactions or, to be more precise, as reactions to boundaries that circumscribe our language- and goal-guided conduct. As such they manifest an ability to incorporate what in a strict (not external) sense is excessive into one's own behavior while abandoning the usual forms of controlled conduct. This ability reveals a basic feature of human existence that in *The Levels* I call excentric positionality—and not in specific reference to laughing and crying, but in view of a whole host of other characteristics of this kind of existence. The notion of excentric positionality allows me to avoid the skewed one-sidedness to which, for instance, depth psychology succumbs when it makes the drive structure alone [xix] responsible for human forms of expression (and not only, and here it is right, for its pathological divergences).

An idea of the human form of existence as a natural phenomenon and product of its history can only be acquired by contrasting it with other familiar forms of existence in animate nature. I chose positionality as the concept to guide this endeavor because I believe it to be a fundamental characteristic by which natural animate entities can be distinguished from inanimate ones. As intuitive as it is, the concept of positionality is broad enough to allow for the modes of being in plant, animal, and human life to be represented as variables without having to resort to psychological categories. The concept of positionality is itself not a construction, but is rather derived from the intuitional structure of the so-called things of perception. I am very aware of the limitations of a study beginning with the question of what conditions must be met when intuiting an occurring entity so that it can be addressed as animate. Not everything that elicits the impression of vitality must, according to the criteria even of common sense, let alone those of biology, "actually" be so. There are possibilities for deception in the realm of the intuitive indicators of the organic. This does not, however, release us from the duty of studying them seriously—or should I rather say, does not make it worthless? The nature lover and the biologist do not normally have to concern themselves with the indicators of vitality, but the question cannot be avoided when a problem, such as in biochemistry, boils down to the questions "Does the behavior of a compound with a known structure satisfy the criteria of vitality, to what extent does it satisfy them—or does it not do so at all?" We ought to know, after all, what we mean when we use the terms "life," "alive," or "animate." The fact that they are used metaphorically and have their origins in a conceptual history comprising religious and metaphysical definitions and, furthermore, that the definition limited to organisms does not have to have priority over the others does not diminish the urgency of clarifying to what state of affairs these terms refer. Nor is this diminished by the fact that views about the number, self-sufficiency, and importance of the characteristic properties of animacy have changed throughout the history of science and will continue to change as research progresses.

This study remains strictly within the framework of outer intuition, upon which the work of the biologist and the ethologist is based [xx]. Wherever it seems appropriate to include theoretical statements from the natural sciences or, in the final chapter, from the humanities, I do so in an exemplifying way. Never do I use such statements to bolster my line of reasoning. In Chapter 3, I write, "What is called for is a development of the essential characteristics of the organic and, in place of purely inductive enumeration, at least the attempt of a strict justification. Our task is an a priori theory of the essential characteristics of the organic," or, to use Helmholtz's term, of the organic modals. "A priori" is meant here in the strict sense of seeking to discover the conditions of possibility that must be fulfilled in order for a certain state of affairs to pertain in our experience. The theory is not, then, a priori by virtue of its starting point, as if it wanted to develop a deductive system from pure concepts and axioms, but only by virtue of its regressive method of finding a fact's internal enabling conditions.

What fact constitutes the starting point for the theory of organic modals? The following answer must suffice for now: it is the fact of boundaries and the independence [*Selbständigkeit*] they warrant for what is considered an animate physical body. My aim is to establish this characteristic of having boundaries, which is found in all organisms regardless of their degree of organization, as the minimal condition that must be given for there to be life. The reasoning of this entire book is based on the hypothesis of this minimal condition. It is important to note that the boundedness of physical bodies is not to be understood in any derived way, but in its visual and tactile intuitiveness. Outlining and contouring indicate this state of affairs but are not identical with it. While outlines and contours can be drawn, the fact of boundaries can only be understood, not drawn. The factors upon which the boundaries are based and that in physics and chemistry are defined as forces of cohesion, chemical bonding, and so forth, have to be left aside in the logical analysis of this state of affairs.

The emphasis in the following is on the relationship of the bounded body to its boundaries. Two cases are possible here. One, the boundary forms only the virtual in-between of the body and the medium bordering on it. In this case, no matter how sharp the contours may be, the body does not have a boundary, or does so only in the external sense that it comes to an end somewhere. In the other case, the boundary really [*reell*] belongs to the body; it is distinct from and over against the medium bordering on it [xxi], regardless of how sharply the contours may be defined by membranes or other surface formations. The boundary is no longer a virtual in-between, but rather a characteristic of the body that ensures its ongoing existence [103ff.].[12] "If we succeed in developing from the tendency given in case II the basic functions whose presence in living bodies is considered to be characteristic of their special status," the real significance of the difference between case I and case II cannot be doubted—despite the fact that it can only be detected in its consequences for the structure of certain phenomena, but not in itself. If these properties that are characteristic of animacy can successfully be developed from this approach, then "the state of affairs presented in case II will prove to be the foundation and principle of the constitutive attributes of organic nature. Case II would then be the ground (not the cause) of the phenomena of life" (Chapter 3).

I arrived at this perspective by way of the argument over vitalism, in which Driesch's claim of the mechanically inaccessible holism of an organism stood opposed to Wolfgang Köhler's view that wholes are gestalts and thus accessible to "mechanical" analysis. Wholeness and gestalt are both more than the sum of their parts. The question is, "Is the body possessing the property of vitality only structured, in this respect, as more than its parts insofar as its characteristic properties and effects cannot be composed out of the same type of properties and effects as its parts"—Köhler—or is the preponderance of vitality based on a mode of ordering that is beyond gestalt—Driesch? If, in any case, wholes cannot be understood as gestalts, Köhler was wrong on the vitalism question. Whether this means that Driesch was right is another matter (Chapter 3).

He is not, in fact, right, because the either/or choice put forward here is not complete. The argument cannot simply be decided in favor of the mechanism undermining the autonomy of every living thing; it must be settled in another way. There are no insurmountable methodological boundaries for the physical and chemical analysis of life phenomena. It was only because his concept of the machine was too narrow that Driesch was forced to suspend the methodological rules of exact analysis and to resort to non-energetic factors. His introduction of entelechy as a fundamentally unmeasurable natural factor is nothing more than an untenable makeshift solution [xxii], a contradiction in itself. Nor did it ruffle any feathers in the scientific community: Hans Spemann's discovery of organizers in embryonic development was already an advance beyond Driesch, to say nothing of the discoveries made in gene analysis and virology with the help of biochemistry. It is only a question of time before someone will use exact analysis to deduce the essential characteristics of the animate from the laws governing inorganic matter.

But this deduction will only mean the dissolution of these characteristics in the operative sense without affecting them phenomenologically. These are phenomena whose qualities can be clearly related to a quantitatively definable chemical and physical constellation, but that nevertheless maintain their irreducibility as phenomena. We find such relationships in the sphere of the inorganic as well. A given quality of color is defined by a given wavelength, but as a quality it only corresponds to this wavelength, even if it appears as this very color only for a seeing subject by means of a functioning retina and a nervous system. The modals of vitality are qualities whose emergence can be analytically understood (and thus made operable) to the extent that this is possible for qualities. A theory of organic modals that does not concern itself with explaining their occurrence but rather with their logical place and their contribution to the phenomenon of life can only be realized in an axiomatic of the organic (not to be confused with an axiomatic of biology, from which, however, it can learn). I refer the reader here to Chapter 3. The development of the concept of positionality in Chapter 4 will help convey the importance of this explanation of the phenomenal autonomy of "life" for the fundamental question of anthropology.

Vitality is a quality of the appearance of certain bodily things—their construction, their behavior in a medium, a milieu, even in relation to a "world." Some of their "essential" characteristics are immediately evident, but they can also feign vitality. In order for true vitality to be present, there has to be in place a certain ensemble of the essential characteristics on which there is general agreement, both scientifically speaking and in terms of common sense. But there are cases scholars argue about, such as viruses. Are these intermediate forms or parasitic molecules, pseudo-forms, precursors, or genuine forms of life? One thing is certain: the more pronounced and obvious the separation and self-sufficiency of a relatively constant form is, the [xxiii] more we tend to think of a physical thing as alive. The form as manifestation of the boundary is an essential index of vitality. It is for this reason that outward appearance [*Aussehen*] becomes ever more important for the organism the higher it is on the scale of primitive to highly organized. Here its appearance not only reflects its vitality, but becomes an organ, a means of its existence. Outward appearance as bait, protection (mimicry), deterrent, or display behavior is built into the circle of life, but, as Portmann writes, the gestalt of the organism becomes the "true phenomenon" as outward appearance and display. "Self-display must be regarded as a basic fact of life on the same level as self-preservation and the preservation of the species."[13] Taking up my theory of the boundary, he writes, "An opaque boundary surface represents a higher level of the 'boundary possibilities' that play such an important role in the higher organism: in the widest relational field of the simpler entity, display by means of shaping its boundary surface is 'unaddressed,' not directed at other forms of life, but rather a simple manifestation in light and space. And yet this kind of display already contains all the potencies that are realized by the directed, 'addressed' display of the higher forms, which are noticed so much more than the primary expressions of undirected being in light."[14]

Finally, an editorial note: I hope my readers will understand that I did not want to weigh down the Preface to the new edition by engaging with schools of thought that have nothing to do with my book. The works of both Sartre—especially in his early writings—and Merleau-Ponty contain some surprising parallels with my own formulations, causing not only me to wonder whether they had perhaps read the *Levels* after all. But I had the same experience with Hegel, whom I should have referenced if I had been familiar at the time with the passages in question. Convergences are not always based on influence. There is more thought in the world than one thinks.

The appendix includes a number of references, corrections, and amendments. The index, I trust, will also facilitate reading.

INTRODUCTION

J. M. Bernstein

Introducing Helmuth Plessner's Philosophical Anthropology

The human lives only insofar as he leads a life. [310]

Levels of Organic Life and the Human: An Introduction to Philosophical Anthropology was originally published in 1928. By "philosophical anthropology" Plessner intended a new paradigm of philosophical thought and understanding—namely, the installation of the human form of life within living nature in a manner that would enable a revocation of mind-body dualism and the displacement of reductive materialism. Plessner's biologically framed understanding of the world is one in which human living exists in emphatic relation to plant and animal living, however much the human becomes a cultural animal whose form of life can no longer be gauged strictly by, say, zoology or evolutionary psychology. What enables Plessner's new philosophical paradigm is a categorical articulation of living bodies that

xxxvii

makes them qualitatively distinct from mere things and hence incapable of being fully understood through the physical and chemical laws that govern the movements of mere matter (even if, as Plessner always insists, every biological action occurs *through* material interactions that are subject to law-like explanation).

In brief, Plessner's new philosophical paradigm aims to challenge both reductive materialism and the panicky metaphysics of anti-naturalism in all its forms—mind-body dualism, theism, panpsychism, vitalism, the culturalism of the linguistic turn, traditional idealism, and on—and, through his philosophical anthropology and the philosophy of life it contains, projects the necessity and unavoidability of pluralism among the natural and human sciences in which no one science can attain to ontological or epistemological self-sufficiency. The human, Plessner argues, including human consciousness, must be understood through its placement in living nature even as it departs from biological determination—we are, Plessner argues, *naturally artificial* beings, beings who can biologically live only through leading a culturally saturated, norm-governed life.[1] The *Levels* is Plessner's masterwork, and it is patently the premier work of European philosophical anthropology;[2] more importantly, its new paradigm should be regarded as belonging to the forward edge of contemporary philosophical thought.

Born in 1892 in Wiesbaden, Germany, Plessner had a not untypical itinerant education, studying medicine in Freiburg before taking up zoology and philosophy in Heidelberg, then moving to Göttingen to study phenomenology with Husserl, and finally, in 1920, writing his *Habititationsschrift* under the guidance of Hans Driesch in Cologne. Having held a professorship in Cologne from 1926, Plessner was forced to resign his position in 1933 in light of his Jewish ancestry on his father's side of the family; he fled first to Istanbul, and then to Groningen in the Netherlands on the invitation of his friend, the psychologist (and phenomenologist) F. J. J. Buytendijk. With the German invasion of Holland, Plessner was forced underground for the duration of the war. In 1946 he was awarded the chair in philosophy at the University of Groningen; and then, in 1951, after seventeen years of exile, wandering, and hiding, Plessner accepted the newly inaugurated chair of sociology in Göttingen. When Max Horkheimer was reconstituting the Institute for Social Research in Frankfurt, and with his friend T. W. Adorno still in Los Angeles, it was Plessner who stepped in to do Adorno's teaching. Such was his regard in those years that Plessner was at different times president of the German Society for Sociology and that of Philosophy. Not long after his retirement, Plessner became the first holder of the Theodor Heuss Professorship at the New School for Social Research—a professorship that was created in order to honor the New School's welcoming of refugee German scholars fleeing the Nazi regime. As even the briefest sketch of his life makes patent, Plessner emphatically lived through the historical storms of the twentieth century; in his case, both the most significant intellectual trends and political events of the century infused his philosophical thought and writing.[3] Helmuth Plessner died in Göttingen in 1985.

In this Introduction, I want to begin by motivating consideration of the *Levels* in relation to the constitutive conundrums of contemporary philosophy before offering a broad account of the founding gestures of Plessner's project. For heuristic purposes, I will begin with what will strike some as exaggerated versions of the positions being criticized in order to bring into view the stress points and divisions that Plessner's philosophical anthropology means to address.

Motivating the Turn to Philosophical Anthropology

Part of the need and indeed urgency for a new philosophical paradigm derives from the broad cultural fact that, however often challenged and criticized, no new conception of reason and knowledge has been able to fully overturn the authority of mathematical physics; as a consequence, throughout the modern age, ever since the undoubted achievement of Newtonian science came to be recognized, the social and human sciences have been devalued, standing under the constant threat of outright delegitimation and epistemic disenfranchisement. As Plessner notes, the grounds of the threat are powerful: if mathematical physics operates through the reduction of the observable qualitative features of the world to different kinds of quantities that can be accounted for through mathematical laws that can be set in logically deductive relations to one another, then the qualitative *lifeworlds* of living subjects are illusory, mere semblance, from a scientific standpoint. If nothing is in itself constituted by qualitative features and properties, then the objects of the human and social sciences are illusory. The final story about the human will be written in the austere language of mathematical physics.

In what can be viewed as the first round of a familiar dialectic, this chilling story faces a quick riposte: however illusory, the qualitative features of experience had to be housed *somewhere* even as they are withdrawn as constituents of material reality. Mind or consciousness or subjective experience thus became the locale for the now materially detached qualitative features of experience. This is the deep logic behind Descartes's division of the world into two ontological kinds, *res extensa* and *res cogitans*; the irreducibility of thinking stuff to extended stuff is what enabled the qualitative emptying of material nature to proceed without hindrance. The persistence of mind-body dualism, no matter how wildly improbable it now seems, should be understood in its dialectical insurgence as the shadow of a mathematically purified material nature.

In a second dialectical moment, the untenability of mind-body dualism launches extremist reductionist claims from both sides of the dichotomy: continuing the project of physicalist reduction, as in the efforts of sociobiology, evolutionary psychology, and neurophilosophy, and, oppositely, the various projects of anti-naturalism that attempt to show that the nature-culture divide is solely a *product* of reason or language or culture or that science is just another social institution governed by all-too-human protocols of social conformity, hierarchy, and power.

Both reductionist programs and the original mind-body divide are prima facie implausible. The *cognitive* achievements of modern science are undeniable in terms of their theoretical complexity, integration, and sophistication; their remarkable predictive accomplishments; and in their now endless technological applications—electric lighting, penicillin, men on the moon, computers, and nuclear weapons. However expressive of deep human interests the forms and structures of natural science might be, however much they may be regarded as products of social and historical formations of knowing and rationality, outright science skepticism appears irrational in the face of scientific achievement. Conversely, if human and animal consciousness contains a dimension of "what having this experience is like," of what being in pain or seeing red or hearing a high C or feeling cold is like—and ignoring the "what it's like" to understand a Greek tragedy or mathematical theorem—then the very idea of reducing consciousness to neuron firings is flatly misguided, a case of reductionist ideology flaunting rational judgment.[4] Putting the matter this way, even if one supposes that there *exists* nothing but atoms in a void in the universe, the scientific laws explaining the movements of those atoms will not *explain*, for example, the great display of plumage by male peacocks.[5] And more generally, we have overriding reasons to believe that the predictive power and explanatory force of lower-level physical theories recede as we move up the evolutionary scale and, even more emphatically, as we engage with purely human phenomena—the causes of the Civil War or the significance of *Hamlet*. What of *cognitive worth* might physics say about such items?[6] Further, we are not going to be able to make sense of the use of the panoply of concepts necessary for scientific rationality—theory, evidence, explanation, falsification, logical and conceptual criticism, modeling, theoretical depth, and scope—in a manner that makes those concepts suddenly cognitively idle once items apart from the most basic come into view; the terms constitutive of cognitive meaning and rational understanding are themselves plural, without a priori restrictions to their range of application. Nor, on the other side of the debate and just as importantly, although the claims of reductive materialism do motivate abreactive claims for dualism, for some idea of mindedness and consciousness as somehow fully independent of their material substratum, even a modest, commonsense naturalism presses toward a compromise position that begins with the thought that the evolution of life must form part of the explanation of the development of animal and human consciousness, and thus that the mind is a biological phenomenon, at least in the first instance.

It is impossible to avoid noticing that the deflating and unwinding of the opposing reductionist views converge on and initially occur in response to biological phenomena, features of this debate that are both philosophically and historically pertinent. It is worth recalling how, under the shadow of Newtonian science with its commitments to atomism and mechanism, for the better part of a century biological phenomena were essentially philosophically forgotten; that forgetting, and all the suppressions that forgetting subtends, became a recurring feature and force in Western rational culture. The third moment of this philosophical dialectic, the one that makes living nature and biology the pivot for a nonreductive form of naturalism, is thus best approached from the moment when biological phenomena were reintroduced into the philosophical conversation, from the moment when Kant argued that it would be "absurd" to hope that someday "another Newton might arise who would explain to us, in terms of natural laws unordered by any intention, how even a mere blade of grass is produced";[7] from that moment there has been a tradition of thought claiming that biological phenomena are denizens of the material world whose characteristic properties appear resistant to explanation through solely mechanical laws. Joining history with theory, it can be argued that for Plessner it was, in part, the account of reflective judgment and organic phenomena in the *Critique of Judgment* that helped open the pathway to philosophical anthropology.[8] Thus, even if Kant's setting up of the issue was peculiar in making purposiveness somehow dependent on intentionality, his ground intuition has proved prescient: there are a range of features of living organisms that challenge the hegemonic claim to universality by Newtonian physics and its even more mathematically powerful successors.

Kant noted that living organisms both appear as organized in a complex part-whole manner and are self-organizing; a living object "is both cause and effect of itself."[9] In consideration of a tree, he notes three features that do not easily accord with mechanical explanation: first, the tree reproduces itself and was produced by a tree of the same species; hence "with regard to its *species* the tree produces itself."[10] Lumps of coal do not produce new lumps of coal to keep the coal line going. Second, the tree also produces itself as an individual: it grows by changing inorganic materials into parts of itself, a change that Kant distinguishes from "any increase in size according to mechanical laws." Third, its parts depend on one another and on the whole tree as the sum of those parts for their continued existence. In an earlier essay, Kant noted a fourth feature—namely, plasticity, the living being's ability to adapt, to change the functioning of its parts, which allows it to maintain equilibrium or continue similar activity despite injury and obstacles. Call this feature of living organisms their capacity for repair and regeneration.[11]

In order to see how proximate Kant's account is to contemporary biological thought, consider, for comparison, Tibor Gánti's criteria for real (absolute) life and potential life.[12] A real living system: (1) must inherently be an individual unit that cannot be subdivided without losing its properties (thus requiring irreducible whole-part relations); (2) must be able to perform metabolism, assimilating material and energy into the system, transforming them into its own internal constituents, and eliminating waste products from the system (what Plessner terms processes of "assimilation and dissimilation"); (3) must be capable of remaining stable despite changes in the external environment (so a certain independence from the material surround and a capacity for persistence through change); (4) must contain an informational subsystem that is essential for its origin, development, and function—a property presupposed by Kant's understanding of the purposiveness of organic behavior; (5) must be regulated and controlled. Because sterile plants and animals, mules for example, and infertile senescent animals are plainly living while incapable of reproduction, Gánti distinguishes these five criteria from three further criteria for potential life: (1) the capacity for growth and reproduction; (2) the capacity for hereditary change and evolution; (3) the capacity for death. On Gánti's account the simplest units of life are cells, while viruses, because incapable of metabolism, are excluded. Further, and obviously, if the units of evolution are taken to be items that can multiply and have hereditary variation, then units of evolution and units of life do not fully coincide.[13] Gánti was a renowned chemist whose primary interest was in constructing the principles for the simplest living system; nonetheless, for that project to be able to pose the appropriate question to chemical understanding, Ganti had first to nonreductively bring *the phenomenon of life* itself into view, just as Kant had before him.[14]

Rather than beginning with consciousness—reason, language, value—where the resistance to reduction is highest once material nature is stripped of qualitative features and consciousness becomes the sole abode for qualitative nature,[15] Kant's original challenge, which itself arose in response to the abyss separating the noumenal domain of freedom and the causally determined world, turned on first profiling the phenomenon of life as a way of marking why mechanistic explanation could not handle the lawlike regularities displayed by living organisms. And despite the fact that biology has progressed by assuming there could be physical and chemical explanations for biological phenomena, increasingly those explanations have come to be understood in a nonreductionist light, much as Kant was arguing they would. The presumption of classical reductionism is that secondary qualities could be reduced to pure mathematical quantities without remainder; secondary qualities would thus disappear from the physical universe. Call this *the disappearance view*. Whether or not one believes the disappearance view is true of physical nature,[16] there is a growing consensus—forever resisted by defiant reductionists—that this view is inapplicable to the biological sciences because the phenomenon of life itself is irreducible.

For this thought to have salience, the presumption is that the actual practice of the science of biology is incompatible with the presumptions underlying the disappearance view. Roughly, the guiding thought here is that "there is not much role for scientific *laws*" in biology; that a large part of biology concerns the search for mechanisms, where the effort is not to reduce the higher-level features to those mechanisms, but rather to show "*how the higher-level features arise from the parts*"; that rather than searching for mathematical laws, biology invests a good deal of its theoretical energy in "modeling, or model building"; that modeling serves distinct cognitive purposes in distinct settings, entailing that models, even powerful ones, operate through simplifications that involve abstraction and idealization; hence, even optimally, science will continue to require "several models for any given system," which is as much as to say that mechanisms and models are not even intended to explain away, *disappear*, the systems they make increasingly comprehensible and controllable.[17] In the philosophy of biology, the explanatory structures of scientific thought are bound to the particular features of biological life they are intended to make cognitively accessible. The phenomenon of life, including the individuals biology attends to—single and multicellular organisms, species, and families—remain in view, remain as *real material inhabitants* of the material world. It further follows from such binding of explanatory structures to biological phenomena that the philosophy of biology is not only concerned with how science works, but equally "to understand something about the natural world itself, the world that science is studying."[18]

The binding of biological explanation to the prefigured horizon of biological phenomena is, however, insufficient on its own for an understanding of biological phenomena as part of a robust, nonreductionist naturalism; there should be, at least, one further constraint on the philosophy of biology—namely, that if consciousness is a biological phenomenon, then bringing biological phenomena into view must minimally not make consciousness under some first personal description incompatible with the best understanding of living systems and ideally will make the very idea that consciousness is a biological phenomenon an at least possible and ideally a fully intelligible furthering of the processes underpinning plant and animal life. Such a constraint is necessary if the intermediate case of living systems is to be a route to satisfying the exigencies raised by the implausibility of the first two moments of the dialectic generally and by the implausibility of mind-body dualism specifically. More precisely, considering the phenomenon of life as a developing anticipation of consciousness inverts or reverse engineers Kant's original puzzle about organisms. Looking for a bridge between self-determining mind and matter, Kant found the defining characteristics of life—self-organizing, self-producing, and reproducing—as mind-like in their implication of a kind of autonomy and self-making and hence unintelligible outside intentional purposefulness. To reverse engineer this claim would be to show how consciousness or mindedness is best considered as a further elaboration of basic life processes: how, say, "the self-organizing features of mind are an enriched version of the self-organizing features of life";[19] or how, contra dualism and panpsychism, sentience "is brought into being somehow from the evolution of sensing and acting [that even single cell organisms evince]; it involves being a living system with a point of view on the world around it."[20] Both these compelling strategies for binding consciousness to life—from Evan Thompson's *Mind in Life* and Peter Godfrey-Smith's *Other Minds*, respectively—inscribe organic life, beginning from single-cell organisms, with sufficiently mind-like (consciousness-like or self-like) features in order to make possible a deflation of the so-called hard problem of consciousness while avoiding outright anthropomorphism.

While Plessner agrees that an accounting of organic life should structurally adumbrate consciousness, his philosophical anthropology works the same set of problems and issues that motivates Thompson and Godfrey-Smith from an adjacent intellectual terrain. Instead of mind or consciousness being the mark of the human that requires adumbration and elaboration within a wider biological framework, Plessner has his sight on the specific characteristic of humans as beings who live a cultural life, a life in accordance with social rules and norms, a life that is everywhere mediated through artificial tools, a life that is *naturally artificial*—that is, a life that transcends itself (is more than life) without departing from itself (it remains a form of living) that now requires adumbration and elaboration within a suitably adjusted biological framework. Plessner's statement that "the human *lives* only insofar as he *leads* a life" [310; italics mine] uses the concept of life equivocally: the first time referring unequivocally to biological living; the second leaning toward social life as the way humans accomplish their individual ends and purposes, for better or ill—as one might say, "I do not know what I want to do with my life" or "My life is in ruins" or, insisting on the equivocation, Ferdinand Kürnberger's lament "Life does not live." Plessner's statement is the primary thesis of his philosophical anthropology and the question it addresses: what must life be such that it becomes categorically intelligible that human beings are living beings who can live—survive and reproduce themselves—solely through leading a norm-governed social existence? How does the latter feature relate to and deviate from the former feature? However one answers these questions, we should now regard it as rationally impossible to conceive of these framework questions as not now guiding all interrogation into the human.

The Phenomenon of Life: Boundaries

Standard analytic philosophy of science operated as an underlaborer to mathematical physics, directing its attention to wholly formal or logical issues surrounding scientific explanation—deduction, induction, probability, the confirmation scientific hypotheses (falsification, Bayesian inference, etc.), the status of theoretical entities, causal versus teleological explanations. Conversely, the philosophy of biology cannot proceed without becoming in part a philosophy of nature, a bio-philosophy that in elaborating the science of biology brings into view the phenomenon of life itself. Plessner, following in the wake of Kant's challenge, argues that in some fundamental way our understanding of life must *precede* biology if the phenomenon of life is going to stand as a conceptual horizon that underlies, enables, and critically limits scientific theorizing; without a preunderstanding of living systems there would be no object for science to turn its attention to [117]. Of course, science can criticize, refine, and explicate prescientific understanding; but to acknowledge the implausibility of extending the disappearance view to biological phenomena is to acknowledge the necessity of there being a *categorical account* of the phenomenon of life [118].

The categories of life cannot intelligibly be either wholly a priori—they are not knowable independently of life, they are not *imposed* on living things by the understanding (by reason or language)—or wholly a posteriori, discovered in the course of experience, like discovering a new continent or a new star in the heavens or the structure of DNA. As the phrase "phenomenon of life" implies, what is at stake is a mode of appearing and hence the condition of possibility for that appearing and *what* appears, the object of the appearing. For Kant, whose transcendental method is a significant influence on Plessner, the categorical conditions of the possibility of experience are simultaneously the categorical conditions for the possibility of the objects of experience. But Kant's own entangling of the categories of cognition with material categories is at one with his idealism; his is an imposition view: the categories are what they are because they are material analogues of forms of thought that originate from the innate structures of human judgment. Plessner's account is more realist than this: the categories of life are not just how life appears, but equally conditions of possibility for an item *being* a living thing. The categories of life must consequently have a material a priori status, both necessary conditions for the possibility of the *phenomenon* of life and necessary conditions for an item *to be* a living one.²¹

In this respect, Plessner's "deduction" [122] of the categories of life is methodologically anomalous, since it must operate from both the subject and object position. Rather than resolve this anomaly, Plessner deepens it by demonstrating that the dual aspects of inner and outer directionality, inside out and outside in, belong to the structure of life itself: organisms have a point of view on the world *through which* it appears; but what appears is a living reality. The fact that the jackdaw can only perceive the grasshopper when it leaps does not make the latter any less nourishing when caught. Living beings are necessarily geared to their living environment, and their living environment imposes itself in harmonious or hostile ways on the organism. Plessner contends that it is only with human perception, in virtue of its forever incomplete break from the organic point of view, that a *further* division between a perspectival subjective point of view and a wholly perspectiveless, objective point of view comes into play. This is precisely what Plessner has in mind in considering human existence as somehow broken, our being simultaneously both centrically anchored in the living world and dislodged from it: human perception involves a division that poses a living, centric perspectival subjectivity against a wholly indifferent outside, an outside whose indifference and thus categorical deadness are extensionally equivalent to its presumptive objectivity. Until that evolutionary moment of breakage, an organism's *positional* orientation on the world was its constitutive access to its objective reality, the actual world as it is for that organism; hence its positional orientation was both the organism's way of separating itself from the world and its mode of access to it. As we shall see in more detail directly, life is *positional* because it is coextensive with boundary regulation; needing to sustain its boundary involves an organism in being positioned—oriented and relationally coordinated—with respect to its environment.

But this precisely begins to insinuate Plessner's novel construction. Rather than deploying and reshaping existing biological concepts, Plessner sought to innovate a categorial frame that could robustly capture life phenomena, as in the Kant or Gánti enumerations, while operating under the dual constraints that life categories must adequately anticipate experiential consciousness and be simultaneously operative in and culturally displaced by human cultural existence.

The phenomenon Plessner focuses on is living beings appearing as wholes with parts, as unities in a way that mere things are not unities. His wedge question is thus: "So how is it possible for something that is verifiably a gestalt to appear as something else—namely, as a wholeness beyond gestalt?" [121] Simplifying greatly, Plessner's question is, under what conditions can an item that *must be taken* to be a whole-with-parts, a necessary unity, implicate a wholeness beyond how it imposes itself within experience? In asking that question, Plessner is maneuvering the transition from appearance to reality through the presumption that the phenomenon of life itself, the way life appears to a living being, implicates a more than psychologically imposed experience of wholeness, an experience whose structural characteristics insinuate irreducibly de re properties. As I will argue, it is not just the case of our having to *consider* properties as functions and parts of a living whole, but rather coming to view those considerations as consequent upon the emergence of a categorical *it* or *self* who has a life, secures its life, develops, and dies.

Introduction xlix

Plessner's innovative and surprising categorial terms for bringing into view the particular wholeness of living organisms are *boundary* and *positionality*. Plessner's guiding thought here is that if organic unity is different from the unity of mere things, then the border of each must mean something radically different: things have *contours* or edges, places where the thing ends and its surrounding medium begins; living things have *boundaries*. For the living being *sustaining* its boundary is fundamental: the boundary separates a complex inside from an outside; the boundary is what relates the outside to the inside *and* relates insides to the thing's surrounding outside. Because the boundary must be sustained in order for the being to remain alive, then *boundary* refers to more than where the individual living being stops and the rest of the environment begins; the living being sustains its boundary by taking a position with respect to its environment. *Boundary* and *positionality* are practical categories; they are phenomena that appear in our engagement with living being, and they are practical components of living beings themselves.

Living things have a position (are positioned) in a place in a living environment and take a position. Plessner thus wants to argue that a boundary is as much a biological action and the result of a biological action as it is a property and hence that we experience the boundaries of living things not as like indifferent barriers, a separating wall, but as components of organic activities, of the active life processes that *thing* itself is.[22] And this means too that the living thing is neither a subject, an auto-poietic, self-creating being, nor an object, something made and created by another or merely given, but a subject-object that has both active and passive aspects; indeed, part of what constitutes the category of boundary is its bidirectionality, its implicating two spaces that are not merely adjacent to one another but are in continual interaction with one another, and where the consequence of that interaction is the establishment of one space being *inner* and the other *outer*.

The boundary thus has a dual function: it is a barrier that must be protected and protect the inner life of the organism, and simultaneously it is a transitional medium that provides access to the outer and brings what is outside in: organisms *reach* out toward the surrounding environment to gather in—nutrients, water, food—what is necessary for their life while protecting that inner life against invasion or untoward interruption. "Skin-like" [123] is one of the terms Plessner uses to capture the idea of a border regime in which there is continual alteration between transition and barrier, between outward movement and inner organization, between taking a position with respect to a surrounding medium and accomplishing inner processes against that medium.

A further way of specifying the categorial status of boundary and positionality is through distinguishing the space and time of mere things in comparison to the spatiotemporal movement of living things. Living things are not merely *in* space, the way a thing is, but *claim* space for themselves, actively situating themselves with respect to what surrounds them, and through their situating themselves shaping that environment with respect to inner needs. Biological adaptation, being positioned, routinely occurs through active adapting and constructing the environment—building a nest or burrow or beehive. In this respect, living things spatialize their environment, giving each item in its region a significance in light of what it spatially is with respect to the organism—too close, too warm, in easy reach, presently out of sight and so to be searched for. Further, living things not only live *in* time, but being in continual movement, their inner processes designate the living being's own "rhythmic" regime of growth, development, and senescence. Claiming space, having a *place* in an environment, and persisting through changes that are its organic life, *living things possess a nonmeasurable spatiality and a nonmeasurable temporality.* The qualitative spatiality and temporality of the living organism are irreducible features of the phenomenon of life that derive directly from the categories of boundary and positionality.

Because the boundary has a transitional character that sets in place the dual aspects and dual movement of the organism, as reaching out and being over and against, Plessner goes as far as to argue that the boundary does not belong exclusively to the body or the abutting medium; the boundary mediates inside and outside. Living things, Plessner claims, "are boundary-realizing bodies" [126]. The thesis that organisms are *boundary-realizing bodies* is the categorial lynchpin of his bio-philosophy.

Plessner's contention in Chapter 4, "The Modes of Being of Vitality," is that if life really does rest upon *the unique relation of the body to its boundary,* on organisms *being* boundary-realizing bodies, then not only must that relation show itself in the ways just noted, but the minimum set of categories necessary for discriminating living beings should dialectically emerge from consideration of the boundary phenomenon. Plessner is not here concerned with *empirical* concepts of the living, but with the categories—"life modals"—that organize and articulate the operation and place of empirical concepts. The boundary condition entails that living beings must satisfy a complex integrating demand: they must be capable of sustaining their internal, functional *organization* of parts, since the continuance of relating of parts to whole is a necessary condition for a being to be alive; insofar as this is the case, the individual forms a whole that is as one in each part. The thought of an organism being one in each part is that *it*, the organism itself, is somehow there in each part insofar it needs the part unconditionally, *has* the part as integral to itself, and must actively integrate the part so that it functions as part of the whole it is. All these familiar locutions have two consequences: first, in respect to its organizational elaboration of parts and whole, it makes sense to say that the organism halts *before* its boundary; that the boundary is an outside with respect to the internal life of the organism; that the organism has something that deserves to be called an *inner* life (because an inner that never becomes or could become outer, a border). Second, all the talk of the organism as *having* parts, as accomplishing *its* vital functions entails that the living body "is thus a *self* or a being that is not only absorbed by the unity of all its parts but is equally posited in the point of unity (which belongs to every unity) as a point detached from the unity of the whole." [158]. Plessner's thesis here is that the kind of part/whole logic of organisms, their being living systems, necessarily begins to imply a self—that is, an "it" *having* these parts, functions, needs, and the like in a way that is more substantial than the grammatical locution through which we are forced to speak of an X that has certain properties.[23] The categorial *it*, the implication that each organism is a self is meant to capture how we experience an organism as more than the sum of its parts, as something that persists through change, sustaining *its* life, protecting *itself* against invaders; all of which implicates each living thing as having a point of view on the world—which primitively is what it means to be a self. Each organism is a self; the vast majority are selves without consciousness; but the self-ness of the organism, its invocation of how things are for it, its having things as for itself anticipate the possibility of consciousness.

But this inward turning, organized, and self-organizing aspect is not sufficient on its own; again, living things require nutrients from the outside, as well as doing those movements necessary to acquire them and, at a higher level, avoiding harms and dangers. An organism must have a position in relation to its environment that permits these accomplishments; in this respect it is "one outside of the unity of its multiplicity." Plessner entitles this explicit demand "positionality." The requirements for integration and oppositional positionality must be set into continual functional relation to one another; in the life of a being there must be a functional interaction between (internal) self-organization ("the self-mediation of the unity of the living body through its parts") and (external) positionality. Sustaining the coordination between inner and outer orientations is a condition of life; the collapse of that coordination, death. The "curve of development" that involves not only growth but "aging and death" [146] describes more than a series of empirical facts about organisms; aging and death are life modals.[24]

From Animal to Human, from Centric to Excentric Positionality

Chapter 5 begins by underlining the forms of unity constitutive of organisms: each is a unity for itself; each is a unity in the diversity or multiplicity of its parts; each is a unity in every part [187]. The third claim is ambiguous, since it can sound as if the thought is that each part is a microcosm of the whole, which is patently false. Rather, the claim is that an organism is a unity in every part mediately: every part is a means, a tool, to the unity that the organism is, thus making the organism a unity of tools. "The whole of the actual body," what the body of a living organisms *is*, "is its own *means*" [189]. This is the dynamic entanglement of means and ends in living systems, each having its unity as end and means at the same time. This, however, also overstates the unity and wholeness of organisms: every organism is a needy and dependent being, its being alive mediated by either inorganic or organic others outside itself; it is a self against and through its medium. As we have already seen, the very idea of boundary is for Plessner a way of thinking through the fact that neediness and dependence on the surrounding environment, as well as protection against it, are constitutive of the kind of beings living ones are.

Having introduced his categorial scheme, Plessner continues by demonstrating how the forms or levels of organic life are further determinations of it. Although there are different ways of distinguishing plants from animals—plants use inorganic materials to nourish themselves while animals feed on organic materials, both plants and other animals[25]—Plessner, staying with the boundary principle, structures the distinction as one between open and closed forms of organization. "A form is open," Plessner states, "if the organism in all of its expressions of life is immediately incorporated into its surroundings and constitutes a non-self-sufficient segment of the life circle corresponding to it" [219].[26] While it is a true and important feature of most plant life that it is in direct physical contact with its surrounding medium and that the relation between part and whole is less integrated in plants than for animals, since plants (can) grow by the multiplication of smaller units capable of reproducing on their own (cuttings, graftings, transplantings), I suspect that for Plessner the idea of an open form of organization took on its particular salience through categorial emphases necessary for comprehending the animal closed form of organization. For Plessner, the animal closed form of organization effectively makes explicit what being an organism is. While Plessner is unequivocal that the levels of the organic are neither a value hierarchy nor a teleology,[27] it is equally the case that the actual unfolding of evolution makes it necessary to consider in what ways the processes of variation and selection are constrained. In effect, Plessner's categorial scheme seeks to finesse the issue of constraint by letting the animal form of organization provide the profile against which both human and plant forms of organization become categorially perspicuous.[28]

Plessner contends that greater closure, together with the introduction of mobility (which may be a condition for greater closure), provides for "true self-sufficiency" [226], which is what he means by the claim that in animal life the organic body attains another level of being [227]. A fundamental feature of this self-sufficiency is the way in which the bidirectional boundary function separates out into distinct sensory (passive) and motor (active) functions. Once the sensory and motor functions undergo radical differentiation, they incrementally come to require case-by-case coordination by a "center"; that center eventually becomes a literal center with the development of a central nervous system. Although he is being slightly hyperbolic, Plessner can now elaborate the distinction between plant and animal, open and closed forms of organization, by contrasting the plant's continuous contact with its surrounding environment with the animal's dual, sensory-motor mediated relation to its environment. Because the animal must scan, monitor, and take note of its environing surround and then act in response to whichever perceptual/sensory taking is most urgent (seeing a threat on the right while smelling a feeding opportunity on the left, for example), it follows that the animal self is in relation to "the medium as the other ... *across a chasm*. . . . Across a chasm it has a sensorial and motor connection to the other" [232] italics Plessner). I take the "chasm" metaphor to be an effort at elaborating the extent and significance of the animal's independence from its surrounding medium and thus to be giving the categorial declension "closed" an existential weight.

By animal closure, Plessner therefore seems to be registering at least the following formations of boundary and positionality: (a) independence from the environment premised on (b) strong differentiation of the sensory and motor functions (including the differentiation of receptors/organs of perception in the sensory system); (c) the consequent requirement for the coordination of the sensory-motor system through a core or center; (d) with a center, under conditions of differentiated sensory and motor systems, the conception of the center as *having* its body as an instrument to accomplish various ends, and as thus being dependent on its body (as if it were not wholly identical with that body); (e) this positing of a center occurring in the bodily living being the animal already is. Although there is an alteration of aspects of embodiment in animal life, Plessner's initial thought about the animal's closed form of organization is that these aspects are fused and bound together: "The living thing is itself—in itself. The position is a dual one: being the body itself and being in the body—and yet it is singular, since the living thing's distance to its own body is only possible due to its complete oneness with this body alone" [237]. In this passage Plessner is pressing three different considerations: First, the *having* of a body, the controlling metaphor of mind-body dualism, is only intelligible as a certain development of *being* a body; that the *having* a body idea, however implicit in earlier forms, emerges as explicit in relation to the emergence of the closed form of organization in opposition to plant life. Second, in the animal form of life, because the coordination between sensory and motor systems is largely automatic, the animal lives out the dual aspects of existence without an experience of distance, without a vital experience *of having*.²⁹ Nonetheless, third, the closed animal form of being and having a body emphatically anticipates human bodily experience.

What distinguishes animal life from the human is that the dual aspects of its bodily life, being and having its body, are not yet present to it: "The here/now nature of the animal is not given to it, is not present to it; the animal is still fully absorbed in this here/now nature and carries therein the barrier, hidden to itself, set against its own individual existence" [239]. Of course, it is precisely the animal's lack of a reflexive relation to its being a "here" for which all else is "there" in this "now" that makes the animal "now" one that is repeating and forever, and its "here" as never another's "there" for it. Pointing forward, one implication of this is that once an animal existence can be aware of its here/now it must lose the possibility of living it fully; to be aware of the here and now is to have one's here and now relativized to other spaces, places, and times. With that difference from the human acknowledged, in the remainder of the chapter on the sphere of the animal, Plessner goes on to add various capacities to the animal as further articulations of its boundary regime, each making animal existence more proximate to human living: frontality, bodily orientation (front/back, up/down, left/right), consciousness of a surrounding field, learning, memory, anticipation, intelligence as the capacity for problem solving. Plessner's method in considering the animal sphere of existence is to obey the dual constraints on his categorial system, showing how the animal form of existence as making explicit what being a living organism categorically is at the same time demonstrates that filling out the categorial scheme anticipates just those features that have traditionally been prized as uniquely human. Only through the constrained categorial elaboration of animal life can philosophical anthropology vertically locate the human in living nature.

That said, the emergence of the human does occur as a radical transformation of the closed animal form of organization. As a transformation in the boundary regime of the animal, it is a transformation from a centric to an *excentric positionality*, an undergoing of a decentering whereby the centering orientation of the lived body is always (existentially and potentially) qualified and relativized. One way of expressing what the transformation to an excentric positionality involves is to note that the human realizes the full reflexivity in relation to the living body that is denied to the animal; the excentric positionality of human thus involves animal existence being raised to the level of self-consciousness. In achieving self-consciousness of its bodily existence that bodily existence is *transformed*: the centric character of animal existence whereby the coordination between sensory and motor functions was all but automatically achieved—with animal learning opening the space of interruption—becoming in the human a permanent, moment-by-moment task for each individual that is satisfied with respect to and through rules and norms particular to some specific cultural regime.

Plessner does not explicitly lay out the conditions for this transformation from closed to excentric positionality, although he takes it to involve: the attaining of the upright posture; the consequent freeing of hands and arms from immediate purposes; the premature birth of the human infant with its consequent insufficient binding of physical coordination to instinct (instinctual impoverishment, as it is sometimes called); physical maturation as requiring for its actuality a work of socialization and social formation; the emergence of discursive language and symbolic (mediated) communication; and, importantly, the human power of the negative, of which Plessner argues all higher animals are incapable [270]. All these conditions and features can be reasonably thought of as contributing to the distinctive character of human existence. For philosophical anthropology, however, what matters is how these items make possible, contribute to, and help solidify the new boundary regime of the human. The emergence of self-consciousness—the guiding theme of modern philosophy—is now refashioned into and becomes a consequence of a novel way of inhabiting the centralized form of organization of animal bodily life. In Plessner's scheme, excentric positionality derives from the unbinding of the instinctual-natural relating of the body I am and the body I have, the lived body and the physical body, and thus possesses the consequent requirement that bodily life be executed, something done and accomplished, thereby decentering human experience. It is the "double aspectivity" of human bodily life then that insinuates the insistence of being self-conscious:

> The excentricity of the living being's position [*Lage*]—that is, of the irreducible dual aspect of its existence as *physical body* [*Körper*] and *lived body* [*Leib*], as thing among things at arbitrary points in the space-time continuum and as a system concentrically enclosing an absolute center in a space and a time of absolute directions, corresponds to the excentricity of that living being's structure. [294]

Although it is tempting to think of self-consciousness as something that could be simply *added* to the centralized form of animal existence, Plessner's hypothesis is that it is the transformation in the boundary regime of the human, in the relation between physical and lived body, that invokes the distance involved in one's awareness *of* oneself, the distance from our own life that reaches its zenith in active self-consciousness. For Plessner, self-consciousness is a further elaboration of our excentric positionality, the requirement for coordination that evidences the gap between one's lived and physical body.[30]

If in animal life each stimulus and each action are processed through instinctually established or learned routines, the coming to be of excentric positionality entails that in principle each and every stimulus is only a *possible* question or address requiring an agential response, where the possibilities of response are, in principle, plural, including the possibility of not responding because the address is irrelevant or unrecognized. To comprehend the impact of this unbinding of sensory experience and motor reaction from instinctual preformation is to give categorial significance to the fact that the interpretation of sensory materials and the decision to act become in the human, to a large extent, conscious events for the self who must decide upon them. The human "I," even the "I think that must accompany all my representations," is the center *as* center and therefore decentered, alienated from the very body it is. Plessner states this paradoxical outcome thus:

> As the I that makes possible the full return of the living system to itself [the I that *accomplishes* the relevant unity of sensory in-put and motor out-put], the human is no longer in the here/now [because the unity is something that must be consciously accomplished] but "behind" it, behind himself, without place [aware of itself as set apart from the positional body it is in virtue of being explicitly burdened by the necessity of interpretation and decision], in nothingness, absorbed in nothingness, in a space- and timelike nowhere-never. [292]

This existential-sounding non-place of the self is the simple consequence of the self coming to consciousness of the body she is in its organizational multiplicity-in-need-of-unification. Although it is true that the sensory manifold requires interpretation, conceptualization, and unification, this is primarily so because of the systematic fragmenting and fracture of the sensory-motor system. In the first instance, it is the body I am and the body I have that require unification; it is the requirement for coordination between the dual aspects of human bodily life (as a rupture of animal bodily life) that categorially underpins all the other anthropological unifying and integrating practices.

Excentric positionality as a transformation of centric animal life is most fully expressed in terms of the becoming conscious and dislocated from one another of two components of animal bodily life. *Ex*-centricity refers to a centered animal existence that is only one component of a dual categorial bodily system whose centric character is one that must be always and again accomplished against a continuing possibility of rupture and dispersal.[31] In a famous passage from *Laughing and Crying*, Plessner states the idea this way:

> ... man has, not a univocal, but an equivocal relation to this body, that his existence imposes on him the ambiguity of being an "embodied" [*leibhaften*] creature and a creature "in the body" [*im Körper*], an ambiguity that means an actual break in his way of existing.... A human being always and conjointly *is* a living body (head, trunk, extremities, with all that these contain) ... and *has* this living body as this physical thing.[32]

The being a body/having a body distinction is rightly regarded as Plessner's signature contribution to the field of philosophical thought. I have delayed its full introduction because I have wanted to underline how much this novel conceptualization of bodily life is bound to the constitutive structures of Plessner's philosophy of biology, how his philosophical anthropology contains and leans on his bio-philosophy, and how without it the temptation to understand self-consciousness in broadly Kantian terms continues to be irresistible

Although there is a rich complexity and layering to Plessner's analysis, the broad idea is familiar enough: being a body is a body as it is lived and experienced from the inside; it is my immanent experience of bodily life. Having a body invokes my distance from my body, which is tokened by the experience of being just one body among other bodies, but equally when my body cannot accomplish what I take myself to be attempting to accomplish with it and through it. Thus one's lived body is the body that immanently and with apparent immediacy inhabits a practically meaningful world whose shapes and appearances are coordinate with one's needs, interests, abilities, the world like a table set for one's use. In the ideal extreme, through socialization and a habit-formed second nature, one can unselfconsciously respond to each object-event as if it were meant for one's bodily use and vital satisfaction. Conversely, in the extreme case of being a physical body amongst physical bodies, other human bodies are exchangeable with one's own, while the world recedes from immediate practical intelligibility and appears neutrally, as just things standing in objective space and time. But these cases are extremes, and the latter, arguably, nothing we ever directly perceive or experience—the wholly objective and objectified world of natural science is an idealized construction that never fully overlaps with even the decentered aspect of perceptual experience.

Hence, getting closer to Plessner's thought, the expansive doubling occurs because human lived experience is *not* natural or given, but rather the consequence of our having our bodies as instruments that enable us to make things practically intelligible and familiar: before you know how to use them, chopsticks are alien items, two narrowed sticks, resistant to manipulation by the clumsy and inept movements of thumb, index finger, and middle finger; things in this condition are so resistant that they are at best only potentially useful and our body an unmastered instrument for engaging those things. While we are doubtlessly geared to the world in a variety of ways from the get-go—the newborn's lips pursed in expectation of a nipple to suck—for humans, the negative haunts every native and learned synthesis of bodily orientation and worldly affordance *as if* in the first instance every item is an alien utensil unsuited to bodily manipulation and our body equally an alien instrument incapable of the manipulations necessary to engage the world; even breastfeeding is already a negotiation filled with moments of breakdown and mutual adjustment.[33] For humans the lived body that is practically open to a meaningful world is a *product* of the learning, socialization, and habituation of the body one has: one works one's fingers until they can hold the spoon, write the word, play the tune, thread the needle; one's voice, mouth, tongue, and teeth slowly mold themselves to the insistences of a mother tongue. If ontogenetically the lived body is practically a product of the socially contoured and regimented manipulations of the body one has, nonetheless that lived body is the *unavoidable* centric idea of body. A favored way of expressing this thought is to say that excentric positionality involves a detachment or disengagement *from* life *in* life, a detachment that has to be lived; so excentric positionality involves a detachment from the body within the body.³⁴ Humans must become the lived beings they already are; in this respect the lived body is both antecedent to and a consequence of the process of making, or, as this idea is sometimes phrased, the lived and physical body are ongoing integral aspects of the "physical lived body" (*Körperleib*) as a whole.

If human bodily life involves a cultural work of coordination, it anthropologically follows that our bodily engagement with the world must undergo an equally radical transformation. After stating that the lived body and the physical body in the material sense are one and the same—even though they do not coincide with one another they are not two distinct systems—Plessner presses how the radical double aspectivity of the body infects the way the world appears and means to us.

> In an equally radical way, surrounding field and outer world cannot be converted into each other, although they do not constitute materially distinct zones, either. Point by point, the surrounding field can be inscribed into the outer world, but it thereby loses the characteristics that define it as the surrounding field. [295]

Plessner is here arguing that the excentric positionality of the human body is coordinate with the world coming to emphatically appear under competing descriptions: the *surrounding world* that is the internal correlate of achieved coordination between sensory and motor functions and the *outer world*, which is point by point the very same material world as the world that is just objectively there without being saturated with the significances things have for agential life. In effect, Plessner is here providing a genealogy of the division between subjective life and scientific objectivity, between idealism and realism, even between the realm of freedom and the realm of causality; all these should be taken as variations on and consequences of the radical separation of surrounding world and outer world that occurs with arrival of the dual aspect structuring of bodily life. Because the distinction between surrounding world and outer world is an unsurpassable consequence of our decentered bodily life, it follows that all the competing human orientations and projects that bind themselves to one or another of those worlds must fail in their effort to reduce threatening others to it. Which is precisely where I began this introduction: with the prima facie implausibility of programs that would attempt to reduce human significances to the determinations of matter-in-motion or the saliences of mathematical physics to just all-too-human social practices of constructing and ordering the surrounding world. For Plessner, because the distinction between surrounding world and outer world is an implacable consequence of the dual structure of human bodily life, then the production of competing and incommensurable accounts of the world, human and inhuman, is an ineliminable and irreducible feature of the human form of life.

Plessner elegantly captures the existential predicament of the human that follows from its excentric positionality thus:

> The human, then, lives only insofar as he leads a life. As such, the life of his own existence is always breaking apart into nature and spirit, bondage and freedom, *is* and *ought*. The opposition persists. Natural law stands against moral law; duty fights inclination. Conflict is the center of existence as it necessarily presents itself to the human in terms of his life. He has to act in order to be. [316–17]

Conclusion: The Project of Philosophical Anthropology

Plessner concludes Chapter 7 of the *Levels* with an outline of the three fundamental laws of anthropology: the law of natural artificiality; the law of mediated immediacy: immanence and expressivity; and the law of the utopian standpoint: nullity and transcendence. The first law has been implicitly present throughout my account of the excentric positionality of the human; roughly, because if the human must act and perform in order to live, then she must fashion herself into what will count as appropriately succeeding in this task.

> As an excentric being without equilibrium, standing out of place and time in nothingness, constitutively homeless, he must "become something" and create his own equilibrium. And he can only create it with the help of things outside of nature that originate from his creative action *if* the results of this creative activity take on a weight of their own. [310–11]

The full weight of the dislocations that mark our excentric positionality converge on the obvious point that we cannot satisfy our vitals through whole naturally endowed means. We depend upon artifice—not just the tools, instruments, and utensils, but even more the whole structuring of human action through social norms and human consciousness through language—for living. Plessner is here underlining the most urgent consequence of this obvious fact: the equilibrium, the how-things-are-meant-to-be for animal organisms is given with and by the centric form of organization. It is just the possibility of there being a *given* conception of equilibrium that disappears with the arrival of leading a naturally artificial existence. The thought that cultural products must "take on a weight of their own" thus expresses the constitutive conundrum of culture: it must from out of its own resources demonstrate how its diverse forms of significance can and should matter, how customs and practices can sustain the meaning they express and project, how utility and meaningfulness coordinate in a sustainable manner. And it is within these efforts of founding, grounding, and making intelligible that the temptations toward reductive answers occur: it is the force of nature all the way up or the lambent inflections of language or reason all the way down. In effect, every culture is a working through and an effort to reconcile the irreconcilable duality between culture and nature.

The second law, the structure of mediated immediacy, is simply the consequence of the living subject having "an indirect-direct relationship to everything" [324]. There are always culturally formed connecting links to objects sustaining human existence; hence, even for the most immediate of connections to the world, that immediacy and its object have undergone a cultural shaping and mediation, even when the mediation is unnoticed. Immediacy here is another term for the achievement of the animal's centric engagement with the world. However, because every immediacy is mediated and executed, performed, then, on the other side, we can become aware of the lack of immediacy. It is in this setting that Plessner rehearses the torsions of self-consciousness I mentioned earlier. So he suggests that rather than considering the self-conscious self as standing-above itself, consider the idea of "standing-behind-oneself." Under this description, "the situation of the human is that of being immanent in consciousness. Everything he experiences he experiences as a content of consciousness and therefore not as something in consciousness but as something existing outside consciousness" [328]. Plessner is here wanting to explain the simultaneous temptation to idealism and realism, the world as immanent to consciousness and the world itself directly appearing to consciousness, as a structural consequence of mediated immediacy: we cannot but be directly conscious of the world, nor can we avoid being aware of ourselves as accomplishing that consciousness and hence being at a remove from the world. And it is because we are indefinitely subject to the tensions involved in satisfying our natural ends artificially or being destined to achieve immediacy only through mediation that the human situation can seem an absurdity pressing for resolution. Hence the desire, the effort, the cultural work of escaping the predicaments of excentric positionality belong to it; the pivot between nullity and transcendence are ingredient in human living—which is the third law of anthropology.

One compelling elaboration of Plessner's philosophical anthropology occurs in his remarkable study *Laughing and Crying*. Again, one way of stating the existential complexity enjoined by the excentric positionality of the human is to say that we are fated to become self-conscious of our having a body, of our being a body, and of the forever incomplete effort to bring those aspects of human bodily life into replete alignment. These different forms of human self-consciousness originate from the fact that the world of humans is really composed of three different, overlapping worlds: the outer world we experience through being a body; the inner world we experience in having a body that remains at a distance from the "I," so condemning the "I" to an interior life; and the social world we share with others that can include the recognition of excentric positionality. Nothing is more human than to repress, forget, or deny the extent to which the forms of human life are formations of our being and having bodies. At the individual level, crises of excentric positionality occur when the complexity of a situation involves contradictory dimensions or when we are faced with a situation in which there is no behavioral answer to the urgent question our sensory system asks. Under these conditions, it is, remarkably, our body itself that responds to the pressure of the occasion in place of the agent self: we laugh uncontrollably at the contradictory situation—the dignity-possessed person in the grip of sheer gravity as he slips on the banana skin—or weep uncontrollably at the pain we can do nothing about, or, worse, weep in grief as we acknowledge the finality of death. Laughing and crying are responses to limit situations where there is no position we can take and hence nothing we can do; laughing and crying are the bodily marks of our outsideness, our unnaturalness. Laughing and crying reveal the construction of a cultural world as a humanly made natural world, our natural habitat a social niche, hence our inhabiting a second nature laid over the living, vital body we are and remain, the categorical insistence of our being boundary-realizing bodies.

In this introduction I have sought, first, to explain the need and necessity of a philosophical paradigm that could place human living within the natural living world, thus avoiding the implausible extremes of materialist reductionism and cultural idealism as well as the impossible compromise of mind-body dualism. Second, I have wanted to outline the broad gestures of Plessner's philosophical anthropology as determined by the categories of *boundary* and *positionality*. The test for Plessner's categorical thesis that living things are boundary-realizing bodies is whether it can bring into view the phenomenon of life in a way that simultaneously can distinguish living from nonliving beings while enabling an unequivocal placement of the human within the organic world even in its mode of departure from a purely biologically explainable mode of living. I have been suggesting that formally the idea of living bodies as boundary-realizing ones powerfully answers this test; in concluding with Plessner's account of laughing and crying as responses to limit situations in which the coordination between being and having a body collapses, I was claiming that even the most fundamental of *cultural* categories—viz., comedy and tragedy as synoptic forms for understanding human existence—remain beholden to our boundary-realizing bodily mode of existence. In his 1931 *Political Anthropology*, Plessner further elaborated the argument of *Levels of Organic Life*, claiming that, appropriately understood, politics is a material a priori of the human form of living: we are collectively responsible for and must collectively determine the social norms and forms that structure our life together. This work includes a powerful argument to the effect that responsibility for life must displace knowledge as the grounding orientation of any shared human world. I take this claim for our collective responsibility for life as a central component of Plessner's own effort to legitimate politics and the humanities against the reigning hegemony of natural science in its reductive self-understanding.[35]

Even the most strenuous defense of Plessner's philosophical anthropology need not claim, however, that it is the only task a robust bio-philosophy needs to accomplish: the anticipation of consciousness, the demonstration and elaboration of the relation between competitive and altruistic forms of behavior in evolution, or a fuller consideration of microbial life and the ubiquity of symbiosis would each press bio-philosophy in a different categorical direction.[36] If these are valid categorical approaches to the phenomenon of life, then there is reason to believe that in exactly the same way in which no one model of a living system can capture all its complexities, so no one categorical model of life can capture the kinds of significances borne by and ascribable to living beings, processes, and systems. Pluralism and a kind of promiscuous realism must belong to the bio-philosophies of the future.[37] Plessner's categorical model is deeply worthy in itself but equally should be taken as exemplary of the kind of undertaking bio-philosophy is and must become. If human beings are living beings, then the fact that the philosophy of biology is a marginal subdiscipline speaks to a massive distortion in contemporary philosophy.

In the Preface to the second edition of the *Levels*, Plessner acknowledges how untimely was its original publication. His book was first received as nothing but an elaboration of Max Scheler's *The Human Place in the Cosmos* and then overshadowed by the impulsive existential drama of Heidegger's *Sein und Zeit*. That Heidegger's masterwork is exactly the kind of antinaturalism that Plessner's philosophy means to subvert has since become achingly obvious. Plessner meant his philosophical anthropology to be a radical, naturalist alternative to the major philosophical positions of recent modernity: Cartesian and Kantian dualism, Nietzsche's will to power, psychoanalytic theories of drive and sublimation, Heidegger's existential phenomenology, pragmatism, and all the variations on reductive materialism. Plessner also notes in his later Preface how the writings of both Sartre and Merleau-Ponty sometimes read as if they had read the *Levels*, although there is no evidence that this is the case.

Perhaps the translation of *Levels of Organic Life and the Human* into English can inaugurate the kind of philosophical reception and conversation the original publication deserved but did not receive. If I am right about the situation of philosophy now, Plessner's masterwork should become a keystone for the philosophical conversations of the future.[38]

Philosophy can never prove a hindrance to the advance of empirical science. On the contrary, she traces every new discovery back to fundamental principles, and thus lays the foundation for fresh discoveries. Should there arise a class of men who regard it as more convenient to work out chemistry in their heads, rather than soil their hands in its pursuit, this cannot be considered your fault, and certainly not that of the scheme of your philosophy. Ought we to decry mathematical analysis, because our millers are able to construct more efficient machinery than any that a mathematician could devise?

—Alexander von Humboldt to Schelling, 1805[1]

ONE

Aim and Scope of the Study

Every age finds its own redeeming word. The terminology of the eighteenth century culminated in the concept of reason; that of the nineteenth in the concept of progress; that of the current one in the concept of life. Each of these periods designates something different—reason stresses the timeless and the universally binding; progress, restless becoming and ascending; life, the demonically playful and unconsciously creative. And yet all periods want to grasp the same thing, and the true purport of these words becomes for them merely the means, if not the pretext, for rendering visible that ultimate depth of things without a consciousness of which all human beginning would be without background and without meaning.

There are reasons why a particular concept and no other becomes the symbol or pretext for a certain time period. A word only becomes redeeming if a period finds in it both its justification and its judge. Rationalist ideology never generated more enthusiasm, never revealed the ground of things more clearly than when the battles for freedom, naturalness, and rationality were still being waged but feudal power had, at heart, already been won over. Never was evolutionist ideology more successful, more fertile for insight and action, than in the transition period of the second third of the nineteenth century, when the patriarchal way of life surrendered to incipient technization, industrialization, and capitalization. The great moment for the ideology of life came with the reaction against blind faith in progress, with civilization fatigue, with despair over the creative potential of socialism. A fundamentally resigned period came to unmask what until then had been thought of as the final unshakeable option—the advancement and progress of all organic being and human activity in the world—[4] as the ideology of expansive high capitalism. This awakening, however, also brought with it a longing for a new dream, a new enchantment.

But by what did this wary, skeptical, and relativistic period still allow itself to be enchanted? People had become too enlightened and conscious for transcendence on a grand scale, too open-minded and adventurous for immanence. The human was seen as phylogenetically and historically contingent. At the same time, however, nature and history had lost their persuasive power over the mind since the alleged discovery that their laws and the broad features of their design derive solely from the creative power of the human spirit.

The only thing capable of enchanting was something irrefutable, to be grasped on this side of all ideologies, on this side of God and the state, of nature and history; something from which ideologies may arise, but that with equal certainty also devours them again: in short, life.

It is in this word that our age apprehends its own strength, its dynamism, its risk-taking, its pleasure in the demonism of an unknown future—as well as its own weakness, its disconnection from its origins, its lack of devotion, and its inability to live. With this new incantation, whose impact since Nietzsche has been growing ever greater, our age follows and hounds itself. A philosophy of life has emerged, originally intended to spellbind the new generation, just as every generation has been caught by a philosophy in the spell of a vision. Now it is called on to lead this generation to knowledge and thus free it from enchantment.

The Development of Intuitionist Lebensphilosophie *in Opposition to Experience*

The discoveries made by the theory of evolution, by genetics, physiology, and phylogenetics, have revealed a new aspect of the way the human and his culture are bound to nature. If in earlier times it was taken as a relative given that the human was part of the animal kingdom, in the new view of nature this became knowledge that "explained"—that is, dissolved—the essence of the human. In order, then, for the spiritual to not become, [5] following the well-known recipe, the simple superstructure of a certain kind of animal existence—thus merely leading to the triumph of a biological form of the old naturalism—the affinity between nature and spirit and the position of the human had to be determined from a new perspective.

Two possibilities for this—the only ones, as it seemed—had already been exhausted: materialist-empiricist and idealist-aprioristic philosophy. If the first theory foundered on the facts of consciousness and necessary sensory laws, which cannot be deduced from the physical world or from sense impressions, the second, conversely, failed to come to terms with the facts of perception and the specifics of physical nature.

The old choice between empiricism and apriorism appeared here as follows: according to one side, the human with all his physical and spiritual characteristics is the final link in the organic evolution on earth. In that case, his consciousness, his conscience, his intellect, the system of forms of his spirit—and thus his culture—are products of nature, the result of cerebral evolution, upright gait, certain changes in inner secretion, etc. How this result came about and how physical facts became spiritual elements remain entirely mysterious, however. Alternatively, his own natural history in connection with the history of organisms is, like all of nature, a construction of the human in accordance with the aprioristic, fundamental forms of his spirit and within the scope of his consciousness. How creative spirit comes to exist concretely "in" a human being, how it becomes dependent on his physical characteristics remain, however, equally mysterious.

These two theories use different arguments but follow the same principle. They posit one sphere as absolute—in one case the physical, in the other the spiritual—and make the other sphere dependent on it, without, however, being able to explain how it is that precisely this sphere comes to be dependent on the other. Either spirit is the flower and product or nature is. In both cases the metaphor is flawed; there is no homogeneous way of bridging the two dimensions (to be able to deduce the one from the other, however, would be an unreasonable demand) or even of mediating between the two aspects while preserving completely the radicality of the dual aspect of body and mind.

[6] This gave rise to the following question: under what conditions can the human be regarded as the subject of spiritual-historical reality [*Wirklichkeit*], as an ethical person with a sense of responsibility, *from the same* perspective that is given by his physical phylogeny and his position in the whole of nature? Or, to be more cautious: can the spiritual history [*Geistesgeschichte*] and the spiritual aspect of the present, as they are intrinsic to the subject of cultural activity, and natural history or the physiological aspect of the human be so combined as to, while avoiding the empiricist and aprioristic errors, nevertheless maintain *one* basic aspect? And this in such a way that confirms the natural, pre-problematic view according to which the human emerged from a prehuman phylogeny of organisms and his spiritual faculties developed in a way that is historically connected, both temporally and spatially, to a vast biological past? Failing to maintain the *one* basic aspect leads immediately to a double truth: the world viewed as consciousness and the world viewed as nature, the human as self, as I, as subject of a free will, and the human as nature, as thing, as the object of causal determination. This puts one in the unworthy and intolerable position, also irresistibly comical, of considering the human as a product of a phylogeny and phylogeny as the product of the human, of a creative spirit somehow become event in the human.

How cautious one must be in establishing the *one* basic aspect is shown by Bergson in his critique of Herbert Spencer. Spencer wanted to, as it were, combine empiricism and apriorism by accepting like an apriorist the given relational forms with which consciousness intuits, perceives, and thinks—the a priori conditions of knowledge, in other words—but explaining them like an empiricist. The a priori forms, the categories, he argues, are the results of adaptation that developed arduously over millions of years, proved themselves, and were passed on as acquired characteristics. The generations who, thanks to the efforts of their ancestors, were freed from the work of adapting to nature must have then come to be aware of primary conformity as a given, as the axiomatic system of their existence, whose inner necessity for experience is explained by the congruence, guaranteed by adaptation, of the bearers of consciousness with the world. [7] (The same principle has been used to attempt a biological explanation of conscience and norms.)

Bergson exposed the circularity of this account while at the same time using it to point to the problem his intuitionist *Lebensphilosophie* purports to solve. He operates against Spencer like a transcendental idealist, a Kantian apriorist. According to Spencer, he argues, the categorical forms came about by means of adaptation, by adaptation to nature. The categories of causality, of substance, of reciprocity must then somehow be present in nature—if not as forms of thought, then at least as forms of being. In other words, nature in this explanation is already presupposed as something that only becomes possible by dint of the categories. Spencer thinks that he is deducing the subjective forms of understanding from objective nature. In actuality, however, he is merely positing the system of categories in another form. The tune remains the same; only the key changes—sometimes the system of categories is called "nature"; sometimes it is called "intellect."

Bergson's aim with this polemic is only to show that using the mechanism of nature as a model for the mechanism of understanding necessarily leads to a circular argument or a *petitio principii*. The idea of the emergence of the categories articulated by Spencer in response to the unavoidable demand raised by the facts of phylogeny must, according to Bergson, be grasped in its entire meaning as the problem of the boundary between the mechanical image of nature (corresponding to the categories) and what is, of course, *itself no longer* a mechanical formation of this image of nature (corresponding to the categories, particularly to that of causality). Understood this way, the idea of the emergence of the categories "in nature" revolutionizes the philosophical method. For it would then no longer be a matter of thinking in the categories of paleontology or zoology and of appealing to the lawful workings of nature and the mechanisms of heredity, selection, and breeding as the foundation for the genetic explanation of the categories—the thinking that in Bergson's view is eminently bound to categories or is mechanical is no longer of any use here whatsoever.

Another way of knowing must intervene, then: the intuition [*Intuition*] of which we as living beings are capable. In pure memory, Bergson writes, [8] we win our freedom from the web of categories. In the river of life, but swimming against it rather than with it, we become released from our practical attitude toward things, thereby getting "behind" the category mechanism of action-bound, utility-oriented thinking. We become existential again. Life grasps life; it understands what it is and what it was. It is in this way that intuition [*Intuition*] grasps the inner essence of development, which paleontologico-phylogenetic thought only comes to know in its outward traces.

It is characteristic of Bergson to contrast (mechanical) thought with (organic-vital) intuition [*Intuition*]. We may thus heartily disagree with his solution to the problem. Of interest here, however, and in a certain sense paradigmatic for all "speculative" philosophies of life, is his approach to his solution and the point at which he introduces the concept of life. This point, this lever, for Bergson lies in the coexistence of intellect and the mechanism of nature. The fact that Spencer failed to recognize the essentiality of this coexistence proved his undoing by ensnaring him in a vicious circle. One can neither get behind the intellect/the mechanism of nature with the intellect nor with the essentially correlated mechanism of nature. Instead, one has to, as it were, transpose one's perspective from the surface to a deeper dimension and strictly separate the new foundational aspect from the previous one. Because, however, this foundational aspect is intended to solve the task of comprehending *from one and the same perspective* the conscious existence of the human in connection with his biological history—broadly speaking, of allowing subjectivity to emerge from physical nature, or of assigning subjectivity a particular place in the whole of nature—the concept of a formative force presents itself here, one that encompasses nature and spirit and creatively permeates being and consciousness: the concept of life.

But Bergson does not appeal to the deeper impossibility of gaining insight into consciousness from insight into physical things. This illegitimate and unfeasible shift in perspective, so readily resorted to by dilettantes of empiricist and aprioristic nature philosophemes, does not figure in his work. His argument against allowing the intellect (subjectivity in a limited sense) to emerge from nature he finds [9] in the intellectualistic character of this emergence and of this nature. He calls instead for a pre-intellectually conceived emergence from nature and thus arrives at the concept of life in which physicality and consciousness, inner and outer sides can indeed be seen from one point of view.

Bergson considers the nature of the natural scientists to be a mere counterpart of the intellectual. The intellectual (along with his "categorized" world) himself becomes simply a manifestation, a spawn of creative life like all the other curious formations of plants and animals. As long as he conducts rational science he does not get beyond his own form of life. The descriptions and theories of zoology, botany, and paleontology only provide the perspective of the intellectual, not the essence of living things. Instead of penetrating into every phase and epoch, the gentlemen's own spirit is reflected in the alien matter of misunderstood occurrences.

This is why the view of life to which Bergson leads his readers, what he calls "creative evolution," has been embraced in cultural philosophy and in the philosophy of history. For it is in these fields that Bergson's initial question regarding the nonmechanical emergence of the world system of mechanism (that is, the conformity system of mechanical nature and intellect) returns in *every* question regarding the emergence of a spiritual world or of culture as such.

Taken in its general sense, this question is the vital question of all historical knowledge. In effect, every historian who wants to grasp the past in its objective essence repeats the demand for a nonmechanical derivation of the mechanical world *if he seeks* to keep clear of his own age's system of categories when describing earlier conditions and their interconnectedness.

According to Bergson, this would not be possible for the empirical historian, who attempts to determine the causes of the later in the earlier and to accentuate the singular, unique, and valuable. The empirical historian, in other words, consciously ties himself to his generation's system of categories and, to speak with Bergson, represents the life form of the intellect in relation to the historical material. Objectivity, the application of the principles of causality and sufficient reason, are, after all, characteristic of this and no other life form. As a result (again, following Bergson's line of argument [10]), the so-called objective empiricist cannot discern the true essence of the past and the emergence, or meaning, of current conditions. Just as paleontology and evolutionary history can only make sense of the traces of early life from the perspective of the mechanical intellect, objective historical scholarship only yields an image of history that conforms to its own precepts.

By arguing for his own boundedness to a particular age and people and by also declaring his quest for knowledge and his means of working to be tied to his particular period, the empirical historian calls himself into question and prepares the way for another mode of knowledge: an intuitive [*intuitiven*] philosophy of history. Historians in the eighteenth and nineteenth centuries did not yet dare to take their relativism this far. At most they held themselves to be bound to their religion, their taste, their sense of justice, but thought of reason and that which it judges as true and necessary to be atemporal and thus absolute. The ongoing discovery of the diversity of human cultures and their views of the world has destroyed this remainder of naiveté and undermined faith in the atemporal nature of the categories of knowledge. Today the historian is unable to find his way out of the absolute historical immanence of his reason and is resigned. One possibility is that the realization of being totally tied to his own time leads him to content himself with living the spirit, the life form of his age to the end, and with giving up on genuine truth (since everything is a construction based on the forms of expression and apperception fate has made available to us). Another is that he considers his work to be provisional and, at best, an accumulation of material for the intuitive [*intuitiven*] philosopher of history. The only way for him to escape the system of conformity of his zeitgeist, the soul of his culture, is intuition [*Intuition*], which forms the bedrock of life and the eruptive source of all zeitgeists and souls of cultures.

In addition to the broadening horizon of factual knowledge about history and ethnography, another important reason for the existence of this so-called historical problem situation and its immortalization by *Lebensphilosophie* (especially in the work of Oswald Spengler) was the emergence of cultural sociology and its insight into the social and economic contingency of the life of the spirit. Since Karl Marx and Friedrich Engels declared the culture of an age to be a superstructure, [11] an epiphenomenon of the material conditions of its people, economic historians, economists, and sociologists have been hunting for the superstructure or conformity laws in the economy, social conditions, and spiritual expressive life (that is, life for its own sake), which they are convinced must determine the structure of an age, of a cross section of history. This in turn led to an engagement with the questions posed by the sociology of art, religion, law, and knowledge, which strengthened awareness of the ways in which spirit—and, in particular, knowledge—is tied to its particular time period.

Spengler put forward his intuitionist *Lebensphilosophie* as a radical prescription for getting out of this predicament: he advocated respect for the closed nature of all world systems (souls of cultures) and their metaphysical equality (no age can sit in judgment of another) as eruptions of the creative ground of life and of the soul, abandoning the idea of a human evolution that passes through all cultures or of one system of values binding for all cultures. Given the predicament in which the philosopher of history finds himself, shying away from taking this last step beyond the notion of "progress" exposes him to the same objection that Bergson raised against Spencer. If he in principle affirms that all categories, including those of knowledge and cognition, are tied to their historical and cultural moment, he of course cannot use these categories to understand other time periods and their transformation into each other. This would be just as foolish as attempting to mechanically understand the emergence of the categories of the intellect and mechanical nature as it conforms to them with the help of these very categories and on the basis of this very nature—which is precisely what Bergson accused Spencer of doing.

Spengler's conception of history and Bergson's conception of nature avoid this mistake; indeed, the avoidance of a circular argument can almost be described as the principle of construction of their philosophies. Just as for Bergson life brings forth worlds that are the systems of conformity between an organism and its world, so for Spengler there is a maternal-creative power from which the soul forms emerge. Every soul form becomes visible in the conformity system of a culture, in the *co-naissance* of human *and* worldview. This means that the way one understands oneself and the way one understands nature vary according to soul form. The notion of time as extending infinitely into the past [12] and into the future, in which, for instance, our ideas of nature and history are rooted, for Spengler not only has no greater claim on reality [*Wirklichkeitswert*] than the notions of time manifest in Greek, Indian, Chinese, or Egyptian culture, but reflects no other reality whatsoever than that of our own Faustian soul. And, just as the fate of flowering, maturity, and decay hangs over every soul of a culture, so it does over ours as well.

Bergson's chief work carries an optimistic-sounding title, Spengler's a pessimistic one. Only biological questions exist for Bergson, while for Spengler even the biological questions appear in the medium of culture. But while their philosophies deal with entirely different areas and objects and emphasize different things, it nevertheless holds that they operate under the same principle. This principle has been called "organicist," although this seems to give too much weight to the notions of flowering and decay that Spengler uses to describe the fateful metamorphosis of every soul of a culture. More to the point is the irrationalism of the grounding, formulation, and function of the concept of life and the oddly vague notion of the creative essence of life conveyed by both philosophies. This in turn is intimately connected to the fact that both oppose naïve and scientific experience—Bergson is against natural science, Spengler against the humanities and, in a broader sense, against experience as such. Bergson reduces experience to the intellect as a manifestation of life; Spengler to the Faustian soul of the West as a manifestation of the ground of the soul.

Intuitionist *Lebensphilosophie* in its entire construction is incapable of considering the human as the subject of spiritual-historical reality, as an ethical person with a sense of responsibility, from the perspective that is given by his bodily nature and phylogeny. It fails to combine the aspect that is essential to the human as spirit with the aspect provided by his physical existence in such a way as to, while avoiding the empiricist and aprioristic errors, maintain the unity and homogeneity of *one* mode of experience. Life intuitionism does take care not to undermine the natural, pre-problematic view according to which the human emerged from a prehuman phylogeny and his spirit developed [13] from out of a vast biological past. But it only affirms this view (in a highly limited sense) by way of the nonrational source of knowledge that is intuition [*Intuition*]—and then renders even this concession illusory by devaluing the intellect as a source of knowledge. Without the intellect, however, there is no true experience, which is only true experience to the extent that it grasps objects in their own existence and essence by means of intuition *and* thought.

A philosophy that inhibits thought, thereby robbing scientific knowledge of its worth, can in no way make the claim of maintaining, enabling, or justifying a unified and homogeneous experience of nature and history. What is an experience worth if its access to the truth is barred? Materialism, naturalism, and empiricism, spiritualism, idealism, and apriorism—they all necessarily fail because they are forced to come to a halt in some way before the double truth of the world's aspect of consciousness and aspect of physicality. And intuitionism, whose aim—all hostility toward reason and science aside—was to avoid this double truth as a principle destructive of the basic experiential position, fails in an even more radical sense by tearing apart truth and experience from its vantage point of intuition.

When evaluating the way intuitionist *Lebensphilosophie* has been received, we must not overlook the fact that, to its credit, it at least affirmed the unity of the foundational aspect and the homogeneity of the mode of experience in the face of a philosophy that, amid the applause of the age, advocated the indissoluble split between several foundational aspects and modes of experience. This philosophy's principles can be traced back to Kant. It predicted that every attempt to sacrifice the duality of experiential positions as reflected in the great disciplines of natural science and the humanities to a monistic ideal would meet with the same fate as have every other rationalist or irrationalist metaphysics that tried to overcome the last differences in the consciousness of the world with *one* principle.

To understand the significance of this objection, we must go back and look in more detail at Kant's critical philosophy. This philosophy is the point of departure for the epistemology [14] and theory of the objects of scientific experience from out of which—quite contrary to its original tenor—a new *Lebensphilosophie* began to develop; a *Lebensphilosophie*, that is, that is neither intuitionist nor hostile to experience and that, from the vantage point of history and the humanities, compels a total revolution of the concepts of existence in all its spheres. It thereby shows the way to comprehending the human as a spiritual-ethical and natural existence on the basis of *one* experiential position.

Lebensphilosophie *and the Theory of the Humanities*

Kant only admits the kind of exact knowledge that is found in mathematics and natural science to be true knowledge. In the preface to the second edition of his chief work, he deplores the condition in which philosophy found itself even in his time: the restlessness of constant new beginnings and the perennial battle against the achievements of the past he finds to be symptomatic of its unscientific nature. It still has no method and no clearly defined sphere; it still strives to find these in endless experiments, just as mathematics did before its revolution brought about by the Greeks and physics did before its revolution brought about by Galileo. What then, Kant asks, would be more fitting than to imitate this revolution in the field of philosophy? We must finally put an end to an anarchy of despotic systems, none of which can tolerate the other and all of which contradict each other to the point of annihilation. A state of law must be created for philosophy that affords every philosopher full possibilities of development by regulating the development of all according to a primary agreement.[1]

The secret of such a revolution lies, as mathematics and physics show, in a particular art of questioning. One should not ask aimlessly, but rather in such a way that the question creates the possibility of unambiguous answers. Rational man, writes Kant, regresses to the level of the child whose teacher dictates all the answers to him if he lets himself be taken in tow by the objects, passively parroting the given instead of posing problems himself. He should rather proceed like a judge, for whom hearing the facts of the case is simply the *sine qua non* of his office. Relying on [15] a rational codification of justice and injustice, he listens to the witnesses, gets an idea of what happened, and delivers his judgment by considering the individual case from the standpoint of the laws. Kant makes law into the model of reason and the trial into the model of the scientific method.

Leaving aside the question of whether this view of exactness in the natural sciences and in mathematics is correct, it is clear that Kant's critique of reason and its revolution in philosophy takes only the exact sciences as its standard and hence also as the starting point of its investigation. The humanities and the systematic and historical cultural sciences are excluded from the beginning. They cannot claim for themselves the validity of science once scientificity has been identified with mathematical verifiability. It goes against the very nature of these disciplines' objects, which are human beings and human works, against the form of their existence, which lies in the past and which cannot be brought forth again the way it was, to approach them in the same way as things of nature. Neither does it make sense to subject human beings, actions, monuments, or documents to mathematical treatment, nor is it possible to conduct experiments with them.

It is undeniable, however, that these objects can be studied in an exact way, albeit using an entirely different approach than the constructive-experimental one, as the emergence of the philologico-historical method shows, in particular since the neo-humanist movement of the early nineteenth century. This method also strives to narrow the question down to alternatives and to decide between them with the aid of documents or monuments as illustrative evidence. But the question itself is arrived at in a far more protracted and uncertain process than is represented by mathematics in relation to the natural sciences. The historico-philological approach does not consider constellations of phenomena, but rather spiritual and psychological dependencies that only exist for spiritual and psychological persons because they reverberate in them. To take a stark example: for someone who has no social needs, the social world, including as history, remains concealed. Writings and memorials that tell of this history would be invisible to someone thus blind to values.

The spiritual world (the term we will use to refer to the objective correlates of the writings and memorials [16]) differs from the physical world in terms of its experienceability, beginning with the prerequisites that must be fulfilled by the knower. Things of nature require sense organs in order to appear. Spiritual life requires resonance and can only be grasped in resonance phenomena. Sensible phenomena simply radiate into the perceiver, while spiritual phenomena only become relevant in the ray that returns from the personality of the knower. In principle, anyone can come to understand and verify a theory in physics, as this requires only a minimum of individual humanity. It is not, however, possible in the same sense to expect general agreement in the understanding of, for instance, a set of historical issues, because a majority of people (quite aside from all the incremental differences in their powers of perception and judgement) will have different responses to them.

Even so: it is but an empty claim that the human exists in infinite variations. There seems rather to be a conjunction of finitude and infinity, limitation and limitlessness, a circumscribed wealth of possible individualities in inexhaustible individuals—which is of immediate significance for the scientific knowability of the spiritual world. One individual constitutes too narrow a base from which to grasp other spiritual worlds. If the historian really wanted to rely only on the reverberations in his own breast, he would have to forgo huge spheres of extinct being from the beginning. The humanities scholar must therefore strictly distinguish between a resonance in his own living individuality and a "resonance" in the layers that form the foundation for understanding other spirits by making "understanding" possible in the first place. The cultural scientist develops skepticism toward himself, his age, and the sphere of what is taken to be self-evident and sharpens his ears to hear the deep differences in resonance. In the primary layer of the human, which is never exhausted by any historical, personal, racial, or ethnic configuration, the historian has at his disposal a wealth of possibilities of interpretation.

This shows that one does not have to be religious to be a scholar of religion, that a Christian can study non-Christian religions. One does not have to share any of Caesar's personality traits in order to write his biography and can be the most unworldly person [17] and still describe the cunning of high politics. It is enough to have what is called a "feel" for the topic, imagination, empathy, and the gift of describing and making comprehensible other ways of life in their otherness and across distances while suppressing one's own lived experiences.

Post-Kantian philosophy had to reckon with the fact that this is possible for human beings in what is clearly a more than amateurish and haphazard way, the fact that in the cultural sciences personal aspects can be overcome where there is the will to do so. Of course, there was no shortage of attempts to integrate the spiritual world into nature, or to at least turn the former into the annex of the latter, in order to clear the way for a single science. But the most significant results of these attempts—Auguste Comte's positivism and Marx's historical materialism—were digested by a new science, sociology, without posing a lasting threat to the autonomy of cultural and historical being. The widening of the concept of nature to include objects whose essence is comprehensibility, uniqueness, the ability to be evaluated, and historicity [*Vergangenheit*] must always remain superficial and leads to the establishment of lawfully determined periods—and thus to cultural predictions that seem to shackle human freedom, but that often enough are refuted by this very freedom. As sociology advances, we can even observe progress in our familiarization with the idea of culture as being subject to certain laws without at the same time needing to deny or even merely narrow the sphere of human freedom.

True fate is something other than determination by the laws of nature. The latter only becomes fate when it turns into a factor with which we have to struggle in the aspect of free choice and whose triumph over our will can be seen as a confirmation or repudiation of meaning: a state of affairs that presupposes the valuation of human life, its position in an overarching totality, values, and standards of value, which in any case assert the claim of objectivity. Herein lies the great benefit in recognizing the uniqueness of the historical world, whose meaningful aspect coincides with the aspect of freely chosen action: the fact that the past must be understood not as a second nature, but from the perspective of what is to come, in the awareness of facing a future—that is, *as existence*.

The question, then, posed by the historical [18] and systematic cultural sciences leads to problems that are not the domain of another empirical discipline, such as, for instance, descriptive psychology, but rather that of the philosophy of human existence. The changed situation created by an age enriched by new branches of research thus quite naturally and compellingly leads to the possibility of natural scientific knowledge acquiring as a counterpart the possibility of knowledge in the humanities. A newly accessed world of experiences demands in Kant's terms that we go beyond Kant and undertake a broadening of his theory of knowledge.

Hermann Cohen's old Kantian school did attempt to confront the starting positions Kant delineated in his system with this new task, of which Kant himself knew nothing, and to strengthen them for battle. According to this school, Kant's critiques of practical reason and of judgment constitute theories of cultural science, and ethics and aesthetics as critical disciplines should at the same time function as inventories of the principles of methodology in the humanities.

The neo-Kantians under the leadership of Windelband, appealing to Hermann Lotze's concept of value, also believed that they could hold on to critical philosophy by an internal reform of its logic. They called for a counterpart to the critique of pure (natural scientific) reason, which they saw as being a one- or half-sided logic, and developed the primacy of the practical as the principle of a universal doctrine of value for all of culture. Science was ranked equally with art, law, the state, religion, and other value systems, and the critical groundwork was laid for the cultural sciences—in particular by Heinrich Rickert and his students—by reforming logic. Their twofold development of a logic of the natural sciences and of the humanities dominated the Baden School's program of theoretical philosophy.

This unmistakably led neo-Kantian value theory to turn its attention to the world of goods and values as they immediately concern the whole human, not just scholars. Indirectly, the question of the possibility of the cultural sciences came to be influenced by the understanding of their object—that is, of culture. Thus the public uproar caused by Nietzsche's critique of values poured into the elaborate, multibranched network of channels [19] that is academic discussion. Unfortunately, the high level of this discourse meant that the gradient was too low for the uproar to have much of an effect on the stagnant waters.

Dilthey was the only academic in this period who understood that the call for a critique of historical reason proclaimed more than a mere expansion of the territory of logic. While the others busied themselves with applying their ingenuity to the construction of a counterpart to the critique of the natural sciences and believed themselves to be formalistically satisfying the demands of a matter that in fact affected the very roots of philosophy itself, this man struggled to overcome the sterile dualism between philosophy as mere theory of science [*Wissenschaftslehre*] and philosophy as free interpretation of life. He did so by starting from the structure of the object of the humanities as it is perceived and experienced.

"From the perspective of Kant's theory of knowledge, the alliance of philosophy and the humanities, the articulation of a critique of historical reason as a counterpart to the critique of pure reason, seems at first to be merely an expansion of the *territory* of the theory of knowledge itself. Thus the theory of knowledge of the natural sciences, as developed by Kant in view of Newton's classical mechanics, is joined by the theory of knowledge of the humanities in reference to the work of the historical school. . . . For Kant, the spiritual realities governing the meaning of human existence fundamentally elude science since there is no mathematics in them; he consigns them instead to the practical position of the person, to being realized through disposition and action. And this, then, is where the alliance of philosophy with the humanities enters the picture, in order to render *positive* that which Kant decided negatively—for the humanities take precisely those realities which give human existence its own content as their object."[2]

This positive solution becomes possible, according to Misch, inasmuch as the logical cast of the humanities is based on the expressive character of its objects: "on the fact that these spiritual objects, whether it be a religion or a work of art, can themselves [20] speak to us. They do not merely carry meaning in themselves in their form-holding being, but rather know of their own meaning and express it so that it becomes audible to those who penetrate the forms of being, forging through to the soul of life which shaped itself in them." Object and knowing subject are in this sphere of one being. "This is why a different kind of behavior underlies the knowing of objects subject to scientific analysis here than it does for objects of nature. Knowledge arises here in the understanding of lived experience, which comes about by an inner connection between one soul and another, one force of life and another.... The consequence for methodology is that hermeneutics comes to replace psychology in the foundation of the humanities; indeed, hermeneutics becomes the heart of general philosophical logic.... Initially it is the matter here of making room in logic ... for the kind of living concepts which in the humanities emanate from its objects trembling at the touch of the word by virtue of their particular expressive character—of making room in the hoary doctrine of the so-called logical 'elements': concept, judgment, and conclusion.... But it is the matter of even more, namely of creating logical foundations that are broad enough to stop the excruciating antagonism between natural science and the humanities (which has, according to Hegel's *Logic*, now opened up in logic itself because the sciences have assumed an existence independent of human life) from tearing apart the theory of science."[3]

This, however, can no longer be accomplished by means of a construct of formal logic. If "understanding as the method of penetrating reality [*Wirklichkeit*] used by the humanities—*cognitio rei*" stands opposed to "the theory of causal explanation—*cognitio circa rem*," philosophy cannot simply make do with natural science and its explanations when it comes to knowledge about the natural conditions and foundations of the human spiritual world. And if this opposition, as Misch says in his essay, is merely temporary and of its time, may we hope that it can be overcome as a result of the transformations taking place in the way concepts are formed in physics?

For Dilthey, of course, the state of the natural sciences of his time was the key issue. But he also recognized [21] that it was impossible for the theory of the humanities to be completely uninterested in nature. He does write somewhere that "this stage of life does not care about the back panels of the scenery." But even on this stage, things take place embedded in the context of nature. "For Dilthey, the stage of life which does not care about the back panels of the scenery, on which that takes places which appears, which in qualitative reality [*Wirklichkeit*] is alive, vibrant, painful, and uplifting, and thus is there for us in such a way that nothing is behind it—that stage is nevertheless constructed on something that reaches from nature into life and points back to nature from life. . . . The human psychological whole is singular, like the whole of the earth, which determines it. . . . The secret of the world, put positively, is individuality."[4] Even the method of understanding remains, like the human, tied to the lived body.

Certain layers of nature with which a living being does not of its own accord come into contact can perhaps be viewed as "back panels" whose configuration is indifferent to what appears on their other side. And yet this indifference only goes as far as that indifference that prevails per se between non-appearing and appearing layers of nature. On the other hand, nature in many of its layers provides the object or the background or the medium or the principle in front of or with which the human leads his spiritual existence. Thus nature and the spiritual world are doubly interlocked, in that one carries the other, determines it, and at the same time receives its qualification and interpretation from the other. The true site of interlocking is the *human*, "who *is* not, but rather lives, and whose true life is lived only historically."[5]

Dilthey tries to combine philosophy with the empirical by dropping the restriction of knowledge to the *ontic* and by opposing the *historicity* of spiritual life to the ontic. From the sterile antagonism of a mere theory of knowledge and a free interpretation of life, he arrives at the level of life, where it is possible—indeed, necessary—to grasp spiritual-historical reality *and* nature in one and the same mode of experience.

Certainly, Dilthey's method of understanding is the method of an empirical science. But, as Misch writes, by "bringing [22] the objects, which have their own self, to express this knowledge they have of themselves, the knowledge of their own life, this—as Fichte puts it—hovering and oscillating of objectifying spirit between the object and itself, which releases historical reality from its phenomenological gestalt of being,"[6] this method lays claim to an experiential position on this side of the opposition between empiricism and apriorism. Both the subject of cognition and its object belong to the same life of the one human sphere, whose objectivations in deeds and works are not brought to this sphere as it were from outside, remaining like alien matter foreign to it, but rather themselves emerged from this sphere, as it belongs to the essence of life to transcend itself and at the same time to draw the results of this self-transcendence back into itself and to dissolve them. What sharply distinguishes this conception from any kind of intuitive [*intuitiv*]-ontological metaphysics of life and identity speculation and constitutes its specific novelty is the *experiential* nature of the concept of life employed here. Life for Dilthey is not an omnipotence beheld when turning away from experience, as it is for Bergson and Spengler, but rather something that can be experienced with intuition and intellect and imagination and empathy and that itself makes the experience of itself possible, forces this experience. All of our powers are called on to investigate the past in its essence, and thus life in its essence, for "life understands life."

Life does not consist in its knowledge of itself, but it becomes completed in this knowledge. This subject-objectivity, however, does not become realized in a speculatively devised system—not even if it is a system, like Hegel's, that understands history as the condition of the possibility of the realization of this subject-objectivity. It is realized only insofar as it *has* itself historically or *experiences* itself. Intellectual, cultural, and political history become the medium of self-knowledge; thus, rather than a thought-up system, experience executes the constantly changing self-conception of the human and his interpretation of life. The task of philosophy consists in comprehending this process of understanding itself and thereby rendering objective life's consciousness of itself.

This task gives rise to problems not yet encountered in the history of philosophy, which since Parmenides has had no higher concept than that of being [23]. It must recast its entire set of instruments accordingly. A hermeneutic—that is, a science of expression, of the understanding of expression, and the contingencies of understanding, which is definitely not, as some authors seem to assume, limited to the domain of language—becomes the heart (following Misch's expression) of general philosophical logic. But this hermeneutic is far from merely being, like every logic, a science of forms, augmented, as it were, by a greater wealth of spiritual forms and their aspects than could be found in the narrower boundaries of traditional logic. It rather continues in that tradition that led from ontology by way of Kant's transcendental logic to Hegel's logic and to the modern study of categories—admittedly by challenging this tradition's ultimate principles. The entire purpose of this hermeneutic makes it in the first place a discipline that is material rather than formal. That which Kant treated from the restricted perspective of the possibility of exact sciences or of the mathematizability of experience—the measurability, calculability, and lawful determinability of real [*wirklicher*] objects independent of and given to consciousness, and thus the possibility of formulating sensible, substantive materials—under the expanded guise of a hermeneutic (as a science investigating the conditions of any kind of interpretation) becomes the problem of the predictability and accuracy of linguistic and, beyond that, of any expressive objectification.

A philosophical hermeneutics as the systematic answer to the question of the possibility of life understanding itself in the medium of its experience through history can only be attempted—let alone realized—on the basis of a study of the structural laws of expression. And this, in turn, is only possible if one remains on this side of any specialist treatment of expressive life and studies it in its original state—that is, the way it lives and not the way it presents itself to scientific observation. If philosophical hermeneutics wants to comprehend the possibility of the experience of life, it cannot, of course, work with experiences and concepts of experience. Phenomenological description must step in here and lead the way to and stay with original intuition (although it must abstain from any ontologization of what it intuits).

[24] From this perspective of a universal science of expression, it in turn appears necessary to seek out and to pursue the questions of a philosophical anthropology, a theory of the human and of the structural laws of his lived existence.[7] These questions concern the essential structure of personality and personality as such, its expressiveness and the limits of its expression, the role played by the lived body in determining the type and range of expression; they concern the essential forms of the coexistence of persons within social ties and the coexistence of person and "world"—that is, the significant question of the human life horizon and its variability, the question of possible worldviews. Thus the idea of establishing a scientific foun-

dation for experience in the humanities compels us to engage with questions that extend into the sensible-material, physical sphere of "life"; compels, in other words, a philosophy of *nature* understood in its broadest and most original sense.

Dilthey knew this as well, and, as Misch says, sought "the Goethian way of science, which was to genetically construct the human from the materials of the entire edifice of nature; only the times did not allow him to take this route." We ought not, then, assess the interest of Dilthey and his students in Goethe's philosophy of nature from a purely historical perspective. This interest, in fact, derived from the clear insight that Goethe and his contemporaries' view of nature sensed much and discerned some of what our era, after a period marked by the uncontested domination of methods of exact calculation in the study of nature, has to develop again from the ground up. This knowledge must not be lost—it contains those elements with which an anthropology as the discipline fundamental for the theory of the human experience of life must begin.

A science of the human person as the decisive bearer of history, as the medium of living engagement in the entire realm of culture, cannot *directly* profit from anatomy, evolutionary history, physiology, psychology, or psychopathology. This insight is beginning to assert itself with force in sociology, ethnology, medicine, and in all of the humanities. The times in which one hoped to contribute to the foundations of ethnology by measuring the reaction times of Fiji islanders [25] are over. One experience might under certain circumstances be of use to another, but an experience remains an experience. It cannot be explained by another experience in the same sphere of being. In a word: if one wants to comprehend the human as he lives and understands himself, as a sensuous-ethical being in *one* experiential position, appropriate to human existence and encompassing "nature" and "spirit," then one must also create the means to do so. These means, however, cannot be derived from the traditional conceptual store of the individual disciplines, as each of these disciplines, whether they belong to the natural sciences or to the humanities, performs a certain reduction of the things it studies, without which it would immediately transcend the boundaries of its particular field. This reduction, of course, is reflected in its concepts.

Precisely in the interest of rendering fruitful the empirical knowledge of the individual disciplines, the theory of the human experience of life must guard very carefully against using such empirical knowledge for its own purposes. Doing so does lead to very interesting books that respond to the need for "synthesis," but it destroys the inner structure of the matter at hand. We can only succeed in surmounting the opposition between how the world is viewed in the humanities and in the natural sciences in a way that is not a mere postulate of a striving for unity, a monistic ideal of order, but rather a fact of the human experience of life (only thus far incomprehensible to us), if we abandon the level on which this opposition subsists. Philosophy has a great systematic task to fulfill here. By formulating the problem of anthropology, it also opens up the question of the human mode of existence and his position in the whole of nature.

The theory of the humanities must be clear about this. Anyone who believes that a philosophy of language or culture is all that is required is seriously mistaken and underestimates the significance of the situation that in Dilthey's work became aware of itself. If we understand this significance to include the summons to liberate ourselves from the dominion of the categories that since Greek antiquity have ruled the interpretation of our thinking, our actions, and our hopes, then there can be no doubt that the work of philosophy must probe anew into the ultimate elements, take hold of them, and transform them.

[26] Working Plan for the Foundation of a Philosophy of the Human

Without a philosophy of the human there is no theory of the human experience of life in the humanities. Without a philosophy of nature there is no philosophy of the human. The aesthesiology of spirit in my book *Die Einheit der Sinne* is guided by this principle, which also guides the current study. In the former book I made an effort to show why the humanities and their philosophical interpretation require a philosophy of nature in distinction to (but not in enmity with) natural science. This is true even if one wants for the time being to hold on to the principle that natural science must at most be grounded in a logic and a methodology but not in a philosophy.

The theory of the humanities needs a philosophy of nature—that is, a consideration, not empirically restricted, of the physical world from which the spiritual-human world is, after all, built up, on which it depends, with which it works, and that it in turn affects. Exact natural science does not provide us with this kind of a consideration of the physical world and its specific ways of appearing. Every attempt to *apply* concepts, theories, and results—from whatever discipline of the natural sciences—to the humanities and render them *directly* fruitful for the latter (as was particularly fashionable in the period of Darwinist-evolutionist positivism but is still on occasion attempted today), must fail in that it puts things of different ways of being, of different modes of intuition—of different *levels* of intuition, as it were—things associated with entirely different experiential positions, into the same pot.

Consider this example from the study of facial expressions: naïvely speaking, it seems to be the case that, from the outset, I perceive the bodily movements of another person as interpretable and meaningful, whether I in fact understand them or not. I am not facing a mere body [*Körper*] whose specific movements I can register, but rather an animated lived body [*Leib*]. This means that, far from being presented in this situation with the task of deducing certain psychological causes from the changes taking place in a body, it is rather the case that the movements of the lived body themselves manifest the situation that is meaningful in and of itself and whose interpretation [27] is tied to certain categories. The bearer of the lived body is apprehended here as neither body nor soul, as indifferent toward this conceptual distinction. Science, however, has led us to disregard the naïve situation in which communication and comprehension in fact take place and to construct instead, with the help of the empirical concepts of the individual disciplines, a new situation. This alone explains the strange efforts to justify our certainty of the existence of psychological motives by pointing to the existence of bodily movements. Attempting to overcome this difficulty, allusions were made to analogical inference, empathy, shared participation in the movements seen performed by the other, and, finally, a gift for perceiving the psychological. Each of these theories is an example of overcoming a self-made obstacle resulting from the distortion of originally intuitive (in this case human and person-bound) things caused by employing the empirical concepts of the individual disciplines.

The human is conscious of nature with the same immediacy and at the same level of life at which he existentially relates to himself, to his fellow humans, and to his era, and in which he expresses himself and is conscious of himself. Nature is thus not merely lived experience, but indeed full reality [*Wirklichkeit*] that *becomes* lived experience for him and that carries him as the foundation and framework of his existence from birth to death. From this sphere of existence all representations and ideas of consciousness, if they are alive, draw their inner life and flow back into it. And it is primarily only the language the human speaks naïvely that is suited to express the content and form of this sphere. Scientific treatments, by contrast, remove him from it and only lead him back to the things of existential reality [*Wirklichkeit*] by the circuitous route of connections hidden from the eye, the ear, the hand. If, then, there is to be a science that comprehends the human's experience of himself in the way he lives and in the way he historically recounts his life to be remembered by himself and posterity, then such a science cannot and ought not limit itself to the human as a person, as the subject of spiritual creation, moral responsibility, and religious devotion, but must rather include the entire *radius of existence and of nature, which lies on the same level as personal life and is essentially correlated to it*. If it does not do so, if it is satisfied with being a philosophy of history or of culture and leaves nature, the sphere of physical being, [28] to natural science, then it violates its own idea in the most inconsistent way and makes the same old mistake of purporting to survey from *one* experiential position things associated with different experiential positions.

A theory of the humanities that seeks to make comprehensible the reality [*Wirklichkeit*] of human life as it is reflected in the human is only possible as philosophical anthropology. Only a theory of the essential forms of the human in his existence provides the substrate and the means for a universal hermeneutics. Philosophical anthropology and its key component, the theory of the essential laws governing the (psychophysically neutral) person is, in turn, only feasible on the basis of a science of the essential forms of living existence and must therefore create its own conceptual framework for the entire sphere, the entire radius, in which the human as a (psychophysically neutral) person is situated. This framework cannot be adopted from any of the empirical sciences, since it is a matter of exploring full lived reality and not objectified, isolated, deep reality, shot through with auxiliary concepts. It is a matter, then, of doing what has, apart from occasional attempts, not yet been done.

This program could not have been carried out in Dilthey's time. The sciences were then under the absolute domination of empiricist thinking.

Dilthey himself (in a curious premonition of Husserl's discovery of the phenomenological approach to research) tried to find a way by means of a close-to-life, descriptive psychology and immediately encountered the self-assured opposition of people incapable of understanding the problem he was trying to solve. It was only with Husserl's conception of a pre-experiential structural analysis, which is per se universally applicable to objects of "intending" [*Meinen*] in general, that the tool for carrying out Dilthey's program was found. This remains true despite the fact that the rationalist interpretation Husserl gave to his own discovery was not consistent with Dilthey's work and that at its heart the tendency of Dilthey's thought could not meet with Husserl's favor.

Today the situation is entirely different. Thanks to Husserl, the attention of philosophy (and also of the individual disciplines, which can benefit greatly from him in relation to their basic concepts [29]) has been brought again to those layers of "being" (in nature, the psyche, and spirit) without which even science cannot form a single concept. And yet, to a greater or lesser degree, it strikes these layers from its terminology and in whole subject areas even discards them completely. It is a matter here of the layers of immediacy, of givens or "phenomena" reserved only for lived experience, intuition, or eidetic insight. Insofar as science, however, is concerned with processing things conceptually, with reducing the manifold to easily assessable elements, with working out uniformity, these layers are left aside. Let us take physical optics as an example. The further it progresses and the more theoretical it becomes—that is, the more it introduces mathematical concepts into empirical observation—the fewer assertions it will make about things in which the eye and the sensation of light still play a role. Other forms of examination come to replace the optical organ. Like physics in general, the more optics comes to resemble its ideal of a strict science, the more it becomes sensorially impoverished in order to finally eliminate altogether the senses and the layers of color and form available only to them, which it will then have—understood.

It was precisely the fact that advances in the empirical sciences of being, in all the natural sciences, in broad areas of psychology, sociology, and economics, even in certain fields in the cultural sciences and history, can only be purchased by moving away from perception, from immediately lived experience, that led in the beginnings of science to a weakening of the awareness of these eliminated phenomena. An overestimation of the conceptual and comprehensible, an underestimation of that which can only be grasped by way of sensation, feeling, and intuition [*Intuition*] was the result. Only once science grew more mature was there a recognition of the particularity and limitations of its mode of comprehension. When the physicist explains "what" the color red "is," when the physiologist measures the intensity of a sound sensation, when the psychologist classifies the perseverative tendency of certain ideas—then in all of these cases what is actually happening is the identification of the quantitatively measurable conditions of phenomena that are qualitatively only available to lived experience. Of course, it then seems reasonable to strip these phenomena of their quality of realness [*Wirklichkeitswertes*] and—precisely because [30] their inner qualitative whatness eludes empirical conceptualization—to make them into "mere" sensations of the subject, to merely subjective appearances conditioned by the organization of the human. The essence of redness, however, is in no way grasped by identifying its wavelength or the processes taking place in the optic nerve, retina, and occipital lobe, or by observing (as centuries of modern science were content to do) that it only be experienced through sensation. It holds in general that that which is phenomenal about phenomena (in whatever zone, whether in nature or in society, in history or in contemporary life, in the psyche or in the mind) cannot be made accessible by empirical conceptualization or by any individual discipline. These disciplines do work with these phenomenal elements—the chemist may need a coloration, the physicist a sound, the psychologist a certain mood as indicators—but the elements themselves are never grasped by the empirical concept.

As long as the sterile dualism of empirical science on the one hand and epistemology on the other persisted, psychology was attributed with the competency of exploring the domain of the phenomenal. Psychology, however, was quick to recognize that it was not up to the task. It has to accept pure phenomena, under the heading of sensations or "gignomen" or elements, as the ultimate data of lived experience; as an empirical science it neither has the right nor the means to work out that which is specific about the phenomenon, its lawfulness in relation to other phenomena. Today there are efforts being made to productively overcome the dualism between science and epistemology. We have the means: phenomenology—as a possibility. It is now a matter of using this means to a necessary end.

The end is creating philosophy anew from the aspect of grounding life experience in cultural science and world history. The steps along this path are laying the foundation of the humanities through hermeneutics, constituting hermeneutics as philosophical anthropology, and doing anthropology on the basis of a philosophy of living being and its natural horizons. An essential tool (not the only one) for making progress on this path is phenomenological description. Let me be clear here: goal and aspect are not one and the same. Philosophy in its concept of the world was the goal of Kant's work, the critique of reason was the path to this goal, and the [31] starting point of experience in the natural sciences was the aspect by which he actually trod this path. I too distinguish between goal, path, and aspect in order to shield my entire endeavor from premature appraisal.

In the age of telegrams, however, abbreviated forms are popular. One browses in philosophical books and looks for so-called findings, as if philosophy were like an individual discipline that generates findings that can be separated from the way it asks its questions. These books are read like teenage girls read novels, eager to find out whether they'll end up together. Hardly anyone bothers going to the trouble anymore of thinking about the framework in which a work is enclosed. This sloppy kind of reading is of course promoted by an approach to philosophizing that is no longer systematically schooled or, conversely, by the premature systematism of little world builders. Patience, empathy, and respect for the intentions of the other seem to be virtues of bygone times.

Constituting hermeneutics as philosophical anthropology and doing anthropology on the basis of a philosophy of living being and the layers of nature essentially correlated to it—these are the next steps, the crucial tasks I referred to earlier as facing the philosopher of today. It is only by way of this new aspect, expressed by the attitude of the human to the world that has come to maturity since the final breakthrough of scientific knowledge and that stands opposed to all subjectivist-idealist objections, that the return to the object, the rediscovery of the great problem of ontology, acquires its seminal meaning and at the same time its place in the program being developed here. But how to begin? It is the aspect, of course, that is decisive.

At its heart stands the human. Not as the object of a science, not as the subject of a consciousness, but as object and subject of his life—that is, in the way in which he is object and center to himself. For it is in this peculiarity: to exist—that he enters into history, which is merely the completed way in which he contemplates himself and knows himself. Not as a body (if by body we mean the layer objectified by the natural sciences), not as a psyche and stream of consciousness (if by these we mean the object of study of psychology), not as the abstract subject to whom the laws of logic and the norms of ethics and [32] aesthetics apply, but as a psychophysically indifferent or neutral living entity does the human exist "in and for himself."

We must begin, then, by treating the human as a personal living unit and the layers of being, of being in general, as essentially coexistent with him. Is the concrete situation in which the human is positioned (not this or that one, not this race or that people, but the human as such) accidental or essentially necessary? Do structural laws link the human to his life horizon, to his environment, which for him is the world? How far does this existential coexistence reach, and where does chance begin?

This question can be approached from two directions: horizontally—that is, in terms of the relationship the human seeks to the world in his doings and sufferings, and vertically—that is, in terms of his naturally developed position in the world as an organism among organisms. By approaching the question from these two directions we can hope to actually comprehend the human as the subject-object of culture and as the subject-object of nature without breaking him up into artificial abstractions. It allows us to preserve the one foundational aspect of the experience of life that the human inhabits in relation to himself and to the world: both bound to nature and free, evolved and made, original and artificial.

The level onto which the human must hoist himself again and again, with all manner of effort and sacrifice, the level of spiritual activity, creative work, the level of his triumphs and defeats—this level intersects with the level of his existence as a lived body. Thus the existential conflict without which the human would not be human also has significance for the philosophical method: it exhibits in the Janus-faced nature of this living being the necessity of grasping (and not sublating or mediating) the dual aspect of existence from *one* basic position.

How is the question posed from the first direction, the one referred to earlier as horizontal? Here we are concerned with the human as the bearer of culture. The objectifications of culture—science, art, language, etc.—thus become the medium in which the human is observed. Nota bene: the observation is of the whole human as concrete living unit. Culture is studied here as the specific expression of this living unit. The type and form of its objectifications [33] are expected to provide insight into the structure of the human system of life in all of its layers.

This can only be accomplished if the question is posed *as broadly as at all possible*. A value analysis of cultural accomplishments will not suffice here; the investigation must also include a consideration of the conditions under which these values are realized. The subject of the analysis will be the sensualization of the spirit and the spiritualization of the senses. It is only in this way that the extreme poles of human existence, that of the body and the senses and that of the spirit, can be seen from *one* perspective by way of an exploration of the system of forms in which this existence expresses itself and their mutual dependence, the essential laws of their coexistence, can be grasped.

Here we have the problem whose solution is sought in the *aesthesiology of spirit* undertaken in my book *Die Einheit der Sinne*. The "horizontal" question is only posed as broadly as at all possible if we think in terms of interrelationships, from the highest layers of spiritual meaning all the way to the lowest layers of sensible matter—that is, if we think in terms of the inner conditioning system that exists between the symbolic forms and the physical organization. This inner conditioning system, however, cannot be discovered with the help of any empirical science, as I have no doubt sufficiently stressed previously. Just as Kant needed his critical method to shed light on the ideational conformity between sensory intuition and conceptual thought upon which the precise use of concepts in mathematical natural science rests, so the system of conformity of sensuousness and spirituality on which the understanding use of concepts rests and that we are considering in terms of the much broader aspect of the experience of life can of course only be discerned with the help of a critical method. That is what is meant here by the "aesthesiology of spirit" as a critique of the senses.

To see this as a rejection of empirical sensory physiology and psychology is therefore laughably ignorant. Was the development of cognitive psychology and the psychology of thought processes stymied by the existence of a critique of thought and of knowledge? Certainly, we have only begun to distinguish between these two forms of inquiry since Leibniz and Kant, and a drawn-out battle against psychologism had to be waged before the distinction between a psychology of understanding and a critique of understanding came to be generally accepted [34]. Today, however, it takes a remarkable degree of philosophical illiteracy to treat the difference between these two approaches as depending on the only area in which it was formerly recognized: that of thinking and knowing. Just as this area is open to empirical and critical investigation, so too the senses and their specific correlates can be subjected to empirical and critical consideration. To assume from the beginning that sensory differences are only of an empirical nature was the decisive error made by Kant and the Kantians. Our age has every reason to submit this assumption to a thorough examination and to remember Goethe and Hegel, who were of a very different opinion on the matter.

In the preface to my book *Die Einheit der Sinne* I already emphasized the fact that a critique of the senses can ground the total relativity of sensation qualities—that is, the ultimate building blocks upon which phenomenal nature is constructed—on the unity of the human person or the apriority of the natural environment in view of its material modes. This claim, which I sought to defend in every chapter of that book, has frequently been misunderstood. It was seen to be either paradoxical or self-evident. Now this claim would be paradoxical if it aimed at affirming that that which the human experiences with his senses is his a priori possession before, and independent of, all experience. It would be banal if it asserted nothing other than that because the human is the way he is, spirit can only express itself through his particular lived body and in the sensible matter associated with it, thus also imprinting the stamp of its essence upon this matter. What is meant, rather, is the inner conformity of our sensory organization (and the sensory elements corresponding to it) with the possible forms and types of spiritual meaning. Many philosophers have claimed that such inner conformity does not exist, which implies that the senses and the lived body are indifferent toward the spirit. The physical sphere would then actually only be empirical, arbitrary, without spiritual necessity. There would be no "matter" or modalities specific to particular senses—in other words, no sensory qualities.

Aesthesiology and its critique of the senses, however, show the opposite to be true, providing evidence for the spiritual necessity, the apriority, of the senses. The claim here is that the [35] structure of the intuitive world, the strange fact of its sensory diversity, contains a law of the senses. Of course, the apparatus of our lived body with which we normally apprehend this differentiation evolved over time and is ephemeral. There is no such thing as a "critique" of the retina or of the organ of Corti (or, to be more cautious: aesthesiology does not aim for such a critique). Here empirical necessity follows the laws of nature. The specific achievements of the senses, however, also exhibit in their typology a meaningful and intelligible necessity over and beyond this, which should not be lumped together with their ends and the causes of their functioning.

There were some who saw in the claim of the sensory qualities' total relativity to the unity of the human person an affirmation of their subjectivity—and discovered there a contradiction to aesthesiology's conclusion that the sense qualities possess objective value. Subjectivity and subject-relativity are two different things, however. Everything that is objective only has the potential of being subject-relative. From this we must distinguish, in turn, personality and person-relativity. That which is essentially necessary to the unity of the person must by all means be treated separately from that which is essentially necessary to the subject. Neglecting these differences gives rise to erroneous opinions about the realness [*Wirklichkeitswert*] of sense qualities, indeed of all phenomenal layers that are only available to lived experience, that only exist for lived experience.

An example of this can be found in the phenomenal reality [*Wirklichkeit*] of the lived body, which is not *objective*—in distinction, of course, to the likewise phenomenally real sense qualities that are. The lived body is only person- (or life-)relative; the sense qualities are person-relative (life-relative) *and* objective.[8] The lived body [*Leib*]—we find this insight already in Hegel—is not the same as the physical body [*Körper*], although they are objectively identical. It is true that when I lift my arm or a child learns to walk, the corresponding muscles are innervated—but this only describes the physical process and not that of the lived body. The process taking place in the lived body [36] is of a different nature. Of course, sensations from the organs and the joints play a crucial role in this process, as do skin sensations, tension, different kinds of tactility. But the lived body is nevertheless more than mere sensation or the awareness of one's own physical body consisting of bones, tendons, muscles, vessels, nerves, etc. It is a living reality. This can be seen precisely in the way that we control it. Walking, lifting, sitting down, standing up, and lying down are all living modes of behavior (which, of course, are mediated by physical functions and may thus also be inhibited by these functions) that in the living position of the individual determine a particular aspect it is essential for him to always consider—an aspect that is thus essentially correlated to the person or to the organism as a living being.

The aesthesiology of spirit in *Die Einheit der Sinne* traces the relations between spirit and nature—that is, it approaches the human as a personal living unit in all the layers of his existence from the direction we referred to earlier as "horizontal." This, however, does not exhaust the possibilities of a philosophical anthropology based on the science of life. The question must also be approached from a "vertical" direction, as dictated by the naturally evolved existence of the human in the world as an organism among organisms. The great problem posed by the psychophysically indifferent unit of the human person as a *living being* must now be addressed. The path of aesthesiology does not lead to this goal; new paths, new methods need to be found. The concrete situation, the life horizon, into which the human finds himself placed cannot be exhaustively plumbed by an aesthesiology of spirit. The phenomenal layers of his environment—that is, the areas of being only available to lived experience, intuition, sensation, and eidetic insight—are too rich and fully formed to fit into the framework whose extreme poles are spiritual meaning and sensory qualities. It is precisely the living modes of being that connect humans with animals and plants and that carry the particularly human mode of being that are indifferent to spiritual meaning. And yet they make up a distinctive phenomenal reality [*Wirklichkeit*] whose study is not within the purview of the empirical natural sciences.

Until one has submitted the human as a naturally evolved living existence to pre-empirical observation (that is, one that is not [37] tied to a specific discipline), one cannot hope to find complete answers to the questions posed earlier: what are the layers of existence with which the human essentially coexists? How does he, as a living unit, have to experience himself and the world? The constitution of hermeneutics as anthropology must be based on a science of life, on a philosophy of life in the literal, concrete meaning of the word. We must begin by clarifying what can be described as being alive before further steps are taken to develop a theory of the experience of life in its highest human layer.

I have now specified the current study's relationship to the aesthesiology of spirit in *Die Einheit der Sinne* as well as its goal of an analysis of the essence of living being. These two works pose different questions and use different methods in the service of the same objective of a philosophy of the human. If this objective in both works was presented in terms of, among other considerations, Dilthey's question, it would nevertheless be unfair to judge them by the sustainability of his attempts at reform. As I already noted in *Die Einheit der Sinne*, the science and philosophy of our time has for a while been struggling to justify and develop a theory of the human person. Dilthey's perspective does not dominate all of these efforts. The great credit due to William Stern, Scheler, and later the pioneers of so-called gestalt psychology (who started from a different set of problems) for their establishment of the idea of the person should be judged as independently of Dilthey as are the achievements of Jaspers and Kraus, crucial in their own way. What remains essential is the consistent tendency toward overcoming the fragmented study of the human in philosophy, biology, psychology, medicine, and sociology. This fragmentation, for which Descartes provided the cue, did not always dominate modern science, but invariably achieved dominance again and again. It is an approach that reified the human with its specialized disciplines and that, as a result of this segmentation into spheres of being, lost sight of the living unit, so that only a pale "subject" remained, a mere wire from which existence, reduced to a puppet, performs its dead movements.

TWO

The Cartesian Objection and the Nature of the Problem

Extension vs. Interiority and the Problem of Appearance

It is in the nature of undisciplined experience to take something to be fundamental as long as it allows this experience to proceed with the greatest possible security and provides it with the best intuitional background for understanding its context. These merits, however, are not adequate for the scientific development of experience. To credit something with being fundamental means more than this and demands more from us. Something can be very important for the evolvement of our insights, fundamentally important, as they say, without necessarily having the nature of a genuine foundation.

A genuine foundation supports without itself being supported. An empiricist dutifully pays attention only to the first function. Once he has become convinced of its existence, he tends to automatically assume that the second also holds. The history of empirical science is full of examples of this.

There is scarcely a significant discovery or theory that has not been fundamentalized in this sense—that is, made, at least to a certain degree, into the object principle, or epistemological principle, of things: Darwin's notion of natural selection, Marx's idea of superstructure, Einstein's principle of relativity, Planck's quantum theory, Freud's concept of repression and sublimation.

Whether something is viable for experience is clearly for the empiricist to decide. Whether, however, it itself requires its own supports can only be determined by way of a philosophical investigation concerned with the essence of foundations and principles. Such an investigation may reveal that something is fundamentally important without being a foundation, or, and this should not surprise us, the philosophical discovery [39] of a foundation may remain without influence on the content of scientific experience.

No one questions the extraordinary usefulness and intuitiveness of the distinction between the physical and the psychic. It is a distinction that captures essential differences in the being of reality [*Wirklichkeit*], as is shown by the progress of the sciences of the body and the sciences of the mind. But the idea that this distinction constitutes a foundation today gives rise to objections among, and is rejected by, not only philosophers, but also all those empiricists who are concerned with the baffling connection between the physical and the psychic in the features of the person and of his achievements.

Broadly speaking, the general opinion that it was Descartes who fundamentalized the distinction between the physical and the psychic (albeit in a somewhat different form) seems to be true. He declared the difference between *res extensa* and *res cogitans* to be fundamental and ascribed to it the character of an exclusive disjunction. While psyche and *res cogitans* are not equivalents, these terms both aim at the same sphere. For now I will refer to this—undifferentiated—sphere using the uniquely German word *Innerlichkeit* [interiority or inwardness], which avoids specifying from the outset what we mean by the psyche, consciousness, and the subject.

Further support for the notion that Descartes fundamentalized the distinction with which we are concerned here is given by the fact that the principle of this fundamentalization establishes measurement or a mathematical-mechanical account as the only method for acquiring knowledge about physical things. By identifying physicality and extension, which also posits an equivalence between extension and measurability, the *res cogitans*–*res extensa* dichotomy invariably leads to a fundamentalization of mathematical natural science.

To be sure, the original separation of all being into *res extensa* and *res cogitans* was meant ontologically. This separation, however, has its own ongoing methodological significance that, in a certain sense, allows it to elude ontological criticism. By equating physicality with extension, nature is made accessible only to measurement. Since the only opposing sphere to extension is *res cogitans*, everything in nature that belongs to the intense variety of the qualities must as such be considered cogitative. Thus there are only two possibilities: either we understand physical bodies' [40] qualitative modes of being and appearance in a mechanical way—that is, dissolve them into quantities—or we avoid this analysis and declare these modes to be the contents of cogitations, the contents and products of our interiority.

The fact that I, as an I with its own particular self-position, belong to this strange interiority, whose specificity Descartes referred to as *cogitans*, lays (still in an ontological sense) the foundation for the subjectification of the nonextended components of nature. Next to extension we have here not intensity but interiority, "thought" or consciousness. There is no other place for the nonextended in this system than the sphere of *res cogitans*. The positive definition of one single opposing sphere determines the entire structure of the whole worldview. There is no need *from this perspective* for the further claim, leading into the idealistic-subjectivist lines of reasoning of post-Cartesian philosophy, that interiority only belongs to me as an I.

This restriction of the existence of the *cogitans* to the scope of our own I was brought about by Descartes's famous epistemological idea that there can be no doubt about the existence of our own I precisely because it is a *res cogitans*. The existence of doubt—and doubt is a *cogitatio*—is immune to its own poison. It has to be, if there is to be such a thing as doubt in the first place.

As *cogitans*, however, the I can only be grasped in the self-position—that is, it can only grasp *itself*. We can only speak of a *res cogitans* because of this turn to the self. Everyone can assume that his own I belongs to the *res cogitans* because he comes to himself as an I, to the I as himself, by turning his gaze toward himself in a way that is exclusively available to him. Thus everyone discovers the existence of the *cogitans* in a mode of perception that is reserved only to himself, that is restricted to himself. Only in himself does he grasp it. Only as such is it unquestionable reality [*Wirklichkeit*]. Other I's, which are encountered in a mode of perception not exclusive to oneself, are thus not safe from doubt. The way in which the ontological conception of a *res cogitans* was arrived at turns it into a methodological conception. The proposition that I, as an I with its own particular self-position, belong to interiority has been turned around to mean that interiority belongs only to myself.

[41] So far, we have arrived at the following important insight concerning Descartes's move: the fact that consciousness, our own I, is able to bear the responsibility for the nonquantitative phenomena of the mechanism of nature is founded on the cognitive character, the interiority of the I, which stands in complete contradiction to extension. It becomes clear that the counter-instance of *res cogitans* serves the purpose of saving phenomena from dissolving into extended being and saving qualities from mechanism.

Two modes of experience that are not mutually convertible have been given the capacity to judge: the evidence acquired by the self via internal experience and that provided by the outside via external experience. To put it simply, this is the breakdown of knowledge of the world into the knowledge of physical bodies and knowledge of the I—in modern terms, into physics and psychology.

Appearance as such remains incomprehensible in this conception. Nineteenth-century thinkers, schooled in the natural sciences, were not able to find a different solution to the problem of the sensation qualities than did Descartes and his successors. Anything that was qualitative was deemed subjective, whether in the crude idea of the nervous apparatus being conditioned by specific sensory energies or in the more subtle conception of emotional and spiritual causes in the qualitative realm of phenomena. Modern physiology and epistemology to the present day are accompanied by the same aporia of the conjunction between physical body and self, between extended and thinking thing as the true essence of sensation quality; by the same riddle of how mechanical stimuli and bodily processes in the nerves and the brain impacting "on the I" bring about the qualitatively shaded image of an object.

Only where philosophy summoned the courage to contest the *exclusive* relevance of exact methods for gaining knowledge about nature has it been able to find answers to the unsolvable questions of the Cartesian dichotomy. While the revival of the philosophy of nature in German idealism never achieved any considerable effect, this was due less to the difficult ideas of the philosophers as it was to the shortcomings of the exact natural sciences of the time. The idea that nature is only to be treated as extension—with all the attendant Cartesian consequences—had to first be implemented with total conviction before scholars, particularly in the fields of [42] biology and psychology, started to become suspicious. This is where we are today. The slogan "Away from Descartes!" would draw a large crowd of supporters from the most diverse branches of organic natural science, medicine, and psychology, if philosophy could finally decide, as have done a few brave thinkers, to tackle anew this whole complex of problems.

Appearance as Originating in Interiority

Today such an undertaking no longer runs the risk of being accused of hostility toward the natural sciences, as it inevitably was when the inorganic and organic natural sciences were still fighting for their right to exist. The undisputed successes of the experimentation and measurement method have long since proven its indispensability in securing empirical knowledge. Our entire practice would break down if the advancement of the exact sciences were to be inhibited at any point. Wanting to enhance our knowledge of being by means other than exact ones is to make a presumptuous claim (at least in terms of that which exists in space and time) that must bear the burden of proof regarding the supposed inadequacy of the exact methods.

What an anti-Cartesian movement must oppose is the identification of corporeality with extension, of physical existence with measurability, which has made us blind to those properties of physical nature that cannot be measured. This has gone so far as to make us not only think of the natural sciences as the only possible way of knowing nature, but of nature as the result of natural science, as the product of its methods—a view espoused of late by neo-Kantians. The principle of the equivalence between extension and measurability—this clarified definition can be given to extension once we are no longer concerned with interpreting the historical Descartes but rather with his basic division of the world—necessarily led, however, to reducing physical relationships to those that can be represented in purely quantitative or numerical terms.

Fundamentalizing the exact method, which is clearly capable of breaking all resistance in the sphere of physical existence [43], following the Cartesian principle, has very specific and significant consequences:

If the essence of physicality lies in extension (for which quantity or measurability can then stand in), then the unmeasurable, qualitative properties of bodies must not belong to the essence of physicality. Who is responsible for them then? In nature as the realm of extension, nothing, of course. Since there only remains the sphere of interiority (for which the self can stand in), there is no other choice than to make it responsible for the unmeasurable, qualitative properties of bodies. Thus, for the sake of the total quantification of bodies, all qualities are *subjectivized* and reinterpreted as mere appearances—furthermore, as sensations. A direct path here leads by way of John Locke to Ernst Mach and is still traversed today by every natural scientist.

As *object* in appearance, the body is a system of qualities. Even that which is quantifiable about it—size, weight, solidity—taken objectively in the appearance is qualitative specificity. The same body in all its qualitative properties, however, can also be mechanically and numerically specified. Since it cannot be a quantitative and a qualitative system in the same sense, it must, for the sake of the identity of the ground of existence of its properties, be qualitative and quantitative in different senses.

A possibility for the body to exist in different senses and yet be the same arises from its *contact or noncontact* with another realm of being: nonextended interiority. *To be with this realm would then mean to appear as an object*; to be without it, to be nonobjective. Appearance is external to an existing entity; it is not necessary for an existing entity to appear. What the body truly, in its core, in its essence, is, is thus not affected by its contact with the *res cogitans*. And yet this contact is not without consequence. This consequence lies in the fact of "having been made into an object," in appearance. That in the body which remains mechanically and numerically incomprehensible, the qualitative nature of its phenomenal properties, is now deduced from the *situation* of appearing, from the body's togetherness with the *res cogitans*, with the aid of the *res cogitans*.

Why are bodies not simply there, and why do they not appear as they are: as pure relations of extension? Why is [44] there something else, which is determined by relations of quantity but is not entirely exhausted by them—that is, the quality of a color, of a shape? Because bodies are given for an interiority and are grasped by it; because their pure essences are broken in this medium that is alien and thoroughly dissimilar to them. How else to explain the motley play of colors, the vibrant abundance of sounds and splendors on the bodies of nature? Their appearance, which sheaths their core and is thus displaced forward from it, can only find its cause in their constant contact with that very "thing" whose whole being is absorbed by thinking, by executing acts, by being interior.

The Cartesian dichotomy explains a basic fact of all experience—that bodies in the totality of their appearances consist of qualities—in terms of the indissoluble contact of bodies with interiority. This interiority's characteristic being, as it were, ignites the being of extension so that it erupts into the intense variety of qualitative existence in a way that is equally incomprehensible to both the principles of *res extensa* as it is to those of *res cogitans*. The conjunction of the two substances, which makes possible the presence of the body for the self, becomes the final ground—not of this or that property, but of the qualitative nature of the properties itself.

The body represents itself qualitatively in its appearance. Its substantive core radiates through its thoroughly qualitative properties to the surface, which belongs to the body and is connected (in an incomprehensible way) to the *res cogitans* in its opposing position. This front-line positioning toward the self, enabled by the conjunction of the two substances, becomes the ground for appearance—not, let us be clear, just for the comprehensibility of appearance. The "front" of that which appears lies "in front of" the deep core, the true center of its being, where it is only that which it "truly" is, revealing and at the same time veiling the central compactness of its being. Equally so, and in the same strict sense, the *res cogitans* must *be displaced forward from* the *res extensa*. The *res extensa* is thus never present as such in its nakedness, but only in the "mantle" of appearance.

Thus (1) the identification of physicality with extension, (2) the either/or alternative of extension and interiority (thought, consciousness), and (3) the identification of the *res cogitans* with "I myself" show the essential connection between the subjectivization of the qualitative side of the physical and the forward displacement of the self.

[45]

The Prior Givenness of Interiority and the Forward Displacement of Myself: The Proposition of Immanence

A thing is an object only if it is an object to someone. Objectivity requires something to oppose, just as a front is only a front against something, against a sphere it faces. It is only possible for a body to be an object in a situation that allows for the presence of the body's existence. Presence—in both the temporal [*Gegenwärtigkeit*] and spatial [*Anwesenheit*] sense—is more than mere existence. Being present is a particular relationship of that which exists *to* something, for which it is present and has presence. Since the *res extensa* is only present as a system of qualities, its only opposing sphere, the *res cogitans*, must come to oppose its presence, must become the (pregiven) reference sphere of its givenness. Thus physical existence only appears to interiority. Interior being is displaced forward from the physical object, given beforehand; the self takes precedence over the phenomenal world of bodies.

Considered *without* the specific methodological consequences arising from the conception of the *res cogitans* as the I, objectivity is simply a relation between the two spheres of being. In this case, the members of the relation face each other as equals. Since they both belong to encompassing being, their mysterious contact with each other, upon which the objective world of appearances is said to rest, initially is no different than any other kind of relation-to-another. The members are still exchangeable in the opposing position; their relation is without direction.

Objectivity, however, is only possible as a particular kind of opposing position or contraposition. The relation between that being which objectively appears or is apprehended and that which it opposes and for which it is objective has an *irreversible* direction, which must be based in the polar opposition of the members of the relation. A thing only looks a certain way "from itself" and not "toward itself." Qualities adhere to the object as properties. The sphere against which they are to be objectified must stand in a polar opposition to their essence for the reason of their *mode of appearance alone*. Interiority, then, cannot simply confront the *res extensa* of the body as *res cogitans* if the body is to appear as a body, to be objective and present. As long as interiority and extension in their opposing position still maintain their common [46] character of being, the requirement of a polar opposition between them is not actually met.

The *res extensa* only appears when it stands opposed to the *cogitatio*. It is only the act of the subject that is absolutely polar to the extended thing. The question is: under what conditions does interiority lose its character of being, thereby fulfilling the condition of a genuinely opposing sphere to objective appearance? The answer is: as soon as interiority remains in its self-position as a center of act execution (*cogitationes*), asserting itself and living purely as an I in relation to being. Only insofar as the *res cogitans* is positioned over against the *res extensa* as a self, as an I, does the *res extensa* have the possibility of appearing. Given the consequences set forth earlier, it follows from this, in turn, that only *I myself* as an I am the decisive link in the chain of conditions for the objective experienceability of the world.

The proposition that the *res cogitans* must be displaced forward from the *res extensa* so that the latter can be given is specified even further (of course I am referring here and throughout to the spirit of the Cartesian dichotomy and not to the historical writings of Descartes). To be given comes to mean to be present to myself. And to say that a thing appears is to say that it is present by virtue of its relatedness to me as a subject. Ergo—and here follows a fateful step—in the self-position reserved only for me, I am the condition under which things can objectively appear. That which appears, then, is content of my self, content of consciousness, representation. What began as the ontic *forward displacement* as the prior givenness of interiority for the sake of rendering possible the phenomenal world has become my own forward displacement. I myself direct my gaze to the things of the world. But that which I, that which my consciousness is able to grasp in these things is always "already" related to myself, is appearance held against essence—that is, essence modified by myself.

The *res cogitans* as I does not directly enter into contact with physical being's pure essentiality but, according to the principle of polarity, only with its appearance, which obscures what it truly is. In this way, interiority becomes a force that transforms the physical world; it becomes the principle of this world's appearance, of its veiling, and thus of its falsification. It is true that interiority's executing act, its gaze, its simple turn toward something may imply direct contact with a body. And yet the contact of the gaze remains subjugated to the situational principle of the body's opposing position [47] to interiority: because it is *given beforehand*, interiority gazes at itself, veils the body's naked existence, renders the body in the *image* of its appearance. While its intention is directed to the body's characteristic being, the gaze of the *cogitatio* only achieves an indirect mediation between the body and the *res cogitans* by way of appearance: according to the law of situatedness, whereby the body only appears through presence.—

Without siding with this line of argument, I would like to note here that objectivist and realist theorists have unjustly attempted to portray as illogical the transition from the proposition of being forwardly displaced to the proposition of prior givenness. First, they say, it is claimed that the *res cogitans* confronts the physical thing in full polar opposition to the *res extensa*—that is, as *cogitatio*, act, pure execution of the gaze. Then, however, this claim is abandoned when the *res cogitans* is put in place of the *cogitatio*, when the transmitter of the gaze, interiority, is put in place of the gaze—which necessitates, of course, treating this something that is displaced forward from the thing-object, this other *res*, as an object, despite its subjective function. This ambiguity compels that which was originally intended to be the receiving and opposing sphere to itself receive and objectify, the I-object (me) to be given to the I-subject (I).

This, however, is not the case. Significantly, what is indeed a fateful step from being forwardly displaced to prior givenness is rather presaged by the subject-object structure of the I itself. The nature of the self-position entails a split into the I that is referred to and the I that refers. Interiority is in the self-position only inasmuch as it can grasp itself, and it grasps itself only inasmuch as it is I. This does not rupture the I's identity, but rather enables it in the first place. Identity as sameness *consists* in fact in moving away "from" something that is supposed to be identical (with "itself") as a return "to" it. In order to be identical, then, the self like anything else must move away "from" itself in a return "to" itself. The self has the dual aspect of moving away "from" itself (act, pure gaze, *cogitatio*) in a return "to" itself (I as a center of act execution, transmitter of the gaze, *res cogitans*). The precision of this determination should not be blurred by conceiving of the split [48] into subject-objectivity only as a mode of perception of the I. The I is a living unity brought about by the contraposition of the self as starting point (act) and the self as return (act center); indeed, the execution of this split *as* its overcoming.

An immediate *grasp of the I* is possible and actual only in mediated form, just as the one-and-the-sameness of *being an I* is so only by virtue of being split. Only where this condition is upheld does an I persist in a genuine self-position. The fact, then, that in accordance with what was set out earlier only the *cogitatio*, the living act, the I as self, not as being, can function as an opposing sphere to the *res extensa*, does not exclude, but rather in fact includes the opposing position of the I as *res cogitans*, as being, as object.

We said earlier that its executing act, its gaze, its simple turn toward something is interiority's direct contact with a body, *and yet* this contact remains subjugated to the situational principle of the body's opposing position to the *res cogitans*: the contact of interiority gazes at itself. Now, however, "and yet" becomes "because." The I in the self-position is only in the "I am" form; I *am* only by virtue of the I in self-*position*. As lived unity of first and third person the I constitutes itself as I; it is *res cogitans* and *cogitatio* in one. The profound teaching of the gospel that only whosoever loses himself shall find himself applies to the I in a descriptive sense. As *cogitatio*, it has lost its substantive nature, its character as *res* (which is why the I believes that in its execution of the gaze it has the thing genuinely and directly). It is only in the *cogitatio*, however, that the substantive essence of the *res cogitans* becomes fulfilled as the specific ontic factor that is opposed to the *res extensa* in a polar fashion (which is why the execution of the gaze achieves only a retroactive and mediated relation of an already subject-related content—that is, of the appearance on the subject).

(It is true that this is where the critical difficulties arise. Does the prior givenness of the I before the object mean the same thing as the givenness of the object? Above all: does it follow from the prior givenness of the I that this I closes itself off from the "outside" world and locks itself into its own sphere of consciousness? Is idealism actually unavoidable, and would the arguments set out earlier have to be disposed of for realism to assert itself? Or is there not a third solution: is the direct connection between subject and object perhaps necessarily an indirect one, the immediate relationship with [49] being only possible as a mediated one, the genuine presence of the real [*Wirklichkeit*] only possible in the image? I am only posing the problem here, not setting out an answer.)—

The famous corollary of the principle of the forward displacement of interiority or of the self is known as the "proposition of immanence": the subject cannot reach beyond itself and its sphere. What is given to it is given in it. Grasping what exists means that what exists is present to him who grasps it, or it is objectified for him. His presence or objectivity is the point of reference for that which appears. Accordingly, that which appears becomes equivalent to the content of consciousness and in a certain sense must be identical to it.

A variety of different philosophies of immanence were born out of the different ways in which philosophers attempted to comprehend this identity, passing through several stages of sublimation from physiologism to psychologism and culminating in transcendentalism. The law of the subject-objectivity of the I means that on each of these levels the position of the *res cogitans* is dual and the I just as much source of illusion as it is of truth.

Thus, to cite an example, from a purely physiological point of view, the positive contribution of our senses and nerves to our grasp of an object carries no more weight than the falsification that the necessary function of our sense organs and nervous system performs on it. Physicists and physiologists start from the physical definition of the object when presenting the dual performance of our sense organs. Certain electromagnetic conditions, they say, are qualitatively grasped as being luminous. Thanks to experiments that have demonstrated the likely existence of specific sensory energies of the nerves, we can credit luminosity's mode of appearance to the optic nerve. The optic nerve, however, plays an indispensable role in our ability to have an immediate experience of such electromagnetic conditions in the first place. The nerve falsifies as much as it ascertains truth, and it must falsify in order to ascertain.

In transcendental idealism, the relationship between I and object is entirely sublimated and significantly altered from the Cartesian starting point. Here I-ness has become the principle of the *res extensa* and, if I may simplify to such a degree, the principle of the possibility of knowledge as measurement in the natural sciences. Modes of intuition and thought [50] constitute appearances in their objective meaning. Nevertheless, Kant took into account the principle of the ambiguous character of the self in his idea of the thing-in-itself (the effect of affection on sensibility, the limited nature of possible experience). Objects precisely do not exhaust being, and not even existence. They are the world in its front-line position to the observer and thus appearances whose own being is hidden, albeit not lost (otherwise they would be mere illusion). While appearances do submit to the principles of measurement, they remain superficial in relation to unknown being as such. Even expanding the area of immanence so that it becomes the quintessence of possible experience does not suspend this law of disintegration into an objective sphere partly dependent on the I and a transobjective sphere of being-in-itself turned away from the I. It is precisely the shift of the boundary of immanence brought about by transcendental idealism, which is a defense of the empirical realism of the natural scientists against Descartes and Locke, that makes the principle of immanence all the more explicit.—

The *res cogitans*, indifferent to the determination of its essence, which has intentionally been left open thus far, takes on the *function of the subject* of its own accord. Initially a zone of being of the same rank as the zone of the *res extensa* and in every sense equal to it, it now shows itself to be displaced forward from the latter in order to ensure the possibility of qualities. This forward displacement designates the *res cogitans* as the self, whose function consists of its turn toward and acceptance of the thing. It is this function that necessitates the self placing itself between itself and things, just as appearance lies between being and the gaze that sees it. The self then only intends to reach beyond its own sphere; in practice it does not. It remains caught in its own sphere and is only connected to being outside of it by means of its own contents. The identification of physicality and extension charges the *res cogitans* with the task of saving appearance, something it can only do as a self, and only at the cost of closing itself off from the physical world.

Extension as Outer World; Interiority as Inner World

Only as a self—more precisely, as myself—can the *res cogitans* fulfill the task of making appearance possible, a task that falls to it as a result of the identification of physicality and extension [51]. And it fulfills this task only at the cost of closing itself off from the physical world, only by dint of a *crack* in the whole of being, which as a sundering chasm creates two *experiential* positions that are not mutually convertible.

Things are different for ordinary intuition. Here extended being and consciousness exist independently of, and unaffected by, me as "thinker." I myself am embedded in comprehensive being, which is broken down into these opposing types. There are other I's in addition to myself, *res cogitantes*, who live their own lives, embedded in comprehensive being. Every I can enter into direct contact with every other I, just as with the physical world. The world is open to the eye and the hand that want to grasp it, and the I is open to the world that gives itself to it. For this so-called natural realism of the naïve view, nature and the shared world are self-sufficient in their being, but not closed off. For the living human being and his naïve mindset, giving up this realism in accordance with the proposition of immanence in fact constitutes an unenforceable demand. Evidence clashes here with logic.

Strangely enough, empirical science has explained away this permanent contradiction, despite the fact that it is this science in particular that holds on to the main features of natural realism—and must hold on to them if it does not want to render invalid its methods, which presuppose a world independent of the subject. Empiricism makes use of the Cartesian principle to the extent that it allows for complete freedom of movement while leaving all of those strange questions concerning reality and the possibility of knowledge about the external world, other I's, and the relationship between body and soul to the philosophers—questions that can only be understood from the perspective of the situation of immanence and of the "crack" between two different kinds of being. At the same time, empiricism, overall, acts in accordance with the situation of immanence by fundamentalizing the double experience of both a physical and an inner world.

Its also methodological interest in the purity of sources of experience does explain the tendency of empiricism to hold on to the Cartesian doctrine of nonconvertible experiential positions, even at the cost of coming into conflict with the natural view of the world. Indisputable successes using this method also justify, to a certain degree, the risk of this conflict. Nevertheless, of course, the requirement remains of finding a balance [52] between the natural realism of the naïve view and its scientific interpretation.

Of crucial importance here is the insight that empiricism interprets our natural existence starting from a tacit assumption: I myself am "in" my body; my body "encloses" my self. Our own body is not thought to fall entirely within the world of bodies, but is rather also treated as a boundary of the I against this world, as the periphery of interiority. The consequences of occasionalism are thus avoided by understanding our own body both as belonging to the physical world and as a gateway to it, as the physical world's constitutive principle. My body is an extended thing and at the same time the bearer of senses by means of which the I comes to know of an "outer" world that transcends it. Whether this is enough to evade occasionalism, however, is highly doubtful. The only thing certain is that there is a fear of it, which is why those famous breaches of immanence are still held on to as problems, although any possible solutions are cut short from the beginning by the Cartesian dichotomy.

So how does this famous localization of interiority within our own body come about? How does interiority become an inner world? One would think that simply pointing to the difference between the *interpenetration* [*Ineinander*] of the manifold, which can be comprehended and lived in the self-position as the abundance of being and acts belonging to the I and the *being-in-another* [*In-einem-anderen-sein*] of this same manifold would be enough to guard against confusion and prevent the spatialization of what is by definition nonspatial. And yet a peculiar intuitional compulsion persists and gives rise to the distinction between inner world and outer world.

The fact that I myself see my own body as my body is grounded in the *here*-position of my self and my body. Regardless of where I find myself in space, this never changes. Seen purely as an object, I occupy an arbitrary point in space; as I-subject, it is my nature to be Here. The self is, as it were, "that" toward which all being converges. While it makes little sense to claim that this per se unextended self can occupy space in any way—that is, to say that it is here or there—, it is just as unavoidable to [53] hold on to the hereness of the I-subject as something belonging to it and constituting its being. As such the self is "the here" and not "in the here." The absolute here-point, never becoming a there, unrepresentable as a specifiable point in space, thus does not verifiably belong to objective space. And yet, as the absolute center, it constitutes the intuitive structure of space, its *enveloping nature* [*Umhaftigkeit*].

I cannot, however, say the same thing of the body. It is not "the here," but is now here, now there; its location has to necessarily be determined in relation to a chosen point of measurement. Only of my own body can I say that it is "always" here or in the here. The place it occupies as my own body, belonging to me, is verifiably in objective space. As such this place can become a "there" that provides space for another body. As *my* body it is always here, but never "the here" as such.

The body thus has a distinguished position in relation to the I, which is generally thought of as the I being enclosed by its body. It is precisely the naïve perspective that distinguishes the pure I "around" which our own body also converges from this body, toward which—"from a subjective point of view"—the external world converges; to possibly be separated from this body is not at all an absurd notion. From an objective perspective, our own body belongs to the outer world and thus to the relative order of objective space. This difficulty of finding unambiguous terms for the total convergence of the given onto "the here" and for the position of our own body in the here that distinguish between spacelikeness [*Raumhaftigkeit*] and spatiality [*Räumlichkeit*], between that which determines space and that which is determined by it, is generally enough to compel the localization of interiority within our own body.

Furthermore, there is the important role played by the internal organization in which the manifold of our own body figures. This order does not coincide with the interpenetrating interior manifold of being and acts, although it too (like the former) can only come into appearance in the self-position. Other bodies do not exhibit an internal organization. There we find the specific space-occupying proximity and interpenetration of the parts that is essential to every spatially extended thing.

Because our own body, and precisely in the self-position, is itself extended, this difference is generally overlooked. Our own body is a fullness of depth enclosed by a uniform [54] surface, which as such is palpable throughout and can be penetrated by impulses. A stimulus exerted on the surface presents itself "here and there" in several fields of sensation—visual, tactile, kinesthetic. The impulse that answers the stimulus attacks first here, then there, on the "inside," in order to achieve the corresponding external effect. This system of organ, movement, and joint sensations, traversed by feelings of pain or pleasure, forms a nonmeasurable but qualitatively extended variety, power, and fullness, held together and sensationally limited by the surfaces of the skin and the sense organs. Thus the more primitive capacity to differentiate—for which measurable and nonmeasurable extension dissolves into one—ascribes the characteristics of an internal space for the interpenetration of psychic variety to the internal organization of our own body.

Finally, a very important factor in the localization of the I in our own body is the fact that we do not use our sense organs when we observe ourselves, reflect, contemplate, or concentrate. Because, in principle, all contact with other things that can be comprehended by the senses can disappear without hindering the I's preoccupation with itself, the notion that interiority is "inside" becomes an irrefutable illusion.

The *convergence* of our own body toward the pure here, the *structural relationship* between the internal organization of our own body and the I's interpenetrating manifold of being and acts on the one hand and spatial divergence on the other, as well as the ability to largely *shut off* the body's external organs when observing oneself, lead to the notion of an inner world spatially enclosed and veiled by the body.

In connection with the principle of the immanence of the I, this gives rise to assertions of the greatest consequence: (1) the segmentation of the world into inner world and outer world; (2) the theory of inner and outer perception and their mutual nonconvertibility; (3) the attachment of outer perception to physical objects, of inner perception to our own self (due to the identification of the outer world with the world of extension, the inner world with interiority); (4) the indirectness of outer perception; (5) the derivation of the outer world from sensory perception—the principle of methodological (as well as dogmatic) sensualism; (6) the problem of the connection between inner and outer world.

[55] Here, at one stroke, we can see the breadth and depth of all the questions arising from the Cartesian dichotomy. Famous problems, which under the influence of the dichotomy tend to be treated in widely different branches of philosophy, now reveal their inner coherence: questions concerning the reality of the outer world and other I's; the possibility of physical and psychological awareness of others; the problem of a tenable comparative psychology/biology that is not tied to introspection in its methods or in its objects and extends, in particular, to animals (the set of issues with which Uexküll's *Umwelt und Innenwelt der Tiere* is concerned); the question of the interplay between environment and organism ("adaptation"); the ways in which physical and psychological being are connected.

The Proposition of Representation and the Element of Sensation

Thanks to the Cartesian dichotomy, the intuitively reasonable localization of the I in our own body has helped make a more or less sophisticated idealism and anthropocentrism as pervasive as it is today. If a certain blindness precisely to the most elementary questions posed to us by the appearance of nature, as well as indifference to the immediate reality [*Wirklichkeit*] of nature, still dominate philosophy today; if even, aside from very few exceptions, there is a complete and conceded inability to interpret the significance and achievements of the history of the philosophy of nature from the documents, then we have here the distinct symptoms of a worldview still determined by the misunderstandings of the Cartesian dichotomy.

Once we have gone over to locating the realm of the *cogitatio*, consciousness, behind the forehead and inside the skull, thus following, despite all learned assertions, the old examples of seeking a place for the subject, then of course, given what we have set out thus far, the cleavage in the constituents of the world must lead to a disintegration into two different worlds. The outer world must confront an inner world, the there-world a here-world. According to the principle of immanence, however, the here-world cannot enter into direct contact with the there-world. This means that the ongoing existence of the there-world can only be ensured indirectly, by the [56] mediation of the here-world. The windowlessness of the subject or the absolute solitude of the I becomes an irrefutable starting position.

The privileged position of our own self, which for the sake of the qualitative givenness of physical things had to become a position of prior givenness, finds in the proposition of immanence more than just the assessment that it is closed off from the physical world. This proposition is not a simple declarative sentence. The immanence of the I expresses more than a state of affairs. It also indicates the principle of the priority of givenness over being, a methodological principle of first examining judgments about that which exists by investigating the accessibility of being. That which exists becomes something one has to get *to* or that must be given, must appear, in order to be assessable. The proposition of immanence means that the manifestation of any existing thing corresponds to an act of subjective attention by virtue of which it exists. Accessibility (givenness) and "knowledge of" become equivalent to each other and necessarily remain intertwined despite their opposing directions.

Closed to other being, every kind of consciousness, as Fichte put it, is self-consciousness, regardless of whether it concerns data of the outer or of the inner world. While the receiving act of the I is differentiated into outer and inner perception, corresponding to the nature of the givens toward which it is directed, outer perception only reaches for the outside in a sense: in terms of value, it too remains inside; it too is "inner" perception, self-consciousness. (This is not yet the place for the significant separation of the psychic as the quintessence of objectifiable [perceivable] inner phenomena from the nonobjectifiable I-subject, the pure here, toward which also the contents of consciousness converge. Initially, the I-subject belongs strictly to the world of perceivable consciousness. Kantian reflections on the theory of knowledge do not yet come into play. The pure I is identical with the real, self-positioned subject, consciousness identical with the subject becoming objective to "itself.")[1]

Outer perception attends only to the outer world, which is identical to the extended world, the physical world of things [57] if only because inner being has become an inner world. In an unshakeable correlation, physical contents correspond to outer perception; psychic/consciousness-related contents to inner perception. Externally there are only physical data; internally only psychic phenomena. According to the principle of immanence, however, physical data are not originary and given in direct contact, despite what perceptive intention and evidence lead us to believe; only as psychic data are they originary and only as contents of consciousness are they given directly. Every element of the outer world is thus psychically mediated, mediated by consciousness, has contents of consciousness as its representatives (thus calling into question this world's independent existence).

The law of the representation or mental image [*Vorstellung*] of the outer world can be readily derived from the principle of immanence: appearance has become a mental image, has received the ontological value and rank of a mental image. In vain the natural attitude in its naïve belief in reality [*Wirklichkeitsglauben*] appeals to visual evidence as a witness against argument. In vain it points to the difference in lived experience between genuine appearance as the thing present in its primary self and a genuine mental image (a recollection, a fantasy). These differences become mere differences within consciousness and retain only the value of different persuasions of the I.

(There are difficulties here, however, that the idealism of mental images cannot solve: the presence of heterogeneous sensory material in the mental images with an external character, as well as the fact that the occurrence of these mental images is dependent upon the body's sense organs. Evidently the purpose of the affinity between the I and its own body is to convey to the I information regarding a world outside of it. Even if one does not want to attribute to mental images the quality of true likenesses [no one has seen the originals]—their quality as indications of this sphere of things-in-themselves cannot be denied. Even if we assume that some form of metaphysical idealism was correct, it would not be able to get around this simple phenomenon. The body as extended thing in this case already belongs to self-consciousness, however—to the sphere of outer perception—but nevertheless evidently conveys to the I, the sphere of inner perception, material for the construction of the latter's mental images. Idealistically or non-idealistically speaking, this [58] dual aspect presents the psychophysical whole self, whose outermost zone of its own organs encloses the pure here of the I in a permanent bond. On the one hand, since our own body partakes both of the outer world and the inner world, it forms the periphery of the sphere of immanence. On the other, the inner world, which as self-consciousness contains the domains of inner and outer perception, encompasses the outer world in the treasure trove of its mental images. But how can a sphere with its boundaries be both contained in the outer world and at the same time contain the outer world in itself?)

The fact that radical dualism, the perpetuation of the dual aspect in the form of occasionalism, destroys itself leads to the admission of the affectability of the I by the senses, the affectability of the world known to us only in its phenomena by the acting I. Philosophy justifies this admission by pointing to the unique position of our own body in relation to our I, its total convergence toward the pure here, and its internal organization. As a system that is palpable from the inside, which can, in varying degrees, be controlled by impulses, which is turned outward and is given in part as an outside, whose center is the I, whose internal variety appears to be totally absorbed into the interpenetration of actions and mental images, and whose boundary surfaces, despite their objectivity (despite, that is, their nature as mental images), from the inside play the role of sensory and movement fields, of inlet and outlet zones for acts of perception and impulse—as such a system, the body reveals itself to be the sought-for bridge from the inner world to the outer world.

Since according to the proposition of immanence any bridge between the I and the outer world has been severed, since extension and interiority are completely heterogeneous, the transition from inside to outside, from outside to inside should not be thought of spatially as a continuous exit/entry. The mode of transition necessarily remains unrecognizable. In this problematic situation, one can only attempt to comprehend this transition in the form of *an element equally indifferent to the outer and the inner world, to the body and the psyche*, out of which both worlds are ultimately constructed. This boundary datum of the inlet and outlet zone of myself as a body among bodies and as an I in this body with sense organs and [59] organs of locomotion is *sensation*, the source material of all mental images.

This term and that which demonstrably corresponds to it represents the most that can be achieved following the Cartesian dichotomy: the first step toward an overcoming of world dualism according to the laws of its own perspective.

Of course, one does not have to see such a step in "sensation," which can also be thought of as belonging to representation [*Vorstellung*] and can be situated, like representation, in the interior. In practice, the natural scientist and the psychologist both operate under this assumption. Sensation is separated from that which is sensed in accordance with the differences found in immediate lived experience. The fact that the natural conclusion is drawn from the external position of the sense organs even rehabilitates the naïve perspective to a certain extent. There is nothing in consciousness that has not somehow passed through the senses. Things, forces, or something unknown to us affect the peripheral organs of our self. This entails their transmission in a medium alien to them (nerves with specific sensory energies); after their transformation into psychic elements they can be experienced as sensations. It is out of these sensations that the sensuously intuitional world is constructed, including my own body with its interiority.

If, however, one wants in the spirit of positivism to escape world dualism according to the laws of its own perspective, the elementary layer of sensations does in fact provide a starting place. "Sensation" must, in this case, be understood as a psychophysically indifferent or neutral element not yet belonging to either the outer or the inner world. Its elementary character, furthermore, is not exhausted by this boundary position: sensation is also neutral in regard to the difference between action and object, indifferent to subject and object. The oppositional polarity between subject and object, which comes up against the indifferent elementary layer, can even, according to neo-Kantian thinkers such as Hugo Münsterberg, intersect with the oppositional polarity between nature and psyche. In any case, positivism has developed toward an attempt to get away from locating sensation in the interior, psychic sphere, seeking to comprehend it—like Theodor Ziehen, for instance, with his concept of the gignomen—as a pure boundary datum.

[60]

The Inaccessibility of Other I's according to the Principle of Sensualism

Even the positivist analysis of the concept of sensation cannot mitigate the deficiency and lack of clarity of the two-world situation. The strange phenomenon of the dual aspect of outside and inside remains, as does the fundamental impossibility of moving from one experiential position to another without an absolute rupture. The contradictions remain between philosophy and the interpretations performed in the natural worldview held by the practical and scientifically observing human—although it is worth recalling that the practitioner and empiricist with their naïve and critical realism solve the difficulties philosophy finds in the realistic view of the world in a completely primitive way or ignore them altogether.

What is doubtlessly significant here is the methodological aspect of the fundamentalization of the concept of sensation, which restricts the given to sensory material. From the abundance of what exists, only those elements remain that can be associated with a particular sensory field. Not only nonintuitive components, upon which the unity of existence, its spiritual physiognomy, its character of value, and its categorical nature precisely rest can no longer be given, but intuitive, albeit formal, nonsensual components, such as gestalt, rhythm, situation ("field structure") are also excluded from that which "properly" exists. Only sensuous elements remain as givens and, due to the elimination of all nonsensuous, complex-forming relations and functions, acquire an atomistic structure.

These unifying functions are needed, however, in order to account for the integrity of appearance. The only possible explanation is that they are added to the atomistically given—that is, they are subjective or psychological. They are thus not given, but are brought about by the subject's interventions and can only be understood as resulting from these. The power of the subject reaches as far as the phenomenal world exhibits a nonsensuous and integrated character. If the task of science, then, is to work out as precisely as possible what is objective in appearance, then this leads to the automatic omission of the nonsensory and integrated aspects as products of the subject and as nonprimary [61] givens. It is the sensory material alone that constitutes the foundation. Modern sensory physiology and psychology have raised objections to the prejudices of sensualism and have forcefully demonstrated the essential connection between sensation atomism and the assumption of unifying functions such as "association" and "apperception," which are intended to restore the integrity of appearance.

It is the principle of methodological sensualism that gives the identification of extended being and outer world, or the unambiguous association of physical nature, mechanism, and outer perception, its significance for the practice of research. Outer perception is related to sensory sensations corresponding to physical sensory fields. These sensations form a completely random variety, into which order is brought in the following two ways: (1) by the subject with the help of its functions, so that we have before our senses the phenomenon, and (2) by the object, or that which is at work concealed behind it, so that the given becomes explicable and transparent to our minds in accordance with mathematical and mechanical laws. Inner perception works in a similar way, but that is not our concern here.

We can now properly understand the fundamental skepticism of empiricists and philosophers regarding the possibility of entering into immediate contact with another psyche. "Something is there" means, to begin with, "I am connected to it by means of perception." Perceiving another's psyche remains impossible because it is a matter here of outer perception, which, according to the dichotomy, only aims at physical entities. The famous theories attempting to explain the consciousness of psychic beings existing outside of myself and my ability to understand them are all responses to this difficulty. The "other" human is primarily a physical object ("outer" in his case is also merely a body)—so how can it be possible for my gaze to breach this physical front and reach the emotional stirrings lying "behind" it? These theories go to the greatest lengths to maintain the position of the psychic as "here" and that of the physical as "there," since the loss of the pressure to bridge this divide would render every theory of the possibility of other I's beside myself irrelevant. As soon as the basic premise is no longer that the psychic is tied to the "here" and the physical to the "there" and outer and inner perception become directions of contact that are freely "movable" [62] in relation to the difference between myself and the other, the problem of the perception and comprehensibility of the other I disappears—without, it must be said, being solved.[2]

The principle of methodological sensualism has conspicuous consequences for comparative psychology. Here the existence of an entire science depends upon whether or not one adheres to the principle. Things are complicated, however, by other fundamental concerns. The foreignness of other organisms to the human species makes it doubtful whether the human is capable of distancing himself from his own humanness to the point of being able to avoid all traces of anthropomorphism in his analysis of the psyches of other species—quite apart from the question of how one assesses the possibility of entering into contact with the inner world of an animal (or even of a plant). In any case, it is the question of the viability of the perception of others that ultimately decides the fate of a (nonbehaviorist) psychology truly aimed at studying other psyches.

Thus until quite recently research in animal psychology suffered from the fact that scholars saw in the discipline's very focus a denial of strict scientific method. Why should we, the biologists argued, allow for a psychological approach to other living beings whose organization and mode of life are not only fundamentally different to those of humans, but with whom communication by means of sounds, signs, or even gestures seems impossible, or at least highly doubtful. Do the experiences of hunters and animal owners with higher vertebrates justify abandoning the established principles of causal research for meaningful interpretation? Ought we to see in the timid glance of the deer, the begging pose of the dog, the angry roaring of the lion more than a metaphor, as compellingly intuitive as they may appear to us? It is true that a certain similarity between the organs or overall type of humans and some higher animals make it seem as if the latter are moved by humanlike affects, drives, and notions. But is it permissible for research to claim more than an "as if" here?

The answer was: since we are refused direct access to the inner [63] life of animals and plants, which verbal communication grants us (poorly enough) to other humans, we must content ourselves here with an *ignorabimus*. Psychology, argued the theorists of the Cartesian school, is only possible where psychological content can be grasped directly—in the end, then, in the introspective self-observation of every human being, who in this reflection discovers phenomena immediately visible to him alone. Researchers such as Theodor Beer, Albrecht Bethe, and Uexküll consequently called for the use of objective terminology in the life sciences, avoiding all psychological concepts; Uexküll in particular argued for replacing animal psychology with biology, a science that is able to determine the objectively controllable correlations between stimulus and reaction in the makeup of a particular animal. The scientific program of the "animal psychologist," according to these scholars, should concern not the eternally hidden inner world of animals with its feelings and perceptions, inaccessible to us, but rather their environment—that is, the various unified forms of the elements that affect them and upon which they can have an effect. Instead of cryptopsychology, what was called for was a phenology of living behavior: the explanation of animal behavior visible to us in factors perceivable by the senses.

The Need for a Revision of the Cartesian Dichotomy in the Interest of a Science of Life

This program, which is strictly adapted to the working conditions of experimental study, leaves aside the big questions (mechanism or vitalism? automatic or spontaneous life processes?) and restricts the goals of comparative "psychology" to knowledge that can be gained by sense perception. This stringency is absolutely necessary if we want to banish from science once and for all the anthropomorphisms embellishing the reports about Wilhelm von Osten's clever Hans, Karl Krall's Elberfeld horses, or Mannheim resident Paula Moekel's dog Rolf, which belong to the repertoire of every animal lover. One must not forget to ask, however, whether Uexküll's is a maximum or a minimum program.

If it were a maximum program, all the questions raised by so-called animal psychology would be resolved in terms of the physiology of irritability or of movement. In fact, [64] however, Uexküll's biology (life plan research) does not pursue its program in this sense. It is instead restricted to the study of the stimuli and reactions characteristic of the organization type of the animal in question—not because it rejects the aims of a comparative psychology (as would dogmatic mechanism), but because it considers them to be unattainable due to a lack of means. It in no way denies that the notion of a "life plan" might contain another side; it only contests the possibility of conducting *empirical* research on it, without, however, negating from the outset the possibility of a *non*empirical approach to the matter.

The idea of a plan implies that it is more than the sum of the factors by which it is realized. This applies not only to plans containing a purpose or goal; as soon as a whole takes shape in the ordering of its elements we can speak of a plan. A life plan as a unity of stimuli to which the organism recognizably responds *and* these responses can thus not be identical to the sum of these perceivable processes. This unity of the sphere that is the given framework for stimuli *and* reactions and that is itself invisible but becomes visible *in* the processes belongs neither to the body of the organism nor to the world surrounding it alone. If such "plans" exist, sense organs and organs of locomotion cannot "precede" the world of things *for* which they are there—and vice versa.

Is there evidence of such life plan forms or vital categories that are "given beforehand" equiprimordially with the organism and the environment relative to it or, in other words, that as organizing ideas put their stamp on both zones of existence? Empirical observation and experimentation of course only extend as far as the particular *delimitation* of the a priori forms of life organization in the animal's body and "its" environment. Thus only an experiment can determine the fact that a pistol fired at close range does not constitute an acoustic stimulus for a lizard, while a quiet scuttling noise, rustling, or swishing does. The significance of this fact, however, is derived from the principle of the biological contingency of the stimulus threshold, according to which only that becomes noticeable to the organism which is biologically meaningful to it in some way. Any empirically ascertainable conformity of the organism to its environment, any adaptedness of the environment to the organism [65] points to overarching laws equally governing living subject and world. The fact that neither of the two members of this reciprocal relationship has priority over the other cannot, however, be grasped empirically. Here the competency of empirical life plan research ends and analysis of the vital categories begins.

In a more comprehensive sense, then, there is also a nonempirical side to life plan research as regards this curious concord between living being and environment. Shall we adopt the hypothesis of a supra-individual soul, as Erich Becher has done recently, or can we get by here without any hypotheses at all if we devote ourselves in all seriousness to the study of the essential laws of the organic? The expression "vital category" should concern us no further; there is nothing special about it. If, however, it turns out that there actually are laws governing the coherence between living being and world, laws of harmony, concordance, and equiprimordial formation that are grounded in the "what" form, the essential structure of life—in other words, material a priori laws—, then it can also be shown that they must have the value of categorical laws.

In philosophical usage, a category is a form to which experience submits but that does not derive from experience; a form whose domain does not end with the subject's sphere of action but extends to the sphere of objects. This is why not only the experience one has of objects but the objects themselves are subject to it. A category, then, is a form that belongs neither to the subject nor to the object alone, but allows these to come together by virtue of its neutrality. Categories are conditions of the possibility of accord and harmony between two essentially different and mutually independent entities, such that these are neither separated by an unbridgeable divide nor have a direct influence on each other.

Of course, reference to accord and coming together concerns, in the first place, the rational mode of knowledge whose elements are intellect and object. But why should it not be permissible, as an experiment, to detach the *function* of categories from their particular role as forms of thought and knowledge and to pose the question of categories or categorical functions belonging to other, more primitive [66] or more fundamental layers of existence? Kant himself speaks of a priori forms of sensibility that come into play in simple, pre-epistemic perception. Like the categories underlying rationality, these forms are still tied to the unity of consciousness. Ought we then to dismiss entirely the notion that there are preconscious a priori forms, categories of existence, vital categories belonging to the deeper layers of existence of the bearers of life—that is, organisms (understood not as existing objects but as living subjects), upon which the togetherness and cooperation of the organism and its environment rest? They would in any case have the value of categorical functions, as they, while neither being taken from the counterworld [*Gegenwelt*] nor applied to the counterworld by the living subject, determine the structure of this counterworld along with the structure of the living subject that fits into it.

The task of a philosophical biology as the science of the essential laws of life, as well as of the foundational discipline of a possible animal "psychology," lies in the systematic grounding of such vital categories. Closer observation will show that the life plan—if it is actually worked out to mean the spheric unity of living subject and counterworld and not only the individual stimuli and reactions by which this unity takes on tangible contours for the sense perception of the experimenter—also represents the foundation of those relationships between subject and world that determine the former's consciousness.

(Psychology, however, is not identical with the science of consciousness, just as psychic being and occurrences in the psyche are not to be confused with the consciousness of them. And yet the fact that both consciousness and inner life are related to the self-position, the interior aspect, entitles us to connect the study of the structure of consciousness with psychology. So it does still make a difference whether we understand the question "do animals have consciousness or not?" to be the fundamental question of all animal psychology in the sense of a study of the life plan forms, that is, the counterworld forms related to subjects, self-positioned beings, *or* to refer to the study of the interiority of all such beings. The only thing certain is that the fundamental question in the first sense has to be decided positively or negatively before the second way of posing the problem can even be addressed.)

We must in any case break with the common view that a decision concerning animal consciousness or lack [67] thereof exceeds the human power of cognition. Underlying this assumption is an untenable concept of consciousness that has done great damage in psychology and psychopathology. Consciousness is thought of here as an invisible chamber or sphere, as a nonspatial counterpart to the spatial brain, with which it shares, nevertheless, an interior existence in the head behind the sense organs. Of course, it would be impossible to decide whether or not such an interior representation is taking place inside a head.

Actually, things are exactly the opposite: consciousness is not in us, but we are rather "in" consciousness—that is, we relate to our surroundings as motile, lived bodies. Consciousness may be clouded, constrained, disabled; its contents may change; its structure depends on the organization of the lived body; but its actualization is always ensured when the lived body effects a unified, two-way relationship (receptive and motor) between the living subject and its environment. Consciousness is only this basic form of, and basic prerequisite for, the behavior of a self-positioned living being in relation to its surroundings. In this respect Erich Wasmann was right to defend, against Uexküll's anti-animal-psychology program, the scientific legitimacy, at the very least, of using terms such as seeing, hearing, touching, smelling in reference to animals. These forms of awareness are forms and conditions of living behavior that bridge the schism between the lived body's own system and the environment.

Indeed, consciousness is *not* necessarily the form of the subject's relationship to the counterworld instituted by the identification of the I with itself, as is essential to the human. Consciousness does not have to be self-consciousness. Nor is consciousness, the spherical unity of subject and counterworld, an entity within the body of the living subject. It only appears this way because of the peculiar mediating role of the lived body, which both separates and connects the subject and objects in a synthetic way. Nor is consciousness something ultimate, which, inexplicably stretched between life and being, represents a mere zone of the imagination and unreality and whose "emergence" from organic material and whose relationship to the body have to be (hopelessly) debated. Consciousness is rather embedded in those spheres of existence whose categories were discussed earlier. The structural laws of consciousness strictly comply with the more [68] comprehensive structural laws of life plans. In the same way, the intuitive-perceptual or the rational-intellectual affinity between subject and object also complies with the elementary modes of harmony between living being and world.

Anticipating the future, Paul Claudel gave these coherencies their classical form in his *Poetic Art*: to know (*connaître*) is to be born with (*co-naître*).[3]

It remains to be seen, of course, whether there are ways to substantiate these claims of equivalence between life plan form and consciousness. But even if we succeed in disclosing such essential relationship types between living subject and world, we must from the outset strongly emphasize the marginal utility of such an investigation.

Consciousness not only refers to the sphere that is given for all concrete lived experience, but also encompasses the entire play of this concrete lived experience and its contents. This abundance of content can also be grasped in terms of essential types even where no community of experience exists by way of a shared language, empathy, or sympathy; if, in other words, it is a matter of nonhuman, animal, or even plant life with which we cannot enter into an immediate "we-bond." We must stringently examine, then, whether or not questions of the animal's mood or the quality of its lived experience can be answered. Whether, that is, they fall within the scope set, on the one hand, by the experimental method that delimits, in a concrete case, what can be considered an organism's contents of consciousness and, on the other, by the philosophical method that determines the essential laws of an organism's *potential* relationship to things.

Accordingly, the young science of animal psychology is not concerned with lived experiences, but is rather interested in developing a theory of animal behavior, of its forms and factors. Its narrow path runs between the Scylla of anthropomorphic descriptions of the intelligence, loyalty, and love in an animal's soul and the Charybdis of Uexküll's program and its attempt to discredit all research on consciousness. In this way it avoids both the "behaviorism" cultivated by some Americans, whose overanxious compulsion to objectivity leads to the stricture of a physiological stimulus-reaction schema (where the objective description of all individual elements of a behavior jeopardizes an understanding of the whole), and the laypersons' uncritically romantic panpsychism and anthropomorphism.

Uexküll was the first to declare the relationship between the organism [69] and its environment to be the domain of an animal psychology (biology, life plan research) brought to reason. This young science has gone beyond him, however, in the sense that (unlike the "Kantian" Uexküll) it strives to understand this relationship in its vitality and intelligibility, no longer identifying the physiological conditions of its realization with the overall habitus of animal conduct. Wolfgang Köhler, thanks to whom we have excellent studies of anthropoid intelligence, David Katz, and F. J. J. Buytendijk seem to be the most decisive in their approach to this kind of animal psychology: they are fully conscious of the purely image-based nature of this science, of the purely phenomenal character of "behavior," "conduct," or "demeanor," which is destroyed as soon as physical/physiological terms are used to describe it. Animal psychology can make no headway without respect for the gestalt nature of behavior, but will rather lose its particular object of study and be left holding only parts—parts that may occupy physics and physiology, but not a theory of behavior.

Living behavior and conduct are only given in the habitus picture. Physiologists and anatomists often cannot measure the minor differences in the habitus picture that tip the scales toward a particular character of behavior, differences that mean everything here, while there they mean nothing. Immediately "comprehensible" habitus pictures are chiefly rendered only by higher animals, those who are morphologically closest to humans. It is not crucial, however, for a habitus picture to be comprehensible. An animal psychology with universal ambitions will also strive to systematically investigate the habitus pictures of those animals that are entirely unlike humans—that is, the lower ones, whose habitus pictures are not immediately comprehensible and are not accessible to empathy or to the interpretation of expressions—always remaining in touch with physiological observation controls and supported by the experimental method, all the while maintaining its focus on the unified gestalt of living behavior.

A Methodological Reformulation of the Opening Question

This in-depth discussion of the question of principle as to whether it is possible to secure access to other consciousnesses without denying the findings of natural science has led us right into the heart of the key problems of the present study [70]. Our discussion thus complements the essay cited earlier, "Die Deutung des mimischen Ausdrucks," which in its positive sections points the way toward a new solution to the problem of the givenness of other I's. The problems addressed there and in the present work—problems

The Cartesian Objection 65

that *came about* as a result of adhering to the principle of methodological sensualism and its premises—cannot be solved *according* to this principle; they require another method and thus a departure from this principle, an approach not derived from the natural sciences. Needless to say, this other method must recognize the findings of natural science in their truth. Philosophy (and that is what the new method is) ought not to overlook or seek to supplant confirmed experience. Nothing in the world contradicts an experience except another experience. The two methods have to find a way to cooperate, because only together, albeit with each maintaining complete autonomy, can they tackle their complex object in its dual aspect of physicality and interiority.

The following is aimed at overcoming not the dual aspect as an (incontrovertible) phenomenon, but rather its fundamentalization, its influence on the way the question is posed. Everything comes down to invalidating this dual aspect as a principle *rupturing* science into natural science, that is, measurement, and the science of consciousness, that is, self-analysis. We cannot say here in a programmatic sense what complex objects appear in the dual aspect, but we can surmise that they are the "animate" things of the world, things that not only belong to being, but also have being in some way as world, that live with and against it. This assumption can only be confirmed by a more precise formulation of the problem.

Strictly speaking, the only beings that are of the dual aspect are those that manifest as self *and* physical thing, those that function as physical things like other things, but at the same time show themselves to be a self and may be conscious of themselves. According to all the principles that have emerged thus far from the Cartesian fundamentalization of the dual aspect, this being is only *myself*. But despite all concerns regarding immanence and representation, I go beyond *myself* and speak of the *human* as a species realized in many individuals whose nature includes existing in the dual aspect. And [71] since it is uncertain where the boundaries in fact are, the sphere of things for whom an inner and outer existence is essential tends to expand beyond the human and absorb animals, plants, and even inorganic bodies as having consciousness, a soul, and a spirit. It is human to be on a first-name basis with our surroundings and to seek what we are in the reflection of the world.

Philosophy has to resist this uncritical expansion of the dual aspect to potentially include all things in the world if only because it has the duty to first determine whether the phenomenon of the dual aspect is in fact fundamental. This, in turn, is only possible on neutral ground. A preliminary decision for or against the fundamental significance of the dual aspect would deprive the entire investigation of its value.

Emphasizing this from two sides is of particular importance today. On one side we encounter a very carefree kind of "objectively" oriented philosophizing, which makes use of the justified opposition to monomaniacal standpoint theorizing and panmethodism (schools of thought in thrall to the primacy of the subject) in order to freely and without reserve steal truth from the immediate: here we have a misuse of intuition [*Intuition*] by an era that has run out of breath. The writing that makes its way into the world from this angle is not necessarily without depth or truth; the writers in question may even have personal knowledge of the truth—but it lacks genuine objectivity. Since today it is not only newspaper hacks, politicians, and literary figures, but even scholars who succumb to this intuitive [*intuitiven*] directness, I want to emphasize here as strongly as possible that, in my view, philosophizing in passing like this spells the downfall of philosophy. In contrast, this current study follows a strict method.

To the idealists and formalists out of epistemological timidity on the other side, let me say that following a strict method has nothing to do with methodism. A significant tradition has led to the habit of treating philosophical problems as epistemological or consciousness problems (even where idealism is no longer the guiding school of thought), which unintentionally leads to the perpetuation of idealism. This is very understandable, as it is precisely those philosophers who take their work seriously and actually grapple with things who develop new insights [72] by engaging with the current discussion, meaning that they automatically have to draw on old concepts. Very often, too, the conceptual tools used to answer entirely new questions still contain elements of idealist and subjectivist theory. Terms like spirit, consciousness, subject, and reason, which may not even be weighted with the teachings of Hegel, Fichte, or Kant, nevertheless deploy the power coming to them from history in an unanticipated way, evoking an idealist position where none exists.

In the current study, the turn to the object is simply compelled by the problem. I am not postulating the primacy of the object, but rather freely taking up the difficulty with which prescientific and scientific worldviews have to struggle and that obliges us to overhaul the *divisio mundi* Descartes put in place with remarkable simplicity.

The overhaul of the dichotomy might not seem such an urgent matter, given the undeniable fact that the traditional classifications established in the seventeenth century have been losing their importance and that the Cartesian dichotomy now only exerts an indirect influence, in particular since the emergence of the empirical humanities and with the empirical use of the concept of culture. Furthermore, the philosophy of our time has already begun to implement a turn to the object, the primacy of the object. It seems to me, however, that it is precisely the beginning resituation and restructuring of the empirical sciences that impress upon us a reconsideration of the fundamentalized oppositions of body and consciousness, outer and inner world, or subject and object, whose consciously or unconsciously recognized validity, as outlined earlier, profoundly affects our worldview.

It is not as if such a reconsideration is taking place here for the first time. In a certain sense, the entire history of modern philosophy with its theoretical and metaphysical concerns is one big engagement with Descartes's dichotomy principle. Up until Leibniz there was the attempt to find an ontological equivalent to *res extensa* and *res cogitans*; then from Kant to Hegel the admirable tendency to objectify the Cartesian principle from the dimension of transcendental lawfulness—perpendicular (as it were) to the level of division into two substances—and thereby freeing philosophy from its principle-izing effect [73]. Only in Hegel was this tendency strong enough to lead to actual victory, and those who came after him were not able to maintain this standard. Hegel's definition of the substance-subject as spirit gave a welcome handle to minor philosophers to rehabilitate the *res cogitans* and thereby Cartesianism, which had never been forgotten in the first place. The tremendous development of the natural sciences seemed to provide living proof of the validity of this step, and only the concurrent rise of the empirical cultural sciences prevented an open "Back to Descartes" movement.

When the new situation could not be comprehended according to the Cartesian dichotomy, one first turned to Kant for help. In addition to producing much that was of immediate importance to the new questions, this also had the advantage of returning philosophy to a high standard, reasserting its distinct position in relation to every empirical science. This made possible the turn that remains associated with the names Dilthey and Husserl, the turn to a new conception of the elementary phenomena and sources of intuition of every possible kind of experience. The core objective of this turn was to understand those cultural and historical formations whose nature does not comply with the separation between physical and psychological. Precisely this failure of the traditional classification regarding the products, situations, and fates of life—life that, naturally born and naturally conditioned, extends beyond its earthly spheres into a nonactual, yet nonpsychic mortal world and becomes eternalized in the work, which preserves history so that it can become historical—this failure, then, only became palpable with the consolidation of empirical cultural and historical science. The battles over methodology in these disciplines as they were most recently waged surrounding Karl Lamprecht and Kurt Breysig are still in vivid memory.

These fundamental questions of category came to be important to historians, cultural scientists, and sociologists—on the quiet, that is. After all, the scheme of the physical-psychological dichotomy does not allow anyone to empirically comprehend historical, social, cultural entities that are made of sensory material, appeal to the psychic, and are permeated with physical elements, that are spiritually meaningful, valuable, or without value, that participate in the spheres of extended nature as well as in those of interiority, and are made up of nonactual content. The state, the economy, mores, art, religion, science, law—the complex as well as the basic factors in these particular zones of culture and history demand to be understood not as conglomerates of physical and psychic content [74] (as well as, perhaps, content of a third category), but as original entities. The logic of this demand points back to the living foundation from which historically dynamic culture derives: to the human.

Since Dilthey, the task of an elementary reformulation of all fundamental concepts with which the bearers of the spiritual world, persons and associations of persons, are treated as members of the real [*wirklichen*] world has become ever more clearly defined. At first the problem was only treated from the perspective of the humanities. The young sciences of history and culture wanted the realities that traditionally were categorized either as physical, psychic, or psychophysical being to meet the requirements of personality, individuality, and vitality—requirements for which the science of being, however, does not provide a handle. The human in his original living unity, which under the dualism of two essentially different methods, natural science and cultural science, threatened to disintegrate into two beings, was to be held together as a unity, as one, by the mode of experience of the humanities. The purely vital ("natural") sides of the human formation were to be reconciled with his cultural and historical sides in such a way that the (never to be denied) rupture in his form of existence would not lead to the breakdown of the identity of individual existence. According to Windelband and Rickert, of course, this identity is only an idea, and the science of the human necessarily inhabits two essentially different standpoints; thus the transition from the human as natural object to the human as the subject of culture and history will always be a *metábasis eis állo génos*.[4] It may look this way from a formal and methodological perspective, but history and life have always refused to accept this.

Subsequently physicians, educators, practitioner psychologists, psychiatrists, and sociologists took up the problem of the theory of the human, of philosophical anthropology. This theory is now beginning to develop as the science of the "person" (Scheler). Insight into new connections between physical and psychological functions, interest in the concrete human as the subject and object of education, medical aid, and political decisions are compelling the development of new anthropological concepts in response to a new way of looking at living things. The human grasped as body "and" soul remains an aggregate of fragments. His original unity, located this side of the [75] separation into body and soul and for everyone a matter of certainty in active life, exists and can only be understood in the reference of all human characteristics to the goals, goods, and values of human behavior. It is only in the value-laden terms of a life laying claim to meaning that nature and spirit find themselves in that personal tension that, precisely because it is a tension, bears witness to this very life that carries and permeates *all* layers of human existence. Craniometry, serum studies, and psychological reaction time tests do not add up to an anthropology, nor do the attempts to incorporate the *Homo* genus into putative family trees on the basis of comparative anatomy, physiology, and evolutionary history.

Experience, in fact, is of no use in the foundation of such a science of the "person." Experience means isolation by means of guidelines—guidelines that are never simply taken from the given nor forced upon it, but are rather based on one-sided observation that renders the given visible in a controlled context. It is for the sake of these controlled contexts that a verifiable empirical approach is necessary. A romantic flight from experience is thus no more appropriate than a positivist overestimation of its methods and results. Experience provides much, but not its own foundation, not its own starting points. The givens underlying and informing our experience disappear in our image of experience, and it could scarcely be otherwise. The result does not contain its own prerequisites as separable moments; they rather disappear in it because it exists "through" them.

Studying the pre-experiential conditions and starting points responsible for experience is the necessary complement to a purely empirical knowledge and perception of the world. The goal of such a study does not yet include a determination as to whether these conditions are subjective or whether, as Kant assumed, they are formal. This present work does not venture to say anything general about the nature of their apriority before examining them in a differentiated way. Conditions of the possibility of experience, in any case, do not have to be conditions of knowledge; we can also argue about the possibility of objects and substrates drawn on by experience.

Nor does the study of starting points and [76] apriorities itself have anything to do with metaphysics. It would thus be a serious mistake when laying the groundwork of philosophical anthropology to call on the manifold speculations about the relationship of thinking being to extended being or to attempt new monistic, dualistic, or trialistic theories in this vein, given that it is precisely a reconsideration of the fundamental nature of this very dichotomy that is needed. The task is not to make the fundamental opposition comprehensible (after one has simply resigned oneself to it) and to explain the manifold forms of its appearance in unified entities, but to determine *whether* it is a fundamental opposition in the first place. It may be possible to show that certain metaphysical efforts become meaningless because the phenomena and the conflicts between them on which they are based do not in fact constitute ultimately indissoluble givens. Their conflicts may surface in the outcomes of experience but will never be resolved on the other side of experience because they cannot be found on this side of experience—that is, at their point of departure. Solutions to problems are worthless as long as the foundation of the phenomena from which the problems arise is not secured.

The Cartesian Objection 71

It is the problem itself that decides how we proceed, a problem posed not arbitrarily but rather by the development of many disciplines in the more recent science of philosophy, a problem that has forced philosophy itself to revise some of its basic concepts: the problem of the fundamental nature of the Cartesian dichotomy. History and the empirical cultural sciences demonstrate the lopsidedness of the dichotomy for the domain of the human, but if they hold a reformulation of the living foundation of culture and history to be necessary, then this of course cannot be limited to the "human." The human is carried by living nature; no matter how spiritual he may be, he remains subjugated to it. From nature he draws the strength and material for any sublimation whatsoever. This is why the call for a philosophical anthropology automatically includes a call for a philosophical biology, for a theory of the essential laws or categories of life.

I am not, then, putting forward the naturalistic argument that an anthropology must be founded on a biology, philosophically as well as empirically, because the human is the most highly developed being in the hierarchy of organisms and attained his current form the most recently and, furthermore, because all of his spiritual expressions of life depend on his physical [77] characteristics. It is rather because the construction of a philosophical anthropology has as its prerequisite a study of those states of affairs that are concentrated around the state of affairs of "life" that I am taking on the problem of organic nature. The initiative to construct a concrete philosophy of nature falls not to the mode of experience of the natural sciences (to say nothing of its absolutization), but to that of the humanities.

The most obvious way to solve this problem would therefore be to start with the mode of experience of the humanities and its objects of study. This is the route I took in *Die Einheit der Sinne*. It was my goal in that book to come to an understanding of the grounding, that is, principle-guided nature of our senses from the perspective of the humanities or, to be more precise, from the perspective of the experience we have of the spiritual formations of culture. For natural-scientific and psychological empiricism, the senses themselves are merely contents of experience, the sensations associated with each of them something ultimate that has to be accepted; their nature cannot be discussed. Seeing and hearing are attributed to the functions of the eyes and ears, whose determinants are analyzed by natural science.

By contrast, the *value-oriented critique* of the senses with its focus on their specific capabilities brought to light aesthesiological laws governing the fundamental significance of the different modalities of sensory perception for the makeup of the person as unique unity of body and soul. This gave new value to elements belonging to the psychophysical vital layer of the human, which under the influence of the natural sciences had been regarded as either physical or psychic characteristics. Suddenly, physico-psychic characteristics exhibited an a priori side.

If I have not chosen to continue on this path of aesthesiology, it is because I am concerned with establishing the most stringent controls possible for the validity of its results. In science there is no stronger criterion for the soundness of a finding than its confirmation by a different method than the one originally used. It is for the sake of such confirmation that a new method must be found.

This method is defined negatively as one that does not start from the mode of experience of the humanities, or, to put it more radically [78], it does not start from experience at all. Aesthesiology had a "critical" approach, working regressively from given objective formations back to the conditions of objectivity. The new method must not proceed this way; must not be a "critique" or a regressive analysis. Positively, the method is of course defined by its object. This object, however, is not set from the beginning. Since it does not belong to experience, there are no concepts for it in place before the investigation, and its contours will only emerge as the initially posed problem is narrowed down.

The problem must thus be given a form that actually leads to the determination of its object. Seen from the perspective of the subject of cognition, the problem is thus: for objects that appear in the dual aspect, does this rupture signify an alternative position of the gaze with respect to the objects or not? Seen objectively, the question is: do objects that appear in the dual aspect only have alternative determinations, meaning that the unity of the object is not given determinately but only given up as determinable, or are certain characteristics of unity immanent to or given beforehand for the dual aspect? Is the dual aspect perhaps even conditioned by such pre-given characteristics of unity; are they inherent in its nature?

In order to avoid preselecting the wrong objects—for whether other things besides myself appear in the dual aspect is precisely in dispute and is vehemently denied by Cartesianism—and, in the interest of the method, to avoid beginning with experience in any way, the investigation will proceed indirectly. It is still undecided whether it is intuitively meaningful to speak of states of affairs belonging to the same sphere of being as body and soul while belonging to neither sphere entirely. This would have to be the case if such states of affairs prevented the disintegration of real, intuited natural things into things that only have an external aspect and things that only have an internal aspect (or are concealed). These states of affairs do not have to be sensuously intuitional. The fact that they have been so adamantly denied in the sciences almost suggests as much. But even that which lacks sensuous intuitability can nevertheless have a functional value for intuition, a functional value that only comes to bear in the domain of intuition. The present study must begin by clarifying this question. Here [79] lies the ground for all that follows if we are to proceed without unfounded assumptions and without borrowing from experience.

Our era has the courage to declare the philosophical primacy of the object, as well as the strength to prove it. But the tendency to remain formal and to leave the material questions entirely up to the individual disciplines, and thus to experience, is still very strong in philosophy—it is not easy to separate oneself from a century-old tradition of epistemological formalism. It is true that philosophy has always had a closer connection to the concrete things of culture and history than it does to concrete nature, due to the fact that the humanities, on the level of their experiences and combinations, themselves take into account material a priori contexts. Without knowing what "character," "personality," "the state," or "economy" are and what possibilities they contain, without intuition of essences and insight into the forms of modulation of essences, there is no history and there are no humanities. The association of the mode of experience of the humanities with philosophy actually goes so far because the primitive articulation of the objects (not to mention of the theories) of the humanities requires an already given background of essential laws and guiding images without which spiritual-historical reality simply does not *exist*. Exact natural science, by contrast, rests on entirely different foundations: autonomous in relation to its objects, on the level of experiences and combinations it is free from philosophy. The connection between philosophy and concrete nature was accordingly lost to the extent that exact natural science took possession of the latter. Conversely, the influence of discoveries in the natural sciences on philosophy diminished the more philosophy came to see the natural scientific method as the only way to know nature. There can, after all, be no concrete philosophy of nature as long as the dogmatic or methodological anthropocentrism remains valid that brings about the dichotomy between the physical perspective and the consciousness perspective—just as this dichotomy brings about anthropocentrism.

The significance of this problem should be clear from all that has been discussed thus far. A whole host of solutions to basic problems in philosophy and the theory of science depend upon deciding whether the Cartesian dichotomy is fundamental or not. The natural form of the relationship between these problems and thus the reach of the question broached here will only emerge, however, in the actual work of the investigation before us.

[80] THREE

The Thesis

The Question

The intuitive habitus of objects that appear in the dual aspect is characterized by a disintegration into inner and outer. Does this disintegration signify an alternative position of the gaze toward these objects or not? Do objects that appear as unities of inner and outer elements only have alternative determinations, meaning that the unity of the object is not given determinately but only given up as determinable in the idea, or are certain characteristics of unity already embedded in the dual aspect or given beforehand for it, given to it? Is the dual aspect perhaps even conditioned by such already given characteristics of unity; are they inherent in its essential structure? Is the disintegration into two aspects that are not mutually convertible still compatible with the intuitive unity of an object, and under which conditions is that the case? Toward what objects is there a convergent gaze onto what are in principle divergent spheres of objects?

Inner and outer as spatial aspects determine divergent, but not mutually inconvertible sides of an object. One can enter into a jug from the outside; the transformation takes place through undivided space. If the sides of the jug are of the same thickness throughout, the spot where they bulge out has a corresponding concavity on the inside. Convex and concave are directionally polar, and yet only one rotation can bring them into alignment. Here the inner can become the outer, the outer the inner, as when a glove is turned inside out, thereby overcoming the directional polarity of "congruent counterparts," as Kant calls them, of left and right.

Spheres that are divergent in principle and upon whose orientation toward each other the unity of an objective structure is said to rest [81] correlate with each other in a polar way like the interior of a space to its exterior. Unlike this last relationship, however, the divergent spheres are not mutually convertible. Gustav Fechner's theory of the two sides, which harks back to Descartes and Spinoza, aimed for a formulation of the relationship between *physis* and *psyche*, between outer and inner, which particularly in its intuitional foundations evokes the spatial relationship between outer and inner. Fechner's examples especially appeal to this relationship. In contrast, it is in the interest of the advancement of the present analysis, which will furnish answers to the questions posed earlier, to avoid inappropriate comparisons with the spatial relationship between outer and inner when determining the relation of inconvertible spheres to each other. I will do so by looking *at spatial objects* in intuition. This will allow the spatially conditioned relations between outer and inner to clearly stand out from those that are not spatially conditioned and those characteristics of unity to emerge in isolation that, apprehended in the convergent position of the gaze (that is, in the pathways of sense perception), also carry the divergent aspect of the object.

The Dual Aspect in the Appearance of Ordinary Perceptual Things

Every thing perceived in its full character as thing appears in accordance with its spatial boundaries as a unity of properties organized around a core.

The tree outside my window is not merely an aggregate of color data held together by a gestalt; when I approach it, it does not merely add new touch and smell data to this aggregate, and the rustling in its branches is not merely the perhaps final addition to the sum total of its appearance. Above all, the tree out there stands on its own [*ist eine selbständige Größe*], as long as it does not explicitly present itself to my gaze as a phantasm or, as sometimes in a dream or in eidetic consciousness as defined by Erich Jaensch, as a pure image halfway between mere mental image and true perception. The properties *in* which the tree itself manifests are attached *to* it. The sense data embedded in the overarching and dominant gestalt characteristics are neither exhausted in the construction of a colorful phantom with a thin and, as it were, flat organization, nor appear as moments pinned on to the outside of a substance that can be arbitrarily removed from it—that is, that overlay it. [82] The independently established tree-thing shows itself in the sense data and as them, so that every human awakened to a full perception of reality [*Wirklichkeit*] must say that the bark of the tree *is* cracked; its leaves *are* green. The sense data belong to this self-existent tree as its determinations.

To express the state of affairs just described from the reverse direction: it belongs to the nature of this structure that the sensuous intuitive data, as the properties "belonging to" this thing, point into it as a thing that is thoroughly bound to the core at its center without thereby making this core appear completely. The leaf has green on its surface, but green does not conversely have the leaf. The dependency of the property on the core substance of the thing, the state of being-carried in contrast to independence, is intuitively expressed in this state of being-had (which means the same thing here as being supported or carried).

The part of the thing that appears as real [*reell*] and can be verified by the senses as a tree or an inkpot is only one of an infinite number of possible sides (aspects) of this thing. For intuition, this real thing [*dieses Reelle*] is certainly the thing itself—but from one side, not the whole thing, which in a real [*reell*] sense can never be verified by the senses "all at once." The really [*reell*] present side only *implies* the whole thing and appears to be embedded in it, even if it is impossible to procure sensory verification either for the whole thing or for the way the appearing side is embedded in it. We can turn the thing over, walk around it, cut it up as much as we like: that which can be verified by the senses remains an extract from a structure that does not itself appear all at once, but nevertheless is given intuitively as the existing whole.

The real [*reell*] (verifiable) phenomenon points to this carrying whole of its own accord; it transcends its own frame, as it were, by presenting itself as something breaking through or emerging into view [*Er-Scheinung*], as an aspect, as a manifestation of the thing. The way in which the real [*reellen*] phenomenon belongs to the whole thing, which cannot be verified by the senses, consists in this transgredience of the phenomenal content. Only because this transgredience contributes to the determination of the real [*reelle*] phenomenon is the latter more than a mere perspective *on* the thing, but rather an aspect, a side *of* the thing.

For the concrete appearance of the thing there are two directions of transgredience that—curiously corresponding to the spatial determinations—essentially belong together, although they never coincide: the transgredience of the phenomenon "into" [83] the thing and "around" the thing. The first direction aims at the substantive core of the thing; the second at its possible other sides. This double direction of the gaze belongs to the real [*reellen*] image if it is to be perceived as a present thing, and it is only in this dually directed giving of the gaze that the spatial sensory phenomenon appears as a unity of sides organized around a core, as a thing.

Kant, Hegel, and in our time Husserl duly emphasized this law of the necessary one-sidedness of the appearance of the perceptual thing, which, thanks to its phenomenal transparency, has an infinite number of sides. The thing's core, the "axis" of its being, is *neither* really [*reell*] immanent in the real [*reellen*] phenomenon—that is, verifiable, demonstrable in this phenomenon, *nor* transcendent or intellectually appended to the phenomenon and thus without bridges to it. In this sense the thing necessarily appears as adumbrated, as Husserl writes. Not because our senses are not everywhere at once and we cannot perceive the total thing with our concentrically focused sensory system does this law apply, but rather because in the essence of the appearance of a thing that is more than a mere apparition there is aspectivity: being-from-one-side. In no way does this mean that aspectivity is subjectivity; rather aspectivity is only the possibility of inhabiting an opposing position to the subject guaranteed by the phenomenon. Aspectivity as the boundedness belonging to the object itself, as the sidedness structurally belonging to it in its appearance, is not to be confused with the image that remains in consciousness as a perceptual or mental image. Those who take Husserl's law of adumbration to be a reversion to subjective-idealist lines of thought are too influenced by Husserl's own interpretation and are not clear enough about this difference between aspectivity and subjectivity.

The perception of the real [*reellen*] phenomenon anticipates the direction into the thing and around the thing. This anticipation could also be described as the thing appearing as a "deep" continuum of aspects. "Into" and "around" seem to be decidedly spatial predicates, however. Are depth and sidedness brought about by the spatiality of the thing, or are depth and sidedness the cause of its spatiality?

The two characteristics, in any case, are not identical—as the question itself shows. To be spatial means having demonstrable boundaries in space. As a spatial [84] formation, every thing has particular dimensions in a particular place, intuitively speaking; it has contours, a demonstrable periphery, a demonstrable center. In spatial terms, one can place one's finger on its center and sides. One cannot, however, place one's finger on the center and sides as characteristics constituting the thing. The attachment to a center of sides carrying properties is not merely a metaphor for the nonspatial relationship between substantive core and property; by the same token, it is impossible to demonstrate the existence of this attachment in space. Thus moments constituting things and spatial moments, while inseparable from one another in intuition, are not identical.

The law that something has properties and only appears in its properties without being completely exhausted by them can also be defended in relation to the nonspatial reality of psychic life: will, feeling, and thought are more than the sides they turn toward a consciousness. Spatial images for this relationship of transgredience between the phenomenon and the core content of a psychic reality probably only have metaphorical value; nevertheless, it is true that the relationship between the phenomenon and the real core content, which gives the determinations of the phenomenon the value of properties and the real core content the value of a substance, is the same for nonspatial reality [*Wirklichkeit*] as it is for spatial reality. This comparison shows that the structure of transgredience—that is, the relationship between the substantive core and the side or property—is indifferent to the distinction between spatiality and nonspatiality.

The colorlessness of this structure of transgredience allows it to join with the particular structural forms of the material in whose formations it can be identified—whether they be spatial or nonspatial, such as psychic realities—into an inseparable intuitive unity.

It is only conceptual reflection that separates the meaningful relation between core content and property from the materially conditioned mode of being. This reflection makes comprehensible or at least expresses that which the thing really reveals in its temporal existence *if it becomes subject to destruction*: the manifest detachability of that which took up space as reality from that which filled it three-dimensionally as form and material. Thus the ash that the cigar becomes documents the impermanence of both the form and the material—that is, the indifference to their function of representing a reality, [85] to their belonging to a reality in their properties. If the earlier phenomenon of the cigar is not to be more true than "its" current one, if it was just as real [*wirklich*] then as it is now, then that which outlasts the vanishing into each other of the phenomena must also stand above them in its mode of being. In the *Phenomenology*, Hegel shows how consciousness here no longer has the possibility of spatially interpreting the core content and of seeing that which is substantive in the actual as its center. When the appearance dissipates, this substance of necessity becomes that which takes up space without filling it—that is, force. The static character of the thing's core makes way for the dynamic one.

For the spatial thing as it is perceived, the convergence of all its potentially appearing sides, or the properties embedded in its sides, onto the ("central") core content is the expression, corresponding to its spatiality, of the nonspatial state of affairs thereby comprehended. Although intuition tempts us to do so again and again, only reflection convinces us of the futility of trying to get closer to the central core content by really penetrating the thing, by peeling off layer after layer. The spatial center is not the "center" of the core, as intuition falsely believes it to be. And the spatial periphery is not the unity of the property-carrying "sides," as perception must suppose. Between the moments of depth and sidedness constituting thinghood on the one hand and the moments of depth and of a closed surface constituting the spatiality of the thing on the other there exists—in answer to the question posed earlier—a relationship not of causality but of a purely reciprocal determination. Thus, in intuition, characteristics conditioned by space essentially correspond to those conditioning space, as do spatial [*räumlichen*] determinations to spacelike [*raumhafte*] determinations.

It is possible that this peculiar law was not without influence on the famous metaphysical disputes on the relationship of space to substance. Of course, there were entirely different ontological issues at stake in these controversies, but the question as to whether space or substance (conceived for instance as force) is prior inevitably points, regardless of how it is answered, to the differences just touched on between spatial characteristics and those of a spacelike nature because these can be made intuitively evident.
[86]

Against the Misinterpretation of This Analysis:
A Closer Focus on the Subject Matter

The fact that the substance of the thing can neither be understood as the quintessence of its properties nor as the quintessence of that to which it can be reduced following the exact method makes it unavoidably clear that it is inappropriate to apply the calculating method to this initially only intuitive state of affairs of substantive core and property. That which intuition experiences as substance becoming manifest in properties while at the same time remaining hidden beneath and behind them defies any scientific dissolution into elements, into electrons and energies. The substance of the thing is not that of which it is composed, does not refer to an inside in the same way as the medullary ray is inside tree tissue, sawdust inside a wooden doll. Just as the naïve approach of breaking open a thing in order to get at its inside as that which it truly is, its essence and its core, is in its tendency a model for the elementary analysis of atomization in the natural sciences, the scientist employing the exact method necessarily fails to capture substantiality. (A correct understanding of the exact approach shows this not to be a deficit. Only a false interpretation of the work and objectives of natural science leads to an interest in denying that even the simple intuition of the perceptual thing has the structure of substance and property and to a conception of its properties as mere sense data following the principle of sensualism. Only the belief that the exact method is the only way to study nature leads to the attempt to remove anything from the object this method cannot explain.)

To what lengths one has gone, calling upon facts of developmental psychology and arguments of epistemology, to invalidate or cast suspicion on the claim that the relationship between core and property is already embedded in the intuited subsisting thing *as it is intuited*. Such a claim, it is argued, is an interpretation based on experience, a particular expectation based on gradually enacted associations, a sediment of intellectual processes, even of judgments, which, thanks to the swiftness with which the mature human reacts, no longer arise separately from the properly sensuous intuitive substrate—although they are in fact separate from it. "Substance," the argument continues, is a concept to which we come late; it presupposes a certain degree of experience with [87] things in order to be understood, let alone put to work against the world of experience. Or it is argued that substance is a category of knowledge, a concept of the understanding, and thus not intuition. Some allow the relationship between substantive core and property its apriority, but only at the expense of its intellectual rationality. Others declare it to be a relatively late product of experience, but emphasize its rational nature as an auxiliary tool in the interest of science.

None of these theories would make any sense if they themselves did not in some way come upon and find offensive the phenomenon of an intuitive order of sensory subsistence, appearing as a substance-property association. The purpose of the theories is, after all, to explain how such a strange construct entered into the consciousness of intuition in the first place. They find it offensive because it is incompatible with the basic principles of natural science, with the methodological principle of sensualism, and are nothing but attempts to interpret it according to or in agreement with the sensualist principle. No theory, no matter how dissatisfied it is with this phenomenon, can deny (without immediately making itself superfluous) the fact that a certain mature perceptual consciousness intuitively grasps, "means" things (hallucinated or real) as substance-property structures.

But we are concerned here only with the structure of the phenomenon, not with its genesis, *not* with its legitimation, and *not* with its truth value. Even objects of illusion and hallucination exhibit a substantive core—otherwise the subject would not have been misled into believing in actuality where there is none. Having a substantive core, which is essentially correlated with the propertiness [*Eigenschaftlichkeit*] of the sensuous and formal determinations of the thing, is, to begin with, only a special structure of the full appearance of the thing. The substantive core is not contained in the real [*reellen*] image of the thing's appearance and cannot be identified in any possible real [*reellen*] appearance of the thing, nor can it be covered by certain gestalt features. Only as support and background does it provide the point of orientation for the transgredience of the possible appearances in relation to the (itself never full, but only perspectively adumbrated) phenomenal, intuitive unity of the thing.

Leaving aside the present analysis, there is no better witness than Hegel in the early sections of the *Phenomenology* to the fact that the preliminary nature of the concept of the thing immanent to intuition—or of that perceptive intention that corresponds to such a concept—does not mean it is untenable [88]. Hegel launches the process of systematic overloading, which drives consciousness from one apparent rest position into another until it has found itself, by having intuition address itself as perception. Intuition thus demands too much of demonstrably real [*reell*] phenomenal objects by addressing them as something they ("in truth") are not yet. From the standpoint of an adequation between conceptual-cognitive and intuitive certainty this may be untenable; perhaps Hegel was also wrong to side against this static principle of adequation and to defend the dynamic side. In descriptive terms, however, he was right: under no circumstances should the ambiguity of an object be invoked against the intuitional intention in fact aimed at it.

Our first task seems to be completed: the spatially conditioned relations between outer and inner of the spatial object of intuition have been differentiated from those that are not spatially conditioned, thus isolating those characteristics of unity that are apprehended in the convergent position of the gaze (that is, in the pathways of sense perception) and that at the same time carry the divergent aspect of the object. For the structure of central core content/property-carrying sides is one of fundamentally divergent spheres of objects that are, by their nature, never mutually convertible. The object is not destroyed by the schism of an interior that never appears—that is, never becomes an outside—and of an outside that never becomes core content. It rather forms itself, as it were, out of this schism into its typical unity as a thing.

The value of this outcome is, however, qualified by the fact that aspect divergence, which was shown to be a prerequisite of every unity appearing as a physical thing, does not itself appear. Intuition only becomes aware of a closed, solid formation with a core whose outer surfaces enclose an interior.

It is only subsequent consideration that analyzes the preconditions of the reach of intuition going beyond what the senses can apprehend without itself becoming perceivable for sensory intuition. Only philosophical reflection brings to light the fact that the relationship between inner and outer is in fact a genuine aspect divergence and not a relation of relative concealment [89] effected by the outside covering up the inside. The dual aspect constitutes the intuitional form of the thing-body, but as a genuine condition it loses itself in that which it conditions. It is precisely the simplicity of the intuitional object that belies the complicated nature of the preconditions.

The outcome obtained thus far will increase in value if we succeed in finding objects that not only appear by virtue of the dual aspect, but *in* the dual aspect—objects, in other words, where the divergence of the spheres determining the object itself becomes the object of intuition. In concrete terms, the next step must be to seek out those perceptual things for which the relationship between outer and inner emerges in the image of intuition as itself objectively determining the object, to analyze them in their intuitive givenness, and, furthermore, to work out the consequences this holds for their nature.

The Dual Aspect of Living Perceptual Things: Köhler contra Driesch

Physical objects of intuition for which a fundamentally divergent relationship between outer and inner objectively figures as part of their being are called *living*. This definition immediately gets to the heart of the difficulties posed by the property of being alive, which, even on the level of appearance, cannot be placed on the same level as the other properties of the same body. It is not overstating the matter to suggest that it is with this problem alone that every theory of life has ultimately been concerned.—

This does not mean, of course, that a thing that appears as alive is completely different from things in general. The essential characteristics of physical things are the same whether they are alive or not. A frog or a palm tree is subject to the same phenomenal laws of thinghood (not to mention the broad zone of continuous physical commonalities) as is a stone or a shoe. It is only that animate things in relation to inanimate ones have the surplus of that mysterious property of life, which, despite its nature as a property, materially changes not only the appearance of the particular thing, but also formally its *mode* of appearance.

[90] We must turn now to the question of whether this change should be classed with phenomena that can be referred to as formations with complex qualities or with a gestalt nature (despite the differences between Felix Krueger's and Köhler's views) in contrast to those that are merely aggregates.

Traits of a purely aggregate nature accumulate or decrease, forming pure "and-connections." An example of this is when a figure drawn in pencil is filled in with different colors. The original figure is enhanced by certain characteristics in the form of "and . . . and . . . ," and the final whole has more properties than the initial whole.

As has rightly been pointed out in modern psychology, this purely additive and-type accumulation of traits (which itself may become the focus of attention) corresponds in lived experience to a qualitative reassessment of the entity carrying the traits rather than a quantitative reshaping, as one would expect in theory. The pencil drawing is a whole, but the colored-in drawing is another whole, which is experienced as having similarities with the original figure, but (despite the de facto existence of continuous characteristics) not as being partially the same. Although it came about by means of aggregation, it presents itself as a whole. The overall gestalt *appears* to us to be primary; it is only by means of isolating abstraction that we become aware that it is made up of components. We are very familiar with the fact that in the case of musical formations, chords, and melodies, or of facial expressions, a purely and-type change in the makeup of the formation corresponds to a holistic change in how it is experienced. Perception always encounters here a phenomenon with its own original character that does not give up its inner coherence regardless of its greater or lesser similarity to other phenomena.

Of course, it would seem reasonable to apply this law to the phenomenon of the living thing. We seem to have the same case here as with the aggregates described earlier, whose appearance exhibits a specifically holistic character: variations in traits induce qualitative overall changes; the addition or removal of an isolatable element is accompanied by an overall variation of the whole formation. The peculiar preponderance of the property of being alive over other properties such as, for instance, form, [91] color, size, weight, etc., in the appearance of the living body, a preponderance that seems to contradict the nature of properties, can perhaps be explained by gestalt theory in this way. Köhler attempted this explanation in his essay "Some Gestalt Problems."[1] Of particular philosophical interest is his attempt here to replace the "mechanism vs. vitalism" opposition with a new position.—

In discussions regarding the causal integration of phenomena of life, the different parties traditionally begin from the following three assumptions: (1) Understanding natural phenomena means explaining them causally; (2) The scientific explanation of biological phenomena aims at tracing them back to chemical and physical relationships and their laws, which are ultimately modeled on the mechanical relations between isolated particles; and (3) Biological phenomena are natural phenomena with a certain surplus of vitality in relation to inanimate phenomena, identifiable as a number of properties (growth, metabolism, reproduction, ability to regenerate, responsiveness to stimuli), which as an aggregate—although not taken individually—characterizes life. Whether or not they believe this surplus can be traced back to chemical and physical factors distinguishes mechanists from vitalists.

Mechanists can point to major advances in identifying as inorganic things previously thought to be vital, without however being able to convince vitalists that these two categories are basically the same or that the organic can be completely characterized in mechanical terms. Conversely, vitalists strive in vain to convince their opponents that the complication given in the vital character of organic processes signifies an essential boundary between living and nonliving nature. Understanding between the two parties was stymied by the claim of the early vitalists that the autonomy of the living is based on the presence of particular substances or forces unique to organisms. According to this view, the living physical thing inhabits a special position because of certain objective properties [92]. The mechanists countered with the logically well-founded objection that from the standpoint of causal research—that is, of experiments and measurements—no fundamental difference can be made between property and property. Any existing substance or force belongs to the nexus of natural events and formally complies with the uniformity of conditions as which we must always treat reality [*Wirklichkeit*], even if it presents us with insurmountable obstacles.

But the situation is not as inauspicious as that. The history of recent biology is, as it were, the history of the vitalists' retreat before the continual physicochemical conquest of the entire territory of life phenomena; it is the final surrender of the property vitalism of vital substances or forces.

Modern vitalism, as represented by Driesch, does in fact recognize the validity of physical principles for the organic world. It is only that physical characterization does not exhaust living being. The energetically conditioned phenomena of such being, furthermore, betray the efficacy of a factor called "entelechy" that operates into space rather than being enclosed in it, which the concepts of physics completely fail to grasp. Phenomena such as autoregulation and restitution, reproduction, and the development from lower to higher degrees of plurality, Driesch argues, cannot be fully understood in terms of energy or of the constellation of final particles. An ideally conceived developmental physiology could, at best, make exact calculations of the material conditions that give entelechy the *opportunity* to intervene. A biology that wants to become a physics of the organic can only ever provide knowledge of the occasions, the opportunities for that nonspatial factor of entelechy—not to be compared with a force or an energy—to take effect.

Unlike traditional vitalists, Driesch does not believe in this or that objective property of the organism, but rather sees entelechy manifested in the peculiar *preponderance* of those properties considered to be the specific characteristics of life (primarily restitution, development, heredity, and activity) over other properties of the living body. For the properties characteristic of life have this strange preponderance over the noncharacteristic ones of color, contour, [93] weight, size, materiality, etc., *because they operate according to laws of wholeness*.

The sober-minded justification of the intention of natural science to study the organic world according to the principles of measurement invokes Kant's *Critique of Pure Reason*, according to which the introduction of an entelechy factor contradicts the principles of pure natural science. Driesch for this reason sought to use Kant's method of the deduction of categories to legitimize wholeness as a category for the investigation of organic phenomena. In addition to summative causality, which is thought to be characteristic of inorganic nature, there is also, Driesch argues, holistic causality as the specific way in which organic becoming creates conjunctions. Organisms cannot be understood as machines, if by a machine we mean, as in analytical mechanics, "the system" with precisely defined connections. For Driesch, this "machine" constitutes the model of physical lawfulness *as such*, and the notion that biological processes can be derived from physicochemical processes must be abandoned the moment it becomes clear that arbitrary "disruptions of 'stages' (that is, of a system) in a series of events nevertheless render the proportionally correct, compound final whole."[2] It was Driesch's own experiments with sea urchins, ascidians, and turbellaria that ascertained this fact in the first place.

It would, then, pose a serious threat to the theory of entelechial holistic vitalism if it could be proven that there are also inorganic wholes. Thus evidence of the holistic mode of response and structure of crystals and colloids has always been an important element in the polemics against Driesch. Now, Köhler's study, which is grounded in psychological gestalt theory, of the occurrence and properties of physical gestalts has breathed new life into the question.

Both "whole" and "gestalt" mean more than the sum of their parts. Following Köhler's definition, we can say that their characteristic properties and effects cannot be composed out of the same type of properties and effects as their so-called [94] parts. For Köhler, a pure sum is a system that can be made out of parts, "taking one after the other without any of the 'parts' changing as a result of being added to the others."[3] Köhler uses examples of electrical or chemical processes and states to show that formations that are more than their parts, and the restoration of these formations after interference, exist in the inorganic realm—and in fact are not that unusual at all. If in a three-way capacitor system with conducting spheres you remove a third of the total quantity of electricity from one of the three spherical capacitors, the gestalt formed by the charge is preserved at two-thirds of the original quantity of electricity. This could be described as physical restitution. The restoration of a chemical balance, the shape of a drop, etc., amount to the same thing.

Is there an essential commonality between physical phenomena of this kind that are more than the sum of their parts and organic wholeness? Driesch denies this. In his view, Köhler *only proved the existence of unities, not of wholes* in the inorganic realm, and these unities, unities of effect, as he calls them, can be understood summatively, that is, "as aggregates based on our knowledge of the parts, including their dynamic potencies. . . . Certainly, if we introduce a new charge into a system of electric charges, all the parts will change. But their *essences* do not change, only their dynamic actuality [*Aktualität*]—an actuality whose strength in the various directions never depends on the parts alone, as is made clear by Coulomb's and related laws. This, however, falls under the auspices of summative unity of effect alone."[4]

Köhler also proposes that none of the physical givens appearing in gestalt form freely develop their own structures, but that it is always only a matter of a complex of fixed conditions "that spatially bind the structural material and at the same time specifically determine its mode of diffusion. We called these conditions the physical topography, also the physical form of the gestalt region."[5] For Driesch, this constitutes a crucial admission of the essential difference between physical gestalt and wholeness. It precisely shows, he argues, that "physically inanimate structures are not wholes *of their own accord*, as a function of their *own* [95] nature. As a function of their own nature they are only unities of effect, and any wholeness they might have . . . is . . . forced upon them by the topography."[6] Only if it were possible to prove that in the apparatuses of physicists and chemists (for example), the physical topography of a spherical capacitor, lightning rod, etc.—that "machine" made by humans—could restore itself could we equate organic wholeness with gestalt. Wherever "energy and electron distribution appears to be holistic, *however*, it is the wholeness of a *given machine* that ultimately is the basis of anything holistic." Thus the wholeness of what Köhler calls "structure" precisely does not arise from inner forces, does not originate spontaneously from an inner impetus.[7]

According to Driesch, the attachment of physical gestalts to a fixed topography distinguishes inorganic from organic wholeness. In the case of organic wholeness, there is no underlying rigid topography or machine imposing its "property" on each part of a variation. "The fact that the properties and functions of a particular part depend upon its position in the whole to which it belongs is a basic property of all . . . entities with which so-called gestalt theory is concerned," writes Köhler. In Driesch's view, this only serves to obscure the essential difference between inorganic and organic functions and properties: in the former, it is the matter of merely quantitative variations of force or energy; in the second, of complicated achievements and associated capabilities. Driesch's statement that the true fate of a cell is a function of its position in the whole must always be understood in the context of the underlying state of affairs: organic gestalt formation does not take place as a result of a machine whole given beforehand. "And even if we were to see the machine only as a very complex system of fixed peripheral values [*Randwerte*]—well, then, in the *living*, the *peripheral values as such* would reconstitute themselves in wholeness after being disturbed, which inanimate entities never do, apart from a few very simple, very specific cases, such as homogenous drop formation, which can in no way serve as analogies for the biological."[8]—Physical gestalt, Driesch argues, differs from physical wholeness in respect to auturgy (Wilhelm Roux)—that is, self-action. Living gestalt formation as autonomous and automorphic stands opposed to dead gestalt formation as heteronomous and heteromorphic.

[96] Is this deep insight not, however, the result of a somewhat unsound line of argument? Does the essential difference between dead and living gestalt, as Driesch himself sees it, not reside on a level higher and on a plane of being of another order than is determined by the being of the gestalt? The polemics between mechanism and vitalism have moved to a higher level because the adversaries have moved closer together than they ever were before. For the mechanist, a "mechanical" explanation must no longer simply adhere to the model of the sum of particles, but can also follow the model of the gestalt: an effective unity as a transposable totality in form. For the vitalist, the uniqueness of living conditions and processes has been reduced to the auturgy and autonomy of their gestalt systems by virtue of which they come to constitute wholes with a spontaneous impetus. Instead of only negatively distinguishing between effective unity and wholeness, does this not make it necessary to positively identify the boundary that must be crossed *in order for* a gestalt to exhibit the specific predicates of wholeness?

The fact that the broken Leyden jar cannot transform itself into two proportionally correct jars is not a valid objection for Köhler; he does not compare the entity "physical topography + physical structure (of an electric charge, for instance)," but only the physical structure with the living physical thing. The *tertium comparationis* is the gestalt nature—that is, the transposability of the structure when there is variation (such as in the chemical components or in the quantity of an electric charge). In response to interference, the gestalt of its own accord reconstitutes itself (within the scope of what is possible), spontaneously restores itself from an inner impetus, despite the fact that its structural material is spatially bound and has a mode of diffusion that is specifically determined by the given topography.

This is, indeed, remarkable, and it is Köhler who first pointed it out. It is conceivable that the structural material "could" react differently to the physical topography, "could" fail to accommodate the wholeness of the physical form. If a ball is covered in wax and then cut in half, the unheated wax on each half does not acquire a gestalt similar to the initial one. And yet the nonheated layer of wax at the outset and upon the intervention constitutes one effective unity. By drawing attention [97] to the spontaneous, inner impetus-driven gestalt-like reactions of certain structural materials, Köhler is not implying that the occurrence of these reactions is irrational. On the contrary: the physically determined nature of the structural material and of the topography in question necessarily leads to the reaction, which for this reason must be understood as gestalt-like. It is a reaction to the gestalt of the given topography spontaneously ensuing from the inner conditions of the structural material; its mode of diffusion is bound by the topography. For Köhler it is only important that some physical materials show themselves to be capable of bonding and of reacting to gestalt, as not every effective unity in inorganic nature exhibits this property.

There is no doubt that the study of these physical gestalts will help resolve organic processes to the present day still considered to be specifically vital into physicochemical processes. Even Driesch admits that the peripheral values in homogenous drop formation as such reconstitute themselves into wholeness after being disturbed. The progress of colloidal chemistry will shed light on restitution phenomena and especially on the basic phenomenon of the self-differentiation of a relatively undifferentiated original system into a highly diverse one—just as contemporary colloid research has begun to shed light on the previously poorly understood causality of form constancy when parts are highly movable and the state of matter lies between solid and liquid.

The fact that systems with their own material holistically reconstitute their peripheral values (that is, to a state similar to the whole at the outset) without a pregiven machine, microstructure, or physical topography outside or within themselves; the fact that they achieve differentiation (such as by means of complicated developmental processes), where "the" parts are reconfigured not only according to their dynamic actuality [*Aktualität*] but according to their "essence," truly no longer lies, in principle, outside of the domain of inorganic possibility. Premature analogies drawn (particularly by the lay public) in response to the excellent observations by the likes of Ludwig Rhumbler, Otto Bütschli, and Emil Lehmann are detrimental, however. We do have to acknowledge, out of loyalty, that such analogies are foreign to Köhler. The particular context in which a certain gestalt law is made out is also crucial to this debate. *Relative to a system* of electrical [98] charges, the change in the dynamic actuality of the parts brought about by introducing a new charge certainly is a change in essence—after all, what sort of essence do parts of an electrical charge have and can they even have? Relative to the kind of system, change in position and dynamic actuality means the same thing as what jumps out at us as the *qualitative* change in, for instance, cells in an advanced cleavage stage—changes in position, plasma granulation, form, and specific material and functional properties. While we must not attempt to grasp the infinitely complex object in analogy to the much simpler process, it is nevertheless permissible to work out commonalities between the two.

The best way to determine who is right in this debate is to examine the question of the essential boundary between (inorganic) gestalt and (organic) wholeness, independently of the psychological arguments against gestalt theory. It seems likely that when characterizing the autonomy of living beings by anxiously seeking to demarcate them from the sphere of summative causality, things that belong to the sphere of physical effective unity—and that thus are precisely not exclusive to the vital—are cited as evidence. Perhaps Driesch is waging his defense of the indissoluble features of the organic with weapons unable to do any serious damage to the theory of physical gestalts. This does not mean, however, that the gestalt theory of vital wholeness phenomena is correct or that the cause to which Driesch rendered the greatest service is lost.

Clearly our starting hypothesis—physical objects of intuition are called living things if they objectively exhibit as belonging to their being a fundamentally divergent relationship between outer and inner—draws the boundary between living and inanimate beings in a different way, and in a way that includes the gestalt moment. The relationship between outer and inner, which manifests in the particular body as an objective determination like its color, shape, weight, surface texture, or degree of hardness, determines the appearance of the thing-body as a whole. Although this relationship—as one property among others—only denotes one summand in the sum of all determinations of the living thing-body, it does not appear to be on the same level as these other determinations, but above it.

[99] Can this higher position still be integrated into the concept of the gestalt, or is Driesch right to see in it a special kind of order (that of wholeness)? Is the body possessing the property of vitality only structured, in this respect, as more than its parts insofar as its characteristic properties and effects cannot be composed out of the same type of properties and effects as its parts, or is the preponderance of vitality (manifest already in intuitive appearance) based on a mode of ordering that is beyond gestalt?

If the latter is true, we can no longer categorize specific wholeness phenomena under gestalt phenomena. Second, we must determine whether or not this supplies the decisive vote for vitalism.

How Is Dual Aspectivity Possible? The Nature of the Boundary

Whether the organic form is based on a gestalt mode of ordering or one that is beyond gestalt can only be answered in relation to the hypothesis that living bodies appear to us as exhibiting a fundamentally divergent relationship between outer and inner as objective determination. Is the gestalt nature of a formation sufficient for it to exhibit dual aspectivity, or does this require another, higher type of order that encompasses the gestalt nature? These two issues cannot be separated, as they both refer to one and the same circumstance of the intuitively and phenomenally grasped living body as whose essential characteristics they appear, independently of whether this body exists only in intuition or also in reality. It is inconceivable that essential characteristics that belong to the same level of appearance would be indifferent to each other. It is more likely that they point to a shared underlying law.

We must thus now turn to the relationship between gestalt and dual aspect. The analysis of this relationship will include a positive indication of the boundary that must be crossed in order for a gestalt to exhibit the specific predicates of wholeness.

At first glance, the gestalt and the dual aspectivity of a formation have nothing to do with each other, even if one only considers what are clearly physical gestalts. Where would it be possible [100] to distinguish between inner and outer in the totality of electrical charges or even in a chemical equilibrium between reactions, apart from the physical form (topography) to which these gestalts are tied? If physical topography is taken into consideration, however, the distinction between inner and outer becomes a spatial and relative one, or it only characterizes the phenomenal law of physical things as such, discussed earlier.

The hypothesis explicitly sets forth that dual aspectivity must manifest objectively in the thing, in the property position, in other words, in order for the thing to be called "living." For intuition this means that the phenomenal totality of the thing-body presents itself as the exterior of an *undemonstrable interior*, which interior—*nota bene*—is not the substance of the thing but belongs to its (otherwise demonstrable) properties. This is why the core of thinghood as carrier of all possible predicates (properties) of the thing never coincides with the centrality from which the specific vital expressions are perceived as originating and as being held.

In order to be able to distinguish between an inward direction and outward direction in a particular formation, it must contain something that is neutral as regards this difference and allows for a tendency into one or the other direction. The two directions collide, as it were, in this neutral zone; they both originate there. It is through this neutral zone that one must pass in order to get from one area to the other. The difference in direction between the two areas remains if the direction of movement is reversed when passing through the neutral zone. The nondirectional zone may not itself occupy an area, as this would nullify the exclusivity of the directional opposition within the formation in question and thus posit a really demonstrable in-between in addition to inside and outside. This makes it a *boundary*.

The proposition that living bodies appear to us as exhibiting a fundamentally divergent relationship between outer and inner as objective determination can therefore be formulated as follows: living bodies have an appearing, intuitive boundary.

Intuitive boundaries are located in all thing-bodies where they begin or end. The boundary of the thing is its edge [*Rand*] where it butts up against something that is not itself. At the same time, [101] this beginning or ending determines the gestalt of the thing, or its contour, whose lines can be traced with the senses. The thing-body is enclosed within its contours, its edges, and is determined as this thing—or, what is the same, its contours and edges determine the thing as this thing. Only as a vague and abstract figure of speech can a contour be thought of as separate from the thing it contours. A boundary can at times stand out as a conspicuous gestalt, but it still cannot be grasped as something distinct from that which it encloses or against which it abuts. For mere intuition this may seem to be possible, such as when contours are represented by drawing simple lines. But there is no separate entity that corresponds to the line. By visually separating the bounded spatial area from its setting, the line simply registers that which in its nature is a pure transition of the thing-body to the medium that surrounds it.

Everyday language does not make a sharp distinction between things that *have* such and such a boundary and those that *exist with* such and such a boundary. It relies entirely on sensory intuition without accounting for the fact that the thing in question does not have its boundary, gestalt, shape as something existing independently from it, but that it is with and in this boundary, as this boundary, which represents the thing's beginning or ending in relation to another outside of itself. Everyday language thus seems to be blind to the conceptual side of the matter.

On the other hand, we must keep in mind that language basically treats each of the so-called properties of the thing-body the same as it does the boundary contour. As ordinary language would have it, surface texture, degree of hardness, weight, color, tonality, depth structure are all just as much the thing itself as they are its properties. We say that a thing sounds and feels a certain way, weighs a certain amount, is a particular color—but we mean (despite the turn of phrase) that it has sound, surface, weight, and color as properties. Like these the boundary contour is also a relative entity, something that exists independently, because it is transposable and can be recreated in other things, in other materials. "The gestalt" can be made larger or smaller, can be reproduced correctly or in a distorted way, replicated, molded, or destroyed. It belongs to [102] the thing with the same characteristic ambiguity that adheres to the position of all properties, which the thing has as much as it is in them, with them, to put it starkly: is them.

In this respect, however, the boundary contour does not exhaust the meaning of the boundary between outer and inner as set forth earlier, does not signify the zone where the fundamentally divergent directions reverse, but only the spatial boundary between the relative opposing directions of an exterior transformable into an interior and an interior transformable into an exterior. The boundary contour belongs as a property to the sheath surrounding the core of the thing, the mantle into which the core emanates. In relation to the interior, which never becomes manifest, the boundary contour belongs to the outer sphere of the thing. As a spatial boundary, the contour is the zone where only a relative divergence of directions begins, while as a property (as part of the absolute aspect divergence of physical thinghood) it constitutes an outer determination.

What is required, however, is a boundary that is a demonstrable, objective property and at the same time the zone where the absolute divergence of directions begins. This boundary must be both contour or spatial boundary (since it must become objectively manifest in appearance) as well as aspect boundary, where the reversal of the two directions, by their nature mutually inconvertible, takes place. The organic shape boundary must, then, *as gestalt*, have a nature that is beyond gestalt, that is not exhausted by gestalt.

This formal consideration alone shows the difficulty arising from such a situation *for the analyst* (see the polemics between Köhler and Driesch) who has to decide whether it is sufficient to characterize an organic shape as a gestalt or not. For it can never be the case that the organic shape appears to intuition other than as gestalt. The biologist, who is tied to sense perception, will thus always ascertain gestalt and gestalt laws. The moment in the sensory shape boundary, in the contour gestalt, that is external to gestalt—its "value," as it were, as aspect boundary (and in fact, as will be shown presently, as aspect boundary that ontically belongs to the living thing itself and determines its mode of appearance)—can itself not be made to stand out in a sensory way from the sensory image of the contour, the typical outline, and be represented as such.

[103] How can a thing satisfy the demand for a unification of the two boundary functions? Put another way: under what conditions does the contour of a physical thing constitute its determining property (thus determining its nature), so that the boundary contour belonging to the thing and its determining significance for the thing no longer cancel each other out as is generally the case with the so-called properties of a thing? What condition

must be fulfilled so that a relative (spatial) boundary exhibits a nonreversible boundary relation between an exterior and an interior?

The answer sounds like a paradox: if a body in addition to its boundary also has the *crossing of this boundary* as a property, then the boundary is both a spatial and an aspect boundary, and the contour, notwithstanding its gestalt nature, acquires the form of wholeness.

It thus comes down to the relationship between the bounded body and its boundary. There are two possibilities here:

1. The boundary is only the virtual in-between of the body and the abutting media, where the body begins (ends) insofar as another ends (begins) in it. In this case, the boundary belongs neither exclusively to the body nor to the abutting media, but rather to both, insofar as the ending of the one is the beginning of the other. The boundary is pure transition from the one to the other, from the other to the one, and actual only as the "insofar" of this reciprocal determination. Here the boundary [*Grenze*] is distinct from the real border [*Begrenzung*] belonging to the body as its contour, and while it does not properly run "alongside" this border, it is external to it. This is because passing over to the other, while made possible by the border, *as* execution does not belong to its essence—that is, is not necessary to the being of the body.
2. The boundary really [*reell*] belongs to the body, which, as bounded, not only ensures the transition at its contours to the abutting medium, but in its boundedness *executes* this transition and is itself this transition. The boundary comes into existence here because it is no longer the "insofar" (pictured as a line or surface and thus in fact distorted) of the reciprocal determination, the empty transition that means nothing in itself, but rather of its own accord fundamentally distinguishes between the formation it encloses [*begrenzt*] as such from the other as other.

[104] The body does not start insofar as the abutting medium ends (or vice versa). The body's beginning or ending is rather independent of that which exists outside of it, *although sensory determination is not in the position to directly demonstrate the existence of this independence by means of sensory attributes.*

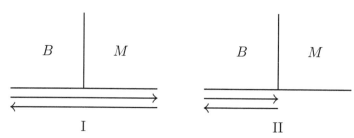

Figure 3.1. B stands for the bounded body; M for the medium bordering on it. Figure I symbolizes the "empty in-between" of the boundary belonging neither to B nor to M/both to B as well as to M. In figure II, the empty in-between disappears, since the boundary belongs to the bounded body itself. The difference between the two cases is expressed by the combination of arrows: the reciprocal delimitation [*Begrenzung*] of B and M in case I is contrasted with the "absolute" delimitation in case II.

As contour (border, gestalt), the shape naturally integrates the formed entity into one intuitive space and thus subjects it to the structure of pervasive reciprocal determination.

It becomes immediately clear that in the second case, the body must exhibit the postulated fundamental dual aspect by virtue of which it appears as a unity of exterior and interior. The dual aspect not only carries the formation, thereby conferring on it the character of thinghood, but also functions as a property, in an essential connection with the gestalt (contour) of the body. As already stated, an aspect attaining the position of a property *formally* rather than *materially* changes the appearance of a living thing-body in contrast to one that is not alive. They need not differ in phenomenal content [*Erscheinungsgehalt*], but must differ in their mode of appearance [*Erscheinungsweise*].

The concept of gestalt does not suffice to characterize the specifically organic form of unity. It has to draw on other concepts to make comprehensible a living thing's groundedness in itself, its self-sufficiency, its being itself and being from out of itself. The notion of gestalt only captures, as it were, one dimension of this multidimensional phenomenon and completely neglects the characteristic autocracy [105] of the living system. Driesch senses this; his arguments betray it, even if he does not expressly shift the emphasis of his arguments onto this point. Whenever he underlines the spontaneity of restitutive, regenerative, evolutive processes or the autonomy of morphogenesis in contrast to the pseudo-spontaneity of inorganic processes, he has it in mind. He reserves the concept of wholeness for this dual-aspect physical system, this unity of spatiotemporal, spacelike, and timelike relations.

The Task of a Theory of the Essential Characteristics of the Organic

Does this wholeness actually exist outside of philosophers' heads? So far, I have developed the law of the boundary based on the proposition that physical objects of intuition for which a fundamentally divergent relationship between outer and inner objectively figures as part of their being are called "living." Proof that this is indeed the case is still outstanding, however. At the moment everything is still hypothetical: *if* it is true that living bodies are specifically characterized in intuition by dual aspectivity, *then* it is also true that, unlike nonliving bodies, they have the relationship to their own boundary found in case II. If this is so, this relationship is also an initial step in "explaining" the property of dual aspectivity.

But how are we to understand the state of affairs described in case II? It was expressly pointed out earlier that sensory determination is not in the position to identify specific criteria for the presence of wholeness. Wholeness does not differ from a gestalt in its characteristics that are verifiable in space and time—indeed, it ought not to, if the exposition given here is justified. From the perspective of empirical natural science, Köhler is right and Driesch is wrong. From the perspective of intuition, which, as is well known, does not entirely coincide with that of empirical determination, Driesch is right and Köhler is wrong. Indeed, even abstract consideration shows the possibility of a form of order beyond gestalt that—as far as the sensory method of demonstration (the tool of empiricism) is concerned—cannot be distinguished from the form of order of gestalt.

[106] The state of affairs described in case II thus indicates (in schematic form) an essential possibility for intuition, but whose immediate empirical verification by intuition itself is excluded from the outset. Does this mean that "wholeness," as it were, is only an impression made on intuition by certain highly complicated physical gestalt systems, perhaps because intuiting

consciousness is not in the position to penetrate into the mechanics of the system? A difference, then, between what the thing itself is and what it looks like? An aesthetic difference?

Clearly, trying to solve the dispute in this manner would completely miss the point. Driesch and Köhler are both concerned with the mode of ordering, with the specific lawfulness of the thing itself. If I nevertheless want to claim here that Köhler is only right in the context of exact determinability, while Driesch is right in the context of full, methodologically unrestricted intuition, then this does not imply a displacement of the difference onto the subjective side of our knowing, but a separation in the object between layers that can be determined and those that cannot, even if they are intuitive.

The legitimacy of this separation will emerge from what follows. *If we succeed in developing from the tendency given in case II the basic functions whose presence in living bodies is considered to be characteristic of their special status* and that serve the vitalists as support for their arguments, there can be no justifiable doubt that the difference between case I and case II is a *difference in being*—that is (and this should be constantly kept in mind while reading this book), *a difference that cannot be experienced in itself, but only in its consequences or its appearance*. If this succeeds for all functions characteristic of life, the state of affairs presented in case II will prove to be the foundation and principle of the constitutive attributes of organic nature. Case II would then be the ground (not the cause) of the phenomena of life.

Granted, this is a lofty goal. Many will take it merely as a sign of an unrealistic assessment of my own capacities and will use it as an opportunity to repeat the old reproaches against the quixotisms of the philosophy of nature. But this [107] should not daunt us. We have no need for subjective sentiments about the restless spirit of research constantly reaching anew for the ultimate things and instead soberly uphold the discipline of questioning. What is called for is a development of the essential characteristics of the organic and, in place of purely inductive enumeration, at least the attempt of a strict justification. Our task is an a priori theory of the essential characteristics of the organic or, to use the term Adolf Meyer borrows from Hermann von Helmholtz, of the "organic modals."[9] For Helmholtz, a modal is qualitatively final such that it cannot be further analyzed by reducing it to other qualities.

A logician of the natural sciences would insert a "for now" into this definition. "Accordingly, an organic modal ceases to exist as such the moment it has been completely broken down into its physical and chemical components, i.e., the moment it has become possible to construe its organic gestalt—for all of these highly dynamic complexes are gestalts—from simpler physical gestalts."[10] I will go even further than the logicians of the natural sciences and claim that not only is it theoretically possible and practically feasible to completely trace back all organic modals to physical and chemical conditions, it is nothing less than necessary. But I am working with a narrower definition of "modal" when I say that in its *quality* it is categorically irreducible and cannot be broken down, which is to say that it never stops existing as such, even when its physical and chemical conditions have been identified precisely.

Inorganic modals, for instance, include qualities of color. Their quality can never be defined electromagnetically. The physicist conclusively correlates a quale to a certain wavelength and speed, without in the least intending thereby to explain the specific color quale in itself—in its particular green, for instance. All he does is correlate this color with the quantitatively identifiable foundations of its being. By no means do the layers of color reality that can be determined in this mathematical way exhaust it[11]: there is also the [108] only intuitable layer of the specific quality, "how it looks," which every empirical conceptualization (whether it be physical, physiological, or psychological) fails to be able to analyze.

One fallacy, with fewer and fewer proponents today, understood inorganic modals such as colors, sounds, etc., to be properties of the senses or the sense organs because they are only accessible to sensory intuition and thus denied their reality. This view does not stand up to criticism. It does not follow from the fact that a whole range of conditions have to be fulfilled not only on the side of the object but equally so on the side of the perceiver that the existence of the quality arises only from the conditions of perception. And even if this were the case, the significant leap from the zone of the chemicophysical being of our sense organs and nerves into the zone of purely qualitative being, pure suchnesses, can no more be ignored than in the case of that other vacuous notion that the givenness of qualities can be explained by the interaction of physical, physiological, and psychological factors.

Every modal is irreducible in its quality, even if the conditions of its manifestation and disappearance can be fully specified. The modals by themselves already designate a thoroughly closed sphere of intuitable, but not directly measurable or quantifiable "being." This means that a theory of modals can only come from an a priori discipline—that is, one that does not work with the tools of causal relations and descriptions of effective reality [*Wirklichkeitsbeschreibung*].[12] A theory of organic modals, tasked with providing a system of the properties characteristic of all life, an axiomatics of the organic (not, to be clear, of "biology"), thus presupposes a "valid a priori theory of the organic that we do not yet have."[13]—

We must avoid the errors of mechanism and vitalism in equal measure. And they are avoidable, as they are [109] errors of one-sidedness. The vitalist has his sights set on the phenomenon of a reality that has not yet been stripped of its intuitable layers. The mechanist, on the other hand, is concerned only with the reducibility of these layers and does not lose any sleep over the irreducibility of the qualities or modals contained in them (justifiably so, as an exact natural scientist). Inorganic natural science is further along than this. If the practitioner of physical optics works with thermal elements instead of eyes, this does not mean that he denies—though of course he does not burden himself with—the necessity of studying the structural laws of the phenomenal spectrum. It is in much the same way that the dispute between vitalists and mechanists must be resolved: their objects simply belong to different ontic levels. Reality, furthermore, ought not to be approached with the preconceived notion that it is only reality insofar as it can be defined by physics and mathematics.

The law of the autonomy of the phenomenon can generally be observed in the relationship between phenomenal and known nature. Something whose inner structure is made up of electrons and energy may appear red or green, low-pitched or high-pitched, hard or soft, smooth or rough. Are the phenomena less real than the dynamic and material constellations underlying them because their appearance is also dependent on the other side, on the perceiving subject? The fact that in this particular case the normal eye perceives green and not gray or red surely has its cause in the overall physical situation. This cause, however, at most serves as the necessary and sufficient occasion of the unique suchness of that which is caused, of its modal character—whose mode of being cannot be deduced from the mode of being of the cause. Empiricism must stop at such *metabasis eis allo genos*. This does not mean, however, that we have here an insurmountable barrier to knowledge as such. It is rather that philosophy is called upon to carry out its own particular work.

If there is to be no doubt from here on forward that it is only a matter of time before all attributes of organic things and processes previously thought to be of a specifically vital nature can be radically explained as deriving from physical and chemical processes, this nevertheless does not mean an affirmation of mechanism. The organic is not dissolved by being explained. Exact [110] biology, as the physics of the organic, shows, if anything, only the system of conditions and occasions for the occurrence of organic modals irreducible in their quality. Vitalism, however, will hardly be satisfied by this possibility of rescue, since its theory—still being propounded today and, at the moment, anyway, irrefutable by facts—that organic processes are "arbitrary," or indetermined (Driesch claims that entelechy can temporarily suspend transformations of energy!), would no longer be tenable. The indeterminacy of living processes in bodies would be limited to the transition from the layer of being that can be studied mathematically to the sphere of the modals. Within the layer of the specifically alive, which in its essence is indeterminate, indeterminacy shows itself in the life process as the principle of transition—just as in the phenomenal spectrum there is an indeterminacy of the transition from one quality of color to its neighbor, despite continuous mediation in the intermediate colors (and despite a demonstrably continuous transition from one wavelength to the next). What is inherently a discontinuous, qualitative transition can clearly not be established by evoking its neighboring qualities.

No matter what direction science takes, vitalism can take credit for having kept alive and honed awareness of the organic modals in the face of hasty identifications of life with the inanimate, particularly during the still uncontested rule of analytical mechanics. Furthermore, even if it overshot the mark here, it has always emphasized the irreducibility of the modals as such, the autonomy of an only intuitable layer of living things.—

These considerations impose upon us the special duty of working out the intuitiveness of the modals. The exposition of the possible relationships between a body and its boundary given in cases I and II may hardly seem intuitive. If, furthermore, we think of what in some cases are quite abstract concepts such as heredity, growth, development, and nutrition that are said to distinguish organic modals, the final elements in a characterization of life, we will be even less inclined to place them in an eminently intuitive sphere, indeed to restrict the knowledge of their uniqueness to intuition. "Heredity," "development," "nutrition," "regulation" [111] do seem to be mere names for categories of what are in part highly complicated and as yet understudied processes we know to mostly take place in obscurity, leading to the assumption that they therefore must elude intuition, if not in fact intuitiveness.

It is true that the organic modals, the essential attributes of life that are initially made available to us by experience, at the same time constitute basic concepts and topics in individual biological disciplines, concealing their elementary intuitional meaning. Nor can it be taken for granted that the intuitional content of these empirically obtained modals belongs to each concept with equal objectivity. But to begin from below, with empiricism, is certain to lead to failure. While experience should never stop serving as our guide, it should not dictate the way. Far from teaching us what really deserves to be called a modal, an essential characteristic, irreducibly final experience unconsciously presupposes it.

Definitions of Life

In order to have a fair grasp of the objects of the current study, those given with the task of an a priori grounding of the essential characteristics of life, we should actually look back at all the different definitions of life and determinations of its essence that in the course of time have shown themselves to be definitive for biological experience. This would also provide some degree of certainty regarding how far we have come in the theory of organic modals, although of course no absolute assurance. A criterion for the completeness of the justification does not exist here either, since the "sum" of the essential characteristics of the determination ultimately only generates a visible unity of what can be called "life" and "living" but is not meant to conceptually define this unity by means of these characteristics. To demand completeness of a justification only makes sense if the number of characteristics composing a concept is set from the beginning.

It will in any case facilitate an overview of the following considerations and make them more comprehensible if we single out two of the abundant attempts to conclusively define life—the two that seem to most carefully maintain [112] discretion concerning the diversity of biological phenomena and to best avoid any form of one-sidedness, despite the strong motives impelling it here. Roux attempted a functional definition of life: "Living beings . . . are natural bodies that distinguish themselves from inorganic natural bodies at minimum by a sum of certain elementary functions directly or indirectly aimed at self-preservation . . . as well as by self-regulation . . . and in their exercise of all these functions. In this process they become very enduring, despite their self-modification—because of it in fact—and despite the complicated and soft structure these functions require." Elementary functions for Roux comprise nine forms of self-action (auturgies): self-modification, elimination, ingestion, assimilation, growth, movement, reproduction, transmission or heredity, and development, which, along with the capacities of self-preservation and self-regulation, form the hallmarks of life.[14] Taking into consideration a whole range of other authors (Xavier Bichat, John Brown, Eduard Pflüger, Carl Hauptmann, Felix Auerbach, Erwin Bauer, Herbert Spencer, Emil Ungerer, Claude Bernard, Wolfgang Ostwald, Johannes von Kries, August Pütter, Hans Petersen, and Armin von Tschermak), Adolf Meyer in his *Logik der Morphologie* then provides a summary of those indicators of life consistently posited by scholars: nutrition (metabolism)—reproduction—development—heredity—growth—stimulability—regulation—movement (energy exchange)—structure. This also supports definitions that include frailness and mortality as essential characteristics of life, such as Bichat's "life is the set of functions that resist death" or Claude Bernard's list: organization, generation, nutrition, development, caducity, illness, and death.

Of course, there has been no shortage of attempts to understand the sum of life indicators from a unified perspective. Driesch's "entelechy" or Johannes Reinke's "dominants" do certainly allow us to grasp what is specifically vital in individual expressions of life. But identifying what is common is not the same as insight into the necessity [113] of differences. Measured against this demand of insight into the necessity of the different characteristics essential to life—that is, of the variety of the organic modals—it would hardly be an exaggeration to compare the current state of research on these characteristics to that on the categories before Kant.

Kant was not satisfied with simply enumerating empirically irreducible forms of being, but set out to discover an order in them and a criterion for their detection. Fichte then accused his deduction from the table of judgments to be nothing but a gleaning of categories rather than an actual deduction from a principle, and Hegel went even further. But Kant had in mind a kind of deduction that was neither rational-emanationist nor metaphysical-teleological; he referred to it as transcendental and deliberately kept it in contact with the open system of experience. The transcendental unity of self-consciousness may be the central point of all categories, but not at the same time its deductive locus, its principle, or the source of its differentiation. Kant explicitly holds on to the irrationality of the categories grounding rationality; the deduction of these pure concepts of the understanding is only concerned with showing them to be principles according to which other synthetic a priori cognitions become possible. The transcendental deduction of the categories thus takes place in light of these cognitions, which are cemented in the exact sciences—(as soon, of course, as the epistemological focus of the theory of categories has been recognized as being one-sided and the actual breadth of the categorical functions as constricting, the problem of the coherence of the categories appears as an ontological problem. This has so far been treated most thoroughly in Hegel's *Logic*).

As much as it would thus be a reversion to the pre-Kantian manner of deduction to deduce the essential characteristics from some concept of life into which one had previously incorporated them, it would be equally wrong to stop with Kant, as it were, and to only admit those characteristics to be essential to life and to gear deduction to them that biology has developed as the "categories" of its empirical work. Essential characteristics in the sense of categories making *biological* knowledge possible are derived intuitionally from objective being and, even if they are [114] first discovered on the occasion of experience, they already guide the biologist's experience as he selects his objects. Kant himself provides this explanation of the categories: "They are concepts of an object in general, by means of which the *intuition of an object* is regarded as *determined* in respect of one of the *logical* functions of judgement."[15] Thus only the intuition of concrete, living reality can ultimately succeed in verifying these categories—that is, come to a fulfilling understanding of what they mean: the categories of empirical biology are rooted in the categories of life itself.

A frequently overlooked distinction needs to be made here between essential characteristics that merely *indicate* a phenomenon of life in the sense of the "habit" of vitality and those whose "complete" occurrence phenomenally guarantees the *actual* presence of something living ("actual" not according to the criteria of empirical natural science, but according to *intuition*). Thus there are certain very characteristic movements that both betray life as well as feign it when the mover is not alive (for instance, a rubber snake). These movements signal the specifically vital type of movement that on its own is an "indicative" essential characteristic. There are also certain rhythms, phenomena of plasticity, or shapes whose high degree of irregularity seems to be subject to a particular rule: these are all cases of essential characteristics of an indicative nature.

Constitutive essential characteristics as categories of life can only be fully grasped (individually and overall) by intuition. These characteristics define life; they never feign it. But they define life as being for intuition, having nothing directly to do with those layers of being conceptualized by physics and chemistry. Indicative and constitutive essential characteristics thus both share the property of intuitiveness, which is also why it is possible to trace the former back to the latter.

A theory of the constitutive essential characteristics or modals of life seeks to grasp their unity and necessity—that is, it is not satisfied with identifying them in their relativity to the intuitive phenomenon of a concrete, living thing as necessary for life, but [115] rather as necessary manifestations of a certain ontological law. It is true that this inevitably removes such a theory from the sphere of concrete sensory intuition in which the essential characteristics of life are embedded (without themselves being of a sensory nature). And yet the theory is based only on truly intuitive [*intuitive*] states of affairs, not on concepts, and by unifying these states of affairs seeks to understand the essential phenomena of life in their differentiation.

Such an a priori theory of the organic, it seems to me, has more in common with dialectics than with phenomenology. It starts with an underlying state of affairs, whose reality [*Realität*] it treats as hypothetical, and moves step by step from one essential determination to another. The essential determinations *follow* from each other, arrange themselves into levels, and reveal themselves to be part of an overall context, which in turn is understood as a manifestation of the underlying state of affairs. Phenomenological study, on the other hand, provides a static description of the essential indicators of the "organic" as revealed by intuition, leaving aside the matter of a theory of these indicators or entrusting it to other disciplines.

It is the nature of phenomenological study to begin with the indicative essential characteristics, and it remains uncertain whether it is capable of penetrating further to those that are constitutive. They certainly belong to its area of research. *Phenomenology is necessarily barred, however, from insight into their categorical character. If we are then to ask the question* of how these constitutive essential characteristics can emerge on their own, in an unforced way, from the problems of a logic of biology or even from the attempt to systematize biology itself, it is not the method of static description we must apply, but we must rather attempt to *guide* the way through the essential layers following a principle or a deduction of the categories of life. As in all scientific matters, whether or not this is successful determines whether the underlying assumption is correct.

The situation is only complicated by the uncertainty surrounding the concept of essential characteristics. The empiricist too is familiar with essential characteristics. Since for the philosopher, however, it is a matter of characteristics in the layer of concrete sensory intuition—assimilation, heredity, regulation, development, [116] aging—it seems reasonable to think here of the empirical entities whose study is the business of the biologist. If he too studies the individual forms and conjunctions of these organic processes, he is ultimately only interested in the general lawfulness to which they are invariably subject: the "nature" of heredity, metabolism, and so on. He thus uses observation and experiment to search for those characteristics essential to, for instance, heredity or metabolism, which at the outset will naturally be of a descriptive nature (for instance, certain universally repeating processes in the nucleus, in chromosome splitting, or in the increase of oxygen consumption with relatively declining CO_2 output). The biologist studies empirical characteristics such as these in order to penetrate to the causes of hereditary or metabolic processes.

This is the path of the empiricist, not of the philosopher. The philosopher must leave the study of the conditions inducing organic phenomena entirely up to the empirical researcher. But there are limits to the empiricist's research as well. He is guided by sensory intuition and separates the different groups of phenomena such as development, metabolism, and regulation from each other. This separation is based on fundamental intuitions distinguishing life from nonlife. His empirical research, then, as concerns the stock of specifically biological categories, rests on premises that can only be analyzed by the philosopher.—

So-called neo-Kantianism, which has become popular among empiricists, understood categories to be forms of thought, modes of judgment, typical concepts. The study of these categories, the empiricists believed, amounted to a specific kind of formal logic and a methodology of (for instance, biological) conceptualization, leaving the sphere of intuition, where the concepts are applied, corrected, implemented, and originally formed, entirely to experience. That is, of course, not at all what is meant here. Categories are not concepts, but rather make them possible in the first place because they signify forms of accordance between heterogeneous spheres, both between thought and intuition and between subject and object. Now no one can say in advance what in the groups of phenomena separated out by empirical research, such as regulation, metabolism, and development, is purely empirical and what is categorical [117], what is an a posteriori and what an a priori determination.

Thus, for example, Helmholtz and Robert Mayer empirically demonstrated the first law of thermodynamics, the law of conservation, despite the fact that it expresses an a priori truth (unlike the second law, the law of entropy). This truth does not furnish us the specific equations. Only measurement can confirm that "our" nature really is a nature whose conditions of possibility include, a priori, (relative) systemic closure. In all instances where empirical conceptualization has achieved a certain maturity it encompasses matters of fact that, when reflected upon afterward, turn out to include a priori components.

It is no different in biology. The empirical researcher who believes (and is supported therein by formal logical positivism or neo-Kantianism) that, for instance, heredity, regulation, or aging may have a certain a priori core at most *as concepts*, but as phenomena are purely a posteriori facts, sees things in a distorted way. His empirical characteristics take on a priori traits, particularly in their objective-intuitive quality, which can only be differentiated in the course of philosophical investigation. Although he does not know it, the empiricist needs these a priori essential characteristics in order to demarcate related phenomena within the scope of the overall entity he is studying. It is based on this that he arranges his concepts. The more he becomes removed from immediate description and applies causal methods to his objects, the freer he becomes, of course, in relation to the initially identified essential characteristics; other concepts that are not immediately verifiable by intuition emerge then, and connections can be grasped that evade comprehension in the "natural" appearance. Research, then, is constantly revising the empirical essential characteristics, changing their boundaries in relation to each other, suspending them. Stimulus processes come to be seen as the result of metabolic processes; growth is attributed to heredity. One "modal" after another disappears as essential characteristics are increasingly classified under common laws and reduced to physics and chemistry.

Empirical essential characteristics are ephemeral; they only have indicative value for another sphere of being whose [118] appearances they are. A priori essential characteristics, on the other hand, are not affected by this ephemerality. They constitute the nonvarying layer of concrete intuitive appearance to which empirical science must return again and again. It is true that they are sometimes referred to by the same names as the empirical modals, the relative essential characteristics, and that this gives rise to misunderstandings. The empiricist will one day be able to declare that there is no longer such a thing as "adaptation" but only "regulations," no more "regulations" but only certain chemical processes: whatever is empirical in the modalities of "adaptation" or "regulation" will then have been determined by means of reduction. The modality of the modal, however, can never be affected by this. As moments that determine what life is as it appears, regulation, heredity, metabolic circulation, etc., are as irreducible as blue, sweet, or rough (and if they can be understood at all, then only philosophically). Compared to these elementary materials of sensory appearance, the modals are at most on a different scale: as constitutive forms of life's phenomenal layer of being, they determine structures that guide both the naïve and the scientific conceptualization of biology.

Nature and Object of a Theory of the Essential Characteristics of the Organic

A systematic development of the organic modals should lead to proof of the wholeness ordering type in accordance with the principle of the really posited boundary (case II). Experience cannot be called upon as a deciding factor here. Every true experience reduces the content acquired purely by intuition to the criterion of observability. Observing or representing a state of affairs, however, means to grasp it in such a way as to make it given in more than one manner.[16] Everyone experiences it being hot in the manner in which temperature is uniquely given. Heat, however, is represented by mercury rising in the thermometer or by a particular substance melting, by [119] evaporation, and so on. Hunger is felt by everyone in the specific manner in which certain sensations, located in particular parts of the phenomenal lived body, a special mood, a tendency, etc., are given. Hunger is represented, however, by providing evidence of, for instance, increased gastric juice secretion. Essential for representation, then, is the translatability of a state of affairs from one mode of givenness into another, or making it given in more than one sensory modality.

Insofar as experience is controlled by such a cooperation of manners of givenness (which, in its inner nature, is the same as an antagonism), it of course leads to a selection of those states of affairs that can only be grasped in *one* mode of givenness. It follows that there is much more to the world than can be observed. Even if it enters into experience and helps shape it, it cannot be retrieved from it again because it does not satisfy the requirements of observation. All content that can only be acquired by intuition is fated to enter into experience without becoming determinable as experience progresses.

All content that can only be acquired by intuition is either content whose mode of givenness is (immediately) given or content whose mode of givenness is not given. The first category comprises those sensations with contents whose mode of givenness is itself expressed, made manifest. The second category comprises essences, ideas, and essentialities corresponding to a so-called intuition, or viewing, of essence. There are two possibilities in this category: either the essences are tied to one mode of givenness, such as in the case of material a priori essential characteristics and laws (for instance, of the optical, acoustic, and tactile sensory circuit), or they are not tied in this way and can rather be made intuitively [*intuitiven*] evident in different manners of givenness—that is, are indifferent to these manners.

Comparing content that can be represented with content that can only be viewed—perhaps a better way of putting it than "can only be acquired by intuition"—allows us to distinguish between the following possibilities in terms of the relationship between content and mode of givenness: (1) Contents of a certain mode of givenness can be represented as long as they can appear in at least one other mode of givenness [120] without losing their identity; (2) Contents can be viewed if they either (a) cannot vary in relation to their mode of givenness because they are inherent in this manner or concern it or (b) do not have a specific mode of givenness. The contents described by (1) *appear* (in one *or* another mode of givenness); they are—in this sense above givenness—the proper objects of *perception*. The contents described by (2), regardless of whether they are hyletic or eidetic, do not themselves appear; they are lacking *that which* could appear: the core that does not appear in itself, that is above givenness, and that can be grasped, identified in this or that mode of givenness.[17]

The wholeness ordering type belongs to the category of contents that can only be viewed. As such it indeed enters into the perception of the organic, but, since it evades all measurement, cannot determine the progress of the empirical knowledge of biology. As essentiality, wholeness does not have a specific mode of givenness. It can just as well appear in a visual as in a tactile way precisely because, strictly speaking, it does not *itself* appear. All that appears is the gestalt of the organic system. The fact that this gestalt can be transformed from one mode of givenness into another gives rise to the confusion between the essentiality that is alien to appearance and the gestalt that is indifferent to givenness. This confusion is further abetted by the fact that, as described in greater detail earlier, wholeness "needs" the gestalt in that the attribute of the boundary, which is characteristic of wholeness, is only observable *as* boundedness (boundary contour) despite the difference in meaning between the two terms.

How can we establish the reality [*Wirklichkeit*] of this wholeness ordering type if the empirical approach is to be closed to us? Reformulating the question [121] may make it appear less formidable, to wit: what conditions must be fulfilled for the state of affairs described by case II to actually figure as a property of a physical body in space and time? We are asking here about the realization of an essentiality; what is given (in formal terms) are the essentialities of "wholeness" and "physical thing."

Formally speaking, then, our goal must be to align these two dimensions with each other so that all determinations of the actual physical thing meet the "demands" of the essentiality of wholeness. In this sense we will proceed in a "deductive" manner. And yet there is no concept given beforehand containing determinations that can, following the schema of analytical logic, be extracted from it—just as there is no entity given beforehand from which, following the schema of emanationist-metaphysical logic, other entities can be developed. Instead, we will follow the principle of the boundary (case II) as we seek to discover the inner prerequisites for the occurrence of the dimension "living physical thing" as it is given to concrete intuition.

It should be clear from the previous discussion that we must also ask the possibility question. I have already shown the obvious *discrepancy* between the phenomenon of the intuition that something is alive and the observable means of the appearance of this life. So how is it possible for something that is verifiably a gestalt to appear as something else—namely, as a wholeness beyond gestalt? Here the answer must not violate this condition in particular: that what is verifiably given is indeed only a gestalt, a (highly complicated) system with a more or less distinct contour in which its physical existence remains enclosed.

We will attempt to arrive at the answer by relying exclusively on the conception developed earlier that the phenomenon of life is based only on the special relationship of a body to its boundary. At the same time, we must determine whether there are essentially original characteristics of life. This automatically leads us to the task of developing an axiomatics of the organic or an a priori theory of organic modals. The question of this discipline: "Are there characteristics uniquely essential to life which—regardless of whether they can [122] be tied to specific physicochemical conditions or not—are qualitatively irreducible?" has to be resolved here. Essentially necessary to life means being the condition of its possibility. Therefore, if it turns out that a physical thing only has the relationship to its boundary described in case II if it takes on the mode of development, stimulability, and reproduction, this proves the modal character of development, stimulability, and reproduction.

Wholeness cannot be realized in an abstract way. To be realized means to become concrete. Wholeness, however, does not become concrete in an immediate way—the thesis of this entire study could be summarized by this statement—but only in the characteristics uniquely essential to organic nature. "In order for" the state of affairs indicated by case II to agree with the conditions of spatiotemporal thinghood or to become actual, the spatiotemporal physical thing must take on the properties of life. Such a deduction of the categories or modals of the organic—*nota bene*, not *from* the fact of boundary realization, which does not in itself exist, but *from the viewpoint* of its realization—makes up the central element of the philosophy of life. It is precisely in this regard that the categories turn out not to be logically justifiable or deducible, but rather to be original modes of realization of a state of affairs that is not realizable for or in itself. Panlogism and a rationalization of categories must thus be left aside if we are to succeed in demonstrating that *life in its essential phenomena is the series of conditions under which alone a gestalt is whole.*

[123] FOUR

The Modes of Being of Vitality

Essential Characteristics Indicating Vitality

If life is really [*wirklich*] based on the unique relationship of a body to its boundary, or, to put it in (even if only approximately speaking) concrete terms, on the "skinlike" relationship (pardon the expression) of the mass of a thing to its form, of the matter to the gestalt, of the "filling" to its "edges," then we may expect living things to reveal this relationship in some way. The intuitive peripheral values of an organic body must differ characteristically from the corresponding peripheral values of an inorganic body. The difference would then be phenomenally identifiable and would not coincide with the empirical differences.

If we follow this unique and—according to our thesis—specific-to-life relationship between matter and form in the individual objects of intuition— matter taken here only as the formed fullness, as the more or less permeable, robust, colored, soft, or hard mass—we will arrive at the indicative

115

essential characteristics I already referred to in a general way and whose deduction I described as necessary. These are the purely intuitional criteria according to which plant and animal things alike are recognized as being alive, without however providing us with more than simply essential characteristics indicative of life. Even in the case of de facto lifeless things, their presence justifies the assumption that their bearer is actually alive.[1]

[124] Everything living exhibits plasticity: it can be pulled, stretched, and bent in a way that brings together the distinctiveness of the whole's boundedness with an extreme shiftability of its boundary contours. Unlike in the case of inorganic bodies, the form here is not the exterior surface of the substance (as a simple expression of the functional unity of its elements), but seems to enclose its actual surface like an invisible skin. The more plasticity—or, to speak with Buytendijk, the distinctiveness of the boundedness—emerges for intuition in the phenomena of development, growth, restitution, or movement, the more the thing appears to be alive.

Everything that is alive exhibits discontinuity in its continuity, regular irregularity, both statically and dynamically. Buytendijk very impressively demonstrated this property in the configuration of organic formations by comparing a variety of shapes: if you hold the outlines of a circle, an ellipse, an egg, and a linden leaf next to each other, the impression of vitality grows with increasing irregularity (despite equally evident regularity) of the lines. This characteristic is of course amplified in real [*wirklichen*] objects, which are entirely different from the schematic simplification of their outlines. It is not as if irregularity alone plays the leading role here; it must rather always show itself to be governed by a (non-isolatable) rule such that even a substantial deformation does not disrupt the overall image but indeed strengthens its effect. This is in fact actually the case.

Regular irregularity, "the erratic form of coherence" (Kries), occurs dynamically in the phenomena of rhythm. Rhythm is tremendously pervasive, making it easy to understand how it could be proclaimed nothing less than the key moment of all life. Here too the hallmark is the relative variability (of the period). A heartbeat can be accelerated or slow, spiky or shallow, etc.; a growth curve can slope upward steeply or gradually; the peristalsis of the gut varies depending on the different stimuli and can be either sluggish or convulsive. Every expression of life is subject to the alternation of night and day, of the seasons, to changes in diet, etc., not in an immediate, mechanical way, but rather in adapting itself to these variations with a rhythm of its own. Strong inhibitions have to be overcome before changes in external factors will lead to an adjustment of periodicity. [125] The processes of life can be slow or quick, full or meager, sure or faltering, tentative or zestful; its rhythms as true gestalts are thus capable of transposition and melody. The forms of becoming also exhibit that "distinctiveness of the boundedness" that prevents living beings from coming to an end where they in fact stop. The erratic form of coherence constantly makes palpable this hiatus between that which is becoming and its rhythm.

This allows us to understand certain characteristics that are particularly distinguishable in the phenomenon of living movement. The hiatus, the intermediate "space" between the process and its rhythmic form given in the distinctiveness of the boundedness, could not emerge so intuitively if it did not appear as a true boundary leading the thing in the process of becoming out beyond itself/into itself. It is precisely thus, if one may put it this way, that the phenomenon of the hiatus, the set-apartness of the formed from its form, comes about. This necessarily reveals *tendency* in a certain direction to be a characteristic that differentiates living movement from dead movement: movements appear to be alive that follow an already given or forerunning tendency and whose real [*reeller*] course is thus given in the character of fulfillment. Dead movements, by contrast, present themselves to intuition as not being grounded in what is to come and as lacking the character of fulfillment. A dead movement presents itself as absolutely determined, "the way it is"; its form coincides completely with the course described by it, *is* this course. Matters are different for living movement. Because it seems to be grounded in and elicited by a tendency, every effectively completed phase of a living movement is characterized by having been undetermined at every point of its course. It presents itself as a movement that could have ensued differently than it actually did. This freedom of form within form belongs to the characteristic of tending in a certain direction. Fulfillment cannot, in the intuitive sense, be forced on a tendency. In the set-apartness of expectation, in the moment of tension seeking a resolution, lies the hiatus of being ahead that can only be bridged by a spontaneous, arbitrary act.

[126] The irrationality and spontaneity of living beings, the tendency to choose the least likely among a given number of possibilities, the property that has been described by the seductive metaphor of "ectropism," becomes

documented in intuition as freedom of form within form. To draw ontological conclusions from this, however, would be a very dangerous place to start. One should never move directly from phenomenological states of affairs to ontological statements. The being that appears is indeed being, but not all of being as it is present and is, to and in itself.

If the investigation begins with the indicative characteristics, the true phenomenological indicators of life, it is only for educational reasons. The discussion of these characteristics may well take on a different significance upon completion of the study. Once we have become convinced that the hypothesis developed in response to a particular problem—namely, that living things are boundary-realizing bodies—is confirmed by the deduction of the organic essential characteristics, the essential characteristics of living being (not restricted to experimental methodology), we may also concede an ontological value to the indicative characteristics of the organic and say that the unique relationship of a closed formation to its boundary becomes manifest in them. The fact that these characteristics in some circumstances feign life can then no longer change anything about their connection with the boundary relationship constitutive of life. Following the law of the autonomy of the appearance, phenomena can occur wrongfully—that is, without a corresponding ontological foundation. But this does not mean that appearance [*Erscheinung*] as such is mere illusion [*Schein*].

There is no need to go into further detail regarding the connection between the essential characteristics indicative of the organic and the boundary structure. It shines through everywhere and is easy to isolate. Incidentally, there are probably more such intuitive indicators, which can be identified with the help of the macroscopic approach applied, for instance, in modern gestalt and complex psychology. Useful insights can also be gained by looking to other subject areas where the notion of the organic is used in a figurative sense, such as in the arts and the social sphere.

[127]

The Positionality of Living Being and Its Spacelikeness

Another simplification is in order here. The states of affairs described by case I and case II can be expressed formulaically, allowing for a clear presentation of the boundary relationships symbolized by the arrows in Fig. 3.1 in Chapter 3. If B stands for the body and M for the abutting medium, then case I can be expressed in the formula B ← b → M. The boundary is between (b) B and M. Case II, on the other hand, has the formula B ← B → M. Here the boundary belongs to the body; the body is both its own boundary and the boundary of the other and is thus over against both the body that it is and the other. I am intentionally avoiding the term "itself" [*sich*] here, as it will take on a special significance further on.

Since its boundaries not only enclose it, but equally open it up to the medium (put it in touch with it), the body that relates to its own boundaries according to the formula B ← B → M is *out beyond the body*. In the other case, however, expressed in the formula B ← b → M, the boundary, belonging neither (both) to the body nor (and) to the medium and thus constituting a purely virtual in-between, the mere possibility of passing over from the one to the other, really [*reell*] belongs to one of the two abutting entities. If the essence of the boundary in distinction to boundedness consists in more than the mere guarantee of the possibility of passing over but rather in this passing over itself, then a thing (regardless of what realm of being it belongs to) with its own boundary must also have this passing over. It would be better, however, to avoid the word "have" here in order to save it for a special case.

The real being [*Reellsein*] of the boundary in one of the two mutually bounded entities expresses itself for this entity as the mode of being out beyond. A boundary establishes an opposing relationship between the entities both separated and joined by it, otherwise it would not be a boundary, and the passing over from the one to the other would be a mere continuation without a qualitative leap, which, as it were, is made of its own accord *and* nullified. As such, the real being [*Reellsein*] of the boundary in the real thing is expressed as the mode of being *over against the thing*.

If a body relates to its boundaries according to the formula B ← B → M such that the boundaries are its own, it must [128] appear as a body that is both out beyond and over against this body. Both of these characteristics ought to get along with the characteristics of physicality despite the fact that they are initially averse to the observable features of a bounded physical thing-complex. For what does it "truly" mean to say that a body is out beyond the body that it is if it measurably ends at such and such a place, or is over against the body that it is if it is verifiably brimming to its boundary contours with genuine being?

And yet, if it is to be ontically and not only logically possible, if it is to really take place, this special relationship a body has to its boundary contours has to be expressed and make itself felt in the real thing in a way that does not run counter to it as a physical thing and is consistent with its "means." A physical thing has as means to express itself and make itself felt only what are generally called its "properties," which in turn make up the whole of its intuitive appearance. As noted earlier, for the sake of its special character as a relationship to the boundary, the real being [*Reellsein*] of the boundary will have to become manifest in the contours of the body. Being out beyond the body and over against it—clearly we have here nothing other than a more precise and intuitive rendering of dual aspectivity—appears (accordingly) as a *peripheral* phenomenon of the physical system.

Thus we can understand the intuitive antagonism of the mutually inconvertible outward and inward directions as a determination of the body's appearance necessarily emerging as a property *of* that body. Furthermore, we can understand this determination as a property that can only be viewed, but not measured, in the sense that the boundary relation in distinction to the boundedness relation cannot be demonstrated (represented), but only intuited (viewed). It is a property that eludes being measured to such an extent that only the demonstrable peripheral properties of a gestalt remain.

As a bodily thing, the living being *is of* the dual aspect of mutually inconvertible and opposing directions: inward (substantive core) and outward (mantle of property-carrying sides). As a living being, the bodily thing exists *with* the same dual aspect as a property, consequently transcending [129] the phenomenal thing in two directions: on the one hand, it posits it out beyond it (strictly speaking: outside of it); on the other, it posits it into it (inside it)—expressions that mean the same thing as those used earlier: to be out beyond it and over against it, to be into it.

The term "to posit"—burdened by the great tradition of idealism—immediately imposes itself here. One should not, however, associate it with the Fichtean sense of an act of thinking performed by the subject. To posit [*setzen*] as to set down [*niedersetzen*] presupposes having stood up, having been lifted up. When attempting to describe the unique complication of the being of the living thing brought about by the being of the boundary, we find ourselves in this very situation. As a physical body, the thing already "is" of its own accord; being does not confront it in any way or set itself apart from it as something that exists. Because the boundary, however, belongs to that which exists, this existent becomes something that passes over in two directions. It is in this sense that it is "lifted up"—we are compelled to follow this image. It cannot remain "lifted up," though, as this would violate the determination that it remains an existing bodily thing despite "passing over."

Only in balance with this determination is the expression "to posit" appropriate with its suggestion of being lifted up, being held in suspense, without thereby losing the moment of rest and stability.

The living organic body, then, is distinguished from the inorganic body by its *positional character* or *positionality*. I mean by this the fundamental feature of an entity that makes a body in its being into a posited one. As described earlier, the moments of "out beyond it" and "over against it, into it" determine a specific mode of being of the living body, which is lifted up as it crosses a boundary and thereby becomes positable. In the specific modes of "out beyond it" and "over against it," the body is set apart from the body and brought into relation with it, or, to be more precise, the body is outside of and within the body. The nonliving body is free of such complexity. It is as far as it reaches. Its being ends where and when it ends. It stops short. It lacks this ease in itself. Since its system does not have the boundary as its own, its being does not include the ambiguous [130] act of transcending. This means that it cannot attain an ambiguous reference back to the system, a self-referentiality of the system (if I may employ this more practicable word here rather than the cumbersome expression "it-relation" [*Ihmbeziehung*]).

In the case of a body with a positional character, the relationship of ambiguous transcending always penetrates that between core and property by virtue of the act of transcending (= dual aspectivity) having the value of a property and appearing as a property.[2]

However, the special nature of precisely this property causes it to predominate over the other properties by penetrating all of them and thus seeming to communicate itself to them from out of the thing's core. The foregoing has shown that the thing's core, while constitutive of the dual aspect in which the living thing appears as thing, has nothing directly to do with the living thing's property of dual aspectivity. It is nevertheless drawn in, because dual aspectivity phenomenally means being out beyond the existing body or into the existing body. The core, which in things not appearing as alive figures merely as a point of orientation, as the X of predication, in living things acquires the character of positedness. Being appears as having passed through (whereby the preposition "through" is only a stopgap in an attempt to do justice to the concept of positedness/*having been* lifted up).

As postulated, the thing's positional character expresses itself intuitively. This becomes immediately evident if we consider that a thing that can be proven not to be alive may undergo a phenomenal transformation to the point that, as in a fairy-tale, it suddenly stands there as alive in full intuition. All at once, a simple thing has become a being [*Wesen*]—that is, an independently existing entity. This being-for-itself or being-for-it—ordinary intuition does not make such precise distinctions—thus forms, as it were, the invisible frame in which the thing sets itself apart from its surroundings with the special distinctiveness of boundedness. A seemingly spontaneous movement of the thing can of course help bring about [131] this phenomenon of positionality. There is a well-known practical joke where a rubber ball is hidden under the tablecloth with a small hose attached to it that the prankster squeezes in order to inflate the ball, thereby making plates and glasses dance. For a few moments, a dancing plate tempts intuition into seeing in it "something akin to life." It does not, incidentally, have to be phenomena of movement that generate the illusion. Any level-headed person must time and again oppose the tendency to imbue a still plant with a soul, just as a child does the bed, chair, room, table, spoon, all the things he comes into contact with. This process of personification clearly shows the initial sense in which that which is "dead" looks at us with different eyes than that which is alive.

A living being appears as being positioned against its surroundings. Its relationship *to* the field *in* which it is starts from it and, in the opposite direction, returns to it. As becomes clear precisely from the way in which we soberly consider a seemingly motionless plant, intuition grasps the positional character independently of any kind of ensoulment or personification. It grasps a thing's positionality in just such a way that it is no longer merely a figure of speech to say that this thing has its parts as properties: these particular leaves, flowers, stems, trunk, and roots.[3] According to intuition, the thing is not exhausted by them, but is something separate as well, because it lives: not a mere thing, but a being [*Wesen*]. It is in this being-for-itself that the thing's set-apartness from the field of its existence lies. It does not merely occupy a position in space, but has a place, or, to be more precise: it claims a place of its own accord, its "natural place."

Every physical, bodily thing is in space, is spatial. As far as its measurement is concerned, its location is in relation to other locations and to the location of the observer. Living bodies as physical things are not exempt from this relative order. Phenomenally, however, living bodies differ from nonliving bodies in that the former claim space while the latter merely occupy it. Every space-occupying entity is at one position. Every space-claiming entity, on the other hand, has a relationship to the position of "its" being by virtue of the fact that it is out beyond the entity (into the entity) that it is. In addition [132] to its spatiality, it is *into space or spacelike* and accordingly has its natural place.

When a nonliving thing breaks to pieces, we may also ask where "it" is now. The scholar of course will tell us that an "it" never existed, but merely a particular constellation of electrons and energy that has now been rearranged. The question, as regards appearance, was nevertheless justified. As concerns a living being, however, we could never be satisfied with such an answer even if we knew it to be objectively correct.

Living Being as Process and Type; the Dynamic Character of the Living Form; the Individuality of the Living Thing

Vitality announces itself with total clarity to intuition only in movement. This is why characterizing positionality in static terms is so difficult and why we are compelled to move on to dynamic territory—not for psychological reasons, not because of the livelier impression something animated makes as compared to something at rest, but for reasons inherent in the matter itself. Life *is* movement, cannot take place without movement. Even if experience had not found this proposition to be consistently true, a priori necessity would affirm it.

A thing of positional character can only be by *becoming*; process is the mode of its being.

It is part of its positional character for a thing to be out beyond the thing, into the thing. In order to satisfy this requirement, the thing must, as it were, be in the position to move away from the thing that it is. Becoming set apart from the thing that it is, easing away its being from this being, provides the only possibility of really having the passage (as the meaning of the boundary) in that thing. However, a thing only ever actually becomes able to stand away in its essence from the domain of its being—that is, out beyond it, into it—if it does not remain within the bounds [*Begrenzungen*] that have been drawn for it (albeit not arbitrarily). Its "being" [*Sein*] is thus essentially destined for passage.

Pure passage is becoming, the unity of not yet and no longer whose empty constancy moves toward the soundness [133] of something that is becoming in it but that still lacks accentuation in the now. "Now" here is the mere limit of passage, the pure in-between of the two modes of not-being. Conceived in this purity, however, becoming as mere passage destroys any bounded entity. The spirit of becoming opposes boundedness of any kind. Passing over, then, must be given a *conditional* mode if it is to remain the mode of a realization—that is, of a concretization within the boundaries of thingness.

Insofar as it passes over, the thing, of course, vacates its boundaries. If it does not posit a persistence of some sort against this (essentially necessary) abandoning of its area, it disintegrates and falls prey to destruction. The reality of its boundary would then be the thing's disappearance, the destruction of the bounding contours and thus the abolition of the boundary. This would mean that the realization of the boundary would show it to be an illusion, and the attempt to demonstrate the reality [*wirklichen*] of case II would have failed.

This, however, is not the way things stand. Besides the moment of passing over, the nature of the boundary includes the moment of standing still, of the unconditional halt. It is these two moments together that determine the essence of the boundary as that which leads to the other and at the same time closes it out. It is thus that the boundary induces boundedness without thereby eliminating the connection between that which is bounded and the other. The isolation achieved with boundedness in fact signifies the *integration* of the bounded entity into the context. Thus the physical thing does not need to specially posit a persistence against the abandonment of its area demanded by the boundary in order to prevent its own disappearance, as

Modes of Being 125

this disappearance is only the effect of a half-determined boundary. Just as the thing's compulsion to pass over derives from the reality of the boundary, so does its power of persistence. Remaining what it is *and* passing over into what it is not (over beyond it) as well as into what it is (into it), must be executed in one move in order to bring about the nature of the organic.

The boundary is a stationary passage, onward as halt, halt as onward. Hence the "product" of the two moments is not simply the boundary-dissolving becoming that contains the new only as limit, but rather the becoming of *a persistence*, the persistence of *a becoming*: the "moments" of standing still and of passing become unified. This unification [134] succeeds in the form of a separation of the two sides of becoming and persistence, which, in distinction to the only abstract—that is, non-self-sufficient—boundary moments of passing over and standing still, can appear as self-sufficient in relation each other, although as conditions of the reality of the living thing they of course only occur in connection with each other. Becoming is essentially actual only in a becoming entity—that is, in contrast to a persistence to which it shows itself to be tied—and persistence is essentially actual only in a persistent entity—that is, in contrast to a becoming to which it shows itself to be tied (which it resists).—The moment of standing still is thus not the same as being persistent; the moment of passing over is not the same as becoming. The side of persistence and the side of becoming are rather each for themselves a synthesis of standing still and passing over, and only the two of them together determine the way in which the physical thing has its boundary as its own.

In order for both sides, whose natures, after all, are opposed to each other, to be together in one and the same physical thing, they have to be correspondingly distributed in this thing. This is only possible if one side recedes in favor of the other. Becoming is determined as the becoming of something (the persistent entity) in the mode of persistence "carrying" becoming; or persistence is determined as the something of a becoming, where the becoming carries persistence. Each of these determinations is a moment of what is called *process*.

In a process, (1) persistence passes over into becoming; something *becomes*, without disintegrating in this becoming or entirely losing its being to it. In the same sense, (2) becoming leads to persistence; something *becomes something*, without being hampered by this something or forfeiting its being to its opposite. A process (as the becoming of something) is such that it no longer contains the now as an empty in-between of the modes of non-being, not yet, and no longer—that is, as the limit of passage—but as the sustained constant of passage.

If process is the becoming of a something, this something reaches through becoming as persistence in a twofold manner: as the beginning-something and as the end-something, as the where-from and the where-to in the modes of emerging and of having emerged of one and the same thing. Something is becoming, thereby moving beyond its prior being. Becoming streams forth from it, without being annihilated in it; the something itself streams forth from becoming, thus [135] becoming something that it was not before, so that *something else* emerges from it.

A bodily thing that realizes its boundary is necessarily engaged in a process; it not only becomes or radiates, but becomes something. Indeed, pure becoming without a persistence aligned with it is no different from a pure passage and, if realized, would abolish the thing's boundedness and thus destroy it as an entity. (This is also to say something essential about the relationship between becoming and thing: it is impossible to conceive of a living thing as something that is *only* becoming *itself* [es selbst]. Pure (non-genuine) being means coming to being in the first place. "Pure" becoming (= passing over) is understood as not having predetermined being, nor being given by becoming. It is thus not suitable for realization. Genuine becoming is a synthesis of passing over and standing still, and only really takes place as a property—of a persisting entity. This essential law already points to the necessity of understanding life as an appended determinacy.)

A living thing becomes something—that is, it changes, becomes other. This change, however, must not call into question the sameness of that which is becoming in the physical thing or of that which yields and carries this becoming. The thing becoming other thus remains the same.

A common mistake is to identify that which remains the same with the physical thing, with its structure of core and mantle, while identifying that which changes with the process. This produces the image of a process that is poured, as it were, over a real bodily thing without the latter itself being involved in it. An easily popularized intuition uses this image in order to present life as a material process (as the secretion of certain substances, intramolecular heat, and the like) running alongside the otherwise given constitution of the bodily thing. The idea of vital spirits, although it operates with an immaterial agent, is based on this model.

The truth is that a process is neither a pure coming-to-being nor an external determinacy indifferently appended to what exists, but that it rather occurs as a genuine property of the thing and thus keeps within the essential [136] boundaries of thinghood. Process, however, means becoming something. In the pregiven essential framework of thinghood, then, the thing must become something or become other. The change must not call into question the sameness of that which yields becoming in the thing (the "it," the subject of becoming). That which results from becoming ("something") must be other than "it" *and* identical with it. *Nota bene*: this requirement derives from process as a property of the thing.

The thing has to fulfill this requirement within the scope of the conditions of physical corporeality, respecting, in other words, demonstrable boundary contours. The process must not destroy the contouring, so that the thing remains the same within its demonstrable boundaries. And yet the thing must change the contouring because the process is the way in which the thing (within its boundaries) *exists* or realizes its bounds [*Begrenzungen*] as boundaries [*Grenzen*].

These contradictory moments have to be correspondingly distributed in order to be unified: the fact that no phase of the process is fundamentally different from the others preserves the contouring of the thing. In the actual contouring, all phases of the process that were ever observable have to exhibit a consistent constant. Or, to put it another way: the actual, observable form of a process phase can only appear as the variable expression of this consistent constant. The identity that is really [*reell*] and consistently maintained throughout all the phases of contouring is determined in relation to all of them—in all of the phases in which it is realized in ever different gestalt—as *their type* or as *gestalt-idea*.

This is why organic form is necessarily a gestalt of a certain type, the manifestation of a form-idea concretely intuitable in an individual gestalt. As such it is *dynamic form*, in which the bodily thing realizes its boundary in the thing that it is.

To summarize: a living thing can exist because it is possible to unify the boundary-conditioned sides of becoming and persisting into a process without thereby giving up the phenomenal corporeality of the thing itself and sacrificing it to the process. The thing does not stand aloof from the process; it rather only remains a thing while engaged in it. How so? It sets apart (1) the thinghood characteristics and (2) the type or the form-idea from the thingly-bodily form factually drawn into the process. For the sake of the constancy [137] of thinghood, which would otherwise be lost in the process, the bodily gestalt of the living thing is typical, or its form is dynamic.

Normally "dynamic form" and "process" are identified with each other despite the fact that they belong to different levels of being. The dynamic form, however, is the condition of the possibility of the process in which the thing is to maintain its identity as thing and gestalt. The necessity of the *modus procedendi*, in turn, derives from the synthetic connection of the sides of becoming and persisting, which, for their part, and each for themselves, are synthetic connections of the boundary moments of standing still and passing over, of "halt" and "onward." Under the compulsion of this necessity, the gestalt acquires the character of the *distinctive* [*ausgeprägten*] type or *shaped* [*geprägten*] form realized in the *individual*, of the gestalt within the scope of the gestalt-idea. The gestalt of the individual thing thereby gives up its mere isolation and appears in the attachment to a formal law of a particular type. This law gives the single thing support and makes it into an individual. The merely single thing must, if it is alive, be a manifestation [*Ausprägung*] of a form-idea or have the character of individuality; the single thing is possible here only as individual.

This allows us to understand the inner reasons behind the law of the type governing all of organic nature. Nowhere does a living body occur, so to speak, as an absolute singularity. It is always a (perhaps the only) case of a type and is subject to a graduation of typical units that are ordered according to grades of kinship. The typicality and thus the possibility of categorizing single living things into particular levels is not a lucky circumstance that happens to suit the interests of systematic biology, nor is it a fact to be explained empirically. While the individual type and its particular level can be explained empirically, typicality and the gradation of the organic world, on the contrary, are the modes essentially necessary for life (as the realization of the boundary of a physical thing) to acquire physical reality.—

But what about the process? This is where the thing changes. This change has to stay within the frame set by the unity of the type and the characteristics of thinghood. It is within this scope that the change takes place. To begin with, change, in concrete terms, means transformation [*Umgestaltung*]. [138] What is there is dissolved, and this dissolution is already the new gestalt. But what is the special character of transformation? Does it always take place in one direction? Upon what moments known from experience does it cast the light of necessity?

The answer is given by the connection of the process with the ontic antagonism expressed in the nature of boundary realization: remaining what it is, passing over into what it is not (over beyond it), *and* into what it is (into it).

Living Process as Development

The individual must become something—that is, always be other. As a merely single thing it would not be able to; as an individual, on the contrary, it has the scope to do so within that which prevents the single thing from changing: that is, within its very boundedness. Because the individual remains what it is under its gestalt-idea, its gestalt can change. Without this setting apart, there would only be a static process, a running in place, as it were, an incessant reshaping on the same level. The totality of all relationships in the diversely structured body would remain the same despite the continuous shifting of the specific relational elements; the degree of diversity would not change. This would violate the stipulation that the individual must actually become other, from the ground up: it cannot lead its own existence alongside the process, but must be really engaged in it.

The stipulation, however, does not have to be violated. Its fulfillment is guaranteed by the requirement that the living thing both remain what it is and *pass over* into what it is *and* what it is not. This gives the process a distinct character. It acquires a direction and a goal. We no longer have to worry that it can only be thought as a static process, as running in place; on the contrary, it is only realizable as movement from a place. The different phases do not simply disappear into each other by taking each other's place and surrendering what has become to becoming. Their disappearance into each other has to distinguish what has become from becoming so that the particular starting point of the process, contained within what has become—it is from here that things will progress—itself moves from its place. It is thus that the process [139] does not merely revolve around its starting point or stand as a whole, but moves forward. The thing actually becomes something that it initially was not. It retains what it has been in the thing that it is, thereby moving beyond it. It really passes over into something that it was not before. The image of the process as progress is not the circle, which represents standing still, but the straight line.

The orientation toward what it is is just as essential to the passage as is the orientation toward what it is not. The progress of the passage, then, has to realize both in each of its steps. From the previous it should be clear that into-it and over-beyond-it cannot be understood as opposite directions in space, as this would transform fundamental divergence into relative divergence. If, then, each step of the process has to realize the absolute divergence or the dual transcendence, to lead the body over beyond the body in opposing directions, the result cannot consist in a spatial distribution onto the exterior and interior of the body. Only the genuine synthetic connection of the two divergent directions guarantees a distribution of the determinations onto the real body in a way that corresponds to its nature. Passing over into what it is must be determined as passing over into what it is not and vice versa. Only under the condition that it is possible to grasp over-beyond-it *as* into-it and that the physical thing succeeds in being the unity of this divergence is the initial requirement satisfied.

Seen from the perspective of the process, the previous can be expressed as follows: the process is not meant to only keep moving forward, thereby bringing the starting point of future becoming to its most recently attained phase. This would mean that the thing engaged in the process would carry what it has been in the thing that it is and—although it is becoming—would merely exist as something that has become (implying that only what is past is properly the subject of the future). The thing would only be that which constantly points beyond the thing. It would, as it were, be divided into what it has been and what it will be. Presence would only be maintained as the constancy of conditions to which the process itself is subject but that are alien to its nature—that is, sustainment of corporeality on the one hand and the form-idea of the type on the other. That which is currently actual in the thing would be pure passage.

[140] But that is only half the truth. The concrete reality of the real is here still evading the other determination of passing over: leading into that which it itself is.

In order to satisfy this determination, the process must run against itself. If it were to do so in a radical way, it would become what it is not meant to be: a closed circle, a stationary process. Running against itself must thus have a different meaning, synthetically combining the determination of "away from it" with that of "against it" into a new meaning. Seen figuratively, this synthesis is only possible if the line of straight progress joins with the line of the closed circle to form the line of cyclical progression, the helix. This, however, is no more than an indication of the new determination of the process, not the determination itself.

In the process, what the thing is is constantly set outside of the thing. If the process were isolated, it would, as pure passage, thus constantly be leading the thing from the mode of the past into that of the future, thereby stripping what it leaves behind of the value of still belonging to what the thing properly is now. It would rid what the thing has been of its essence [*entweste sein Gewesensein*], leaving it behind as mere residue, the trace of erstwhile life. The thing would no longer be present in what is physically there. If this were actually how matters stood, the living body would be its own bereaved. It would be the house of the dead from which life has escaped; its life would only be dying and death not the end of life, life itself not actual.

If this non-genuine realization of the living process is to be avoided and the body is to hold on to its life, the process must just as continuously set the body into the body as it sets it over beyond it. *With a certain limitation*, that which was fended off previously as the sole essential law of processual becoming now takes place *in* the process: the beginning-something becomes the end-something; the body, engaged in the process, has "itself" as the result. But the meaning is new: now it is a matter of the unification of the essential trait of staying what it is with the opposite, equally required trait of yielding something *other* than what it itself is. This synthesis takes place as a specifically directed form of the *modus procedendi*: as *development*. Development is the becoming of what already is without becoming turning into mere coming-to-being. At the same time, it remains what it is by becoming other. Although [141] development as process continuously leads what the thing is out of the mode of the past and into that of the future, the process as development is deprived of being *as that which becomes*. In this becoming there is presence because the thing *is* only to the extent that it *is coming*. It is *in* becoming and *despite* becoming under one condition: that as goal it is *ahead* of the process.

But what in the actual thing can be ahead of itself? Only that which is identical with what it is becoming—that is, the other. It is in this that the beginning-something coincides with the end-something: they coincide in the form-idea. The form-idea, then, is ahead of the developing thing itself. It is necessarily the goal of the development.

Is the goal of development attainable for it or unattainable? If it were unattainable, how could we speak of development? If it were attainable, the development process would be an idealization process: the form-idea would, in the end, be actual. The end would differ from the beginning of the development as ideality differs from reality—if the end of development were not in fact real after all and thus no longer ideal. (It may be possible to realize an idea, but never its ideality.) The difference between the beginning and the end of the development process would thus be lost in this case as well; the form-idea would remain equally unattainable for every phase, and nothing would have come to pass in the course of the process that would justify referring to it as development. The thing would not have changed even with the form-idea being ahead of it.

What does it mean, then, to say that a development has reached its goal and realized it when an idea is in the goal position? What condition must be fulfilled in order for the thing with the form-idea ahead of it to actually become other? Since it is constant and fulfills its meaning only as an absolute constant, it is impossible for the form-idea to contain this condition. We must thus look for the condition in the fact of being ahead. What does it mean in respect to the actual determination of the thing that something belonging to its essence is ahead of it? Surely only that it is still lacking something, an incompleteness that in the course of the process, "in time," can be compensated. Perhaps—this remains open for now—this incompleteness will never [142] be fully redressed, the thing will never in fact become what it is "supposed" to become, but with an essential determination ahead of it, it cannot but become *ever more complete*.—We should not allow ourselves to be bewildered here by lazy ideas. The image of something that merely hovers before the process, which the process chases after, is just as inadequate as the notion of the process overtaking itself. In the first case, that which is

Modes of Being 133

ahead is completely separate from that which is engaged in the process; in the second, on the contrary, it belongs to it completely, coincides with it, and thus, of course, is no longer ahead of it. Spatiotemporal metaphors inevitably miss the heart of the matter.

Crucial here is the fact that that which is ahead belongs to the essence of the thing it is ahead of. This defines the thing as effectively incomplete and each next step of the process as one that is executed in order to compensate for this incompleteness. The unity of the moments of belonging to that which already exists *and* to that which is ahead gives the process—which initially could simply be understood as a purposeful sequence, as an end-oriented course of events always taking place on the same level—the significant *declination* according to which each of the consecutive phases takes place on a higher level than the one before. To remain just as incomplete as it was before the beginning of the process or in one of its previous phases would mean that the thing was no longer engaged in a process that is anticipated by a determination of its "what" [*Wasbestimmtheit*].

This process derives its slope only from itself—that is, from the conditions to which it itself owes its existence. It does not require a factor guiding it from the outside; it guides itself. This self-guidance is not immediately *a tergo* but mediately *a fronte* and therefore may not be ascribed to anything not belonging to the thing engaged in the process, to a factor coming from the outside, to vital forces or an entelechy.

As compensation for an incompleteness, the process exhibits an ascending course toward the goal, the form-idea ahead of it. This inevitably makes the form-idea into the ideal of the process—that is, into a benchmark of approximation that, for the sake of its ideality, remains infinitely far away while at the same time allowing itself to be approached. How is this possible, given that to truly approach and to be infinitely far away [143] are mutually exclusive? The process must exhibit a change in the *second* dimension. Only if the process rises *higher* can the approach meaningfully realize an approximation to the ideal. This is why the process appears to be conditioned by that to which it itself leads. In its inclination toward the goal, the goal becomes the inducement of the process. The process thus satisfies the demand of going against itself specified earlier; otherwise, the thing would transcend itself in only one direction. The form-idea as that which is ahead of the thing engaged in the process necessarily takes on the characteristics of the final cause as whose motivated effect the development of the thing manifests.

If, in view of the fact that the process in every phase is to be a shaped thing, the dynamic form, itself the condition of the possibility of a bounded process, takes on the character of the type, the outline of the thing thereby splitting into concrete gestalt and gestalt-idea; if, for the sake of the genuineness and actuality of the process, the gestalt-idea shifts into the mode of being ahead of the proceeding thing—that is, stamps the thing as one that is being perfected, refines the process into a development, and itself takes on the essential features of the ideal—, then the derivation of these "categories" of the vital, *including* the famous *causa finalis*, shows how helpless the argument between preformationism and the theory of epigenesis is as regards the phenomenon of development.

Both are just as much right as they are wrong. If one wants to do justice to the *phenomenon* of evolution and takes the side of the epigeneticists, who since Driesch's classic experiments refuting August Weismann's ideas about germ plasm have taken the lead as the true party of strict researchers eliminating every final cause, one cannot get around the assumption of a guiding factor. If one sides with the preformationists as the party of teleologists, one comes into conflict with the facts. The first party can only rescue the phenomenon with respect to its causal exploration by vitalistically assuming a *causa finalis*. The second party attempts to harmonize the phenomenon and its causal foundation at the expense of the facts. Both are impossible. The inexorable progress of research presents evolution as a causal-epigenetic [144] course of events. Here as in all cases the phenomenon in its specific quality slips through the fingers of natural science's explanations.

Preformationism and the theory of epigenesis would not help us even if we conceived of their differences in another way, regarding the preformationists as the representatives of strict causal determination (consider especially Weismann's view), the epigeneticists on the other hand as the representatives of vital indeterminism à la Bergson's creative evolution. It is impossible to unambiguously assign these theories to either *causa efficiens* or *causa finalis*, effective determination or final determination, determination or indetermination. Preformation can mean a machine-like initial constellation from which the development [*Entwicklung*] automatically follows as processing [*Abwicklung*] (or unfolding [*Auswicklung*]), or it can mean a constellation of a machine (or matter) and guiding factor. And epigenesis as gradual complication [*Verwicklung*] (the term Uexküll once suggested in place of development) can just as well be causally/finally determined as indetermined. At bottom the two positions do not contradict each other. Their original dispute has long since been resolved by the progress of experimental biology. In the past fifteen years, developmental physiology and modern genetics have made significant advances in elucidating strictly causal conditions, and it does not look as if these disciplines will have to stop at the boundaries defined by vitalism. But it is something different entirely to point out—and this is how vitalism's boundaries get their due regardless of the progress of research—that these explanations contribute nothing to our understanding of the phenomenal character of the development process. Like all qualities, this one too calls for an investigation into its structure of meaning—that is, a nonempirical "what-analysis."

Summarizing the results of our deduction thus far in a convenient formula, we could say that preformation and epigenesis equally determine the essence of development—with the significant qualification, however, that they do not appear as isolated sides of development that can be set apart from each other, but only come into their own in a relative sense in certain developmental moments.

If a physical body satisfies the condition, set out at the beginning, of realizing its boundary, it is developing. [145] Evolution is the necessary mode of being of the body that is ahead of itself in the succession of the process. It thus belongs to the body's inner being to be phenomenally caused by a goal, without however owing its existence (including its being) to final causes separate from it—that is, conditions not belonging to the system of the developing body itself. Separation from the gestalt-idea standing in the goal position, which is thereby rendered as final cause, is the body's own work. The body anticipates this work. *The entelechial agent intervening "from the outside" is the mode in which the peripheral values of the body acquire boundary value.* Insofar as the body only achieves this development by complying with the basic condition of boundary realization, and only because it as a body of a specific nature brings with it matter, substance in gestalt, this development does not, at bottom, bring anything about that was not already given at the outset. In relation to the quintessence of all given substantive and gestalt conditions, the becoming-other occurring from phase to phase is only (so to speak) of a formal character. This formality, however, reaches into the furthest depths of life.—

The a priori characterization of development would be incomplete if we did not take into account and substantiate two moments here: *growth* and *differentiation*. From out of itself, a body can only become more than it is in two regards: in its size and in its variety. A simple increase in size with the same level of variety would not qualify as development; the body in this case would not be raised above what it was. In a thing only quantitatively increasing, it is precisely the quale, the "what" as constant, that sets itself apart from variation. This quale must be *affected* by the process for the thing to change in its being. As a result, all of the relationships contained in the thing whose "constellation" does not depend on its mere "mass" are drawn into the process. The structure along with the gestalt of the thing then suffers a change in relation to the quantum—taken, *nota bene*, in the intuitive, not the numerical (mathematical, physical) sense—in relation to the (phenomenal) mass or fullness. This change not only occurs in its peripheral values but according to the quintessence of all the relationships belonging to it. A mere [146] regrouping of the relations would not satisfy the requirement of the process leading higher. The next level is supposed to be situated above the previous one—that is, not simply leave it behind, but leave it below. "Above" and "below," of course, only mean a difference in the degree of diversity here.

To say that development leads higher means that from phase to phase, the process yields an increase in structure, which, in combination with an increase in fullness—this too must change—constitutes the state of affairs of growth or self-differentiation. Driesch was entirely right to oppose premature analogies between organic growth and inorganic "growth" processes with a description of this essential characteristic of inner differentiation into qualitatively, morphologically, and functionally different elements, parts, and components (cells, tissue, organs) and to identify conditions passing from lower to higher degrees of diversity as specific to development. This qualitative enrichment seemed to him to contain an indication of having been caused by an intensive natural factor, and, with the explanation given earlier, we can agree.

Modes of Being 137

We do not, however, accept the blind factuality of this causal factor as a natural factor *competing* with the observable and calculable factors of energy. As the previous has shown, this notion can be dismissed. Entelechy as natural factor can be replaced by entelechy as a mode of being corresponding to the boundary condition, which can be understood despite the fact that it cannot be characterized physically ("explained"). It is this boundary condition that harmonizes relative auto-causality with the phenomenon of the autonomy of the living system and that renders entelechiality itself comprehensible and necessary.

The Curve of Development: Aging and Death

Are the reasons for the special curve described by development, with its ascending branch called "the period of maturity," its descending branch "the period of aging," only external to it, alien to its essence? Does development carry the law of its own inhibition within itself, or is it inhibited from without, perhaps by the inability of physical thinghood to [147] provide adequate material for the realization of an infinite ascent? Clearly the question being raised here is the question of the necessity of death, upon which all of the following depends.

In formal terms, there seem to be two possibilities here: (1) Death and aging are the result of the competition between life and corporeality, with life as the carrier of an infinite tendency to develop leading to ascent, and corporeality as the carrier of a tendency to draw down leading to descent, ultimately triumphing over life in death. The struggle between two "forces" foreign to each other would, in this view, become visible in real development. Life *qua* life is understood to be infinite in the sense of endless progress. It becomes exhausted only as a function of the finite systemic conditions of the body, so that the characteristic curve of actual development is seen as a gradual process of exhaustion. Contrary to the parabolic course of youth, maturity, and old age, life in its purity must be thought only as a tangent. Accordingly, there is a connection between this view of the absolute alienness of death to life and the notion of immortality beyond actual physical life as guaranteed by the essence of life itself. The hour of death becomes the hour of liberation from the body and entrance into true, eternal life continuing on in the soundness of its own essence.

(2) Death is essential to life; life is finite in itself. What plays out in real development is a struggle posited by life itself between two opposing tendencies, of which the ascending, "positive" one initially, the descending, "negative" one finally has the upper hand with a necessary transition phase in the middle in which they balance each other out. Development in each of its steps is, in this view, *both* living and dying, ascent and descent, even if the emphasis varies from step to step. This assumption corresponds, as the reader will remember, to the notion already fended off earlier, according to which the body, as it were, never attains its life but is constantly running after it (the idea that the process only transcends the body in *one* direction, only posits the body out beyond the body). If the body, however, can never fully satisfy the requirements life places on it, it would not be able to die either. Life and death would then only be aspects providing [148] the body's intermediate state with the hues of reality [*Wirklichkeit*] as if refracted by a hidden sun.

These two possibilities are not exhaustive. There is yet a third: death is immediately external and unessential to life, but, due to life's essential form of development, mediately becomes life's unconditional fate.

And this is in fact the case. As set out previously, the process must run counter to itself in such a way that the directions of "away from it" and "toward it" can be combined synthetically to create a new meaning. When the line of straight progress adds its essential moments to those of the line of circular closure we have a helix, and so it is the helix that symbolizes the curve of true development. Every point located on this line lies above the previous one and does not form a straight connection with any other point in the direction of the line. Without really implicating the opposite direction, the process necessarily leads *in* the opposite direction and turns back without directly turning back on itself in the same dimension. It thereby leads to a point that necessarily lies *above* the starting point. Depicting a particular curve segment on a flat plane (omitting the moments illustrating the process's upward movement) renders a parabola with an ascending branch, vertex, and descending branch. This representation of a continuous direction, without at any point immediately countering itself, at every point (with the exception of the pivot point) determines a counterpoint. Just as the living thing as mere gestalt must submit to the form-idea, which as type grants it the scope of individual variability, the development process submits to the formal law of ascent, height, and decline. Life is not death, its own decomposition, its self-negation, but rather, in its development from stage to stage, moves *toward* death. And because the developmental stages of maturity are not mere compromises between two tendencies of life running in opposite directions alongside each other, life grows toward death as its unconditional destruction and ultimately falls prey to it. Reinterpreting this law to mean that death is an immediately essential moment of life, that dying is [149] identical with living, would be entirely unjustified tragicism. This would amount to rationally ridding death of its sting, and, even in the hour of death, life as mere self-negation would triumph not only over itself but also over the meaning of its negation. Such an abolition of abolition [*Aufhebung der Aufhebung*] is only possible for spirit—that is, for the human in his particular form of life.

Death wants to be died, not lived. It approaches life, which is naturally inclined toward it *and yet* must be overwhelmed by it in order to die. Only this is the true meaning of death: that it is the beyond *of* life and *for* life, the negation of life that is *separated from life*, but at the same time is compelled by it. Empirical theories of death only show half of the picture. Death does remain—this is shown well enough—foreign to life, irreconcilable with it, and incapable of synthesizing with it. As a force that is indeed blind (just as the first thesis conceives of the nature of corporeality), life can neither comprehend nor bear it. But the empiricist reading of death's blindness and transcendence is one-sided. It misses the fact that development anticipates the end, begins to lapse, and ripens toward destruction. Death is made possible by development. Youth, maturing, and old age are a priori to life as it develops. Living beings can and must die of their own accord. It is possible for them to die a natural death.

To make something possible, however, is not the same as to make something actual—and this, in turn, is overlooked by the aprioristic theory of death. The *nothingness* but not the *nonbeing* of life is anticipated. Development brings the possibility of passing away into life (and not just closer to life) in the mode of aging and decrepitude. But life cannot truly pass away of its own accord. *To truly pass away means to execute an absolute boundary crossing into something qualitatively other*. Life—which, after all, *is* this constant absolute boundary crossing—could only do this *as* life. But this would defer the moment of expiration, and life would continue. Life prepares its own transcendence in the developmental mode of decrepitude, but the fulfillment of effectively being lifted out of itself is only brought about by the power of death that is separated from life by an absolute chasm.

[150] Does life create its own death? No—consistent with its own essence and by means of development it only creates the conditions for the onset of death. Thus there is such a thing as a natural death—and an unnatural death. If natural death expresses the apriority of decrepitude, the final phase of aging, unnatural death documents the power separated, ontologically and essentially separated from life that gives rise to destruction.

Philosophically speaking, this is extremely strange. We can see that empiricist and aprioristic theories of death are incorrect in their one-sidedness. Empiricism overlooks the apriority of stages of aging, while apriorism overlooks the essential alienation and pure facticity of the act of death. Herein lies the true difficulty of the problem of death, the solution of which is likely to shed light on other parts of this study as well.

It is evidently possible to understand the a priori necessity of the mere fact (of death) without thereby robbing the fact of its facticity, without relativizing death and turning it into a self-abolition [*Selbstaufhebung*] of life, into a simple negative, a counter-projection of life against itself. This would be the typically dialectical account, which was, generally speaking, developed in its classic form by Fichte and Hegel in order to do equal justice to both the positive and the negative, being and nothingness. Effectively, however, they deprive both terms of their heft by mediating the opposition in a third, in life, the ego, spirit.

Fichte's ego *posits* the non-ego as an essential counter-projection. It is true that the non-ego that exists *hic et nunc* is posited *by means of a* hiatus irrationalis. In order to secure the absolute foreignness of the individual material things in their self-groundedness (otherwise acting would not be *genuine* overcoming!), Fichte abandons the apriority of their material existence and suchness, and his apriorism remains formal. But the *hiatus irrationalis* is only relative to the ego that confronts the individual things, to the ego that is itself a limitation of the absolute sphere *of the pure ego* in which subjectivity and world both securely rest.

Things are reversed, as it were, in Hegel. The apriorism here is fully saturated with matter (it is in this sense that Richard Kroner is right to refer to Hegel's irrationalism). As a result, the *hiatus irrationalis*, if you [151] will, disappears by becoming the principle of the constitution of every determination. The dialectical reversal in Hegel has sublated within itself the rational moment of the essential implication of the other side (of that which the thing is not by virtue of being *such*) and the irrational moment of total essential discrepancy. The result is that a *medium* of mediation of anything and everything heterogeneous was able to remain, whose designations such as substance-subject, spirit, or concept are not even as significant as the whole great foundational conception of a continuous homogeneity and sound mediatedness of all contrasts, contradictions, and oppositions of this world. For Hegel, the negative, lack, pain, and destruction may well be a power equal to the positive, but their security in the world and their spiritual nature remain unshakeable. There are no *intermundia* here; there are no genuine rifts as there are for instance in Leibniz, no *hiatus irrationales* spanned by a world.

Without wanting to draw premature conclusions here, the "law of the hiatus," which has been repeatedly touched upon in the course of the current deduction of the essential characteristics of life, therefore deserves special attention for general philosophical reasons as well. In order to make it possible for a thing to really become engaged in process, the thing's outline becomes the individual manifestation of an idea; the thing becomes subsumed *under* the gestalt-idea, taking on the value of a typical form. Thus, while life creates the condition for the occurrence, for the concrete manifestation of the type, it does not create the type itself. The type materializes "occasionally." Manifestation [*Ausprägung*] and defining [*prägende*] form-idea remain absolutely separate from each other.—The same law becomes evident in the developing individual's relationship to death. *Under* or in the forms of youth, maturity, and old age, life matures toward death. They too create the opportunity for death, which remains meaningfully and ontologically separate from life as its absolute other.

"Type" and "death" are only to be understood as necessary possibilities. They are to be conceived in the strict sense as opportunities that life sets out of itself according to its own law in order to receive a type, to receive death. It remains inconceivable, however, how particular fate can be fulfilled unless, like the dialectical monists, one were to wrongly reconcile the two components that determine fate—that is, to bring the form-idea, [152] or death on the one hand, and life on the other onto the same level of being.

Here I would like to make a remark that deserves careful consideration: traditional philosophy's method of working with a common denominator for all being (*sic*!), everything that is given (!), must not mislead us to believe in the reality of this denominator sight unseen. Otherwise what are unavoidably preconceived fundamental categories of philosophizing may destroy the possibility of solving the problems that could only be identified with the help of these categories in the first place.

Must the form-idea already "exist" *before* becoming visible in a living manifestation? *In* its visibility, the form-idea is what cannot be produced by life, but only accepted and suffered; what is separate from life and independent in itself, persisting in itself, what has always been and will always be. The existence of ideas is—at least something to be discussed, so that the possibility of a connection between the form-idea as something already existing for itself and the physical thing ought not to be dismissed outright. Because the final causality of the form-idea is an essential condition for the process of life, which has to be of a developmental nature, it is only the *mode* in which the contact between idea and body takes place that can be made comprehensible. It remains incomprehensible, however, how the contact between idea and body—consider Plato's doctrine of mimesis, Kant's schematism—itself takes place.

The same is true of the relationship between death and the living thing. Here too we can understand the mode in which death occurs. Here too its occurrence can only be understood as the intrusion of a force not produced by life, separated from life, a force that can only be accepted and suffered. In its visibility, death announces itself as genuine being, as contact *per hiatum* with an absolute other. Does this justify attributing being to death independently of dying life? Surely we may only hold on to a contact between them—similar to the one between type and body—as long as the actual phenomenon of dying demands it.

A being of death, a being-for-itself, an existence before and after and outside of dying is out of the question. Death is only the destruction of life, even if the mythical imagination with a proper feeling for death's transcendence of life has given it the appearance of an angel with a sword or of a skeleton with a scythe [153]. Thus that which was obscured by the previous formulation remains open here: how is it possible that something that transcends life, that can act upon it, does not exist in any way apart from this effectivity? How is it possible for there to be contact with the absolute other that is not contact but rather two entities setting themselves apart from each other, leaving a space in between, not touching—two entities of which only one, the living thing, *is* certain, while the other (the form-idea, death) "is" not? It is not in fact even correct to speak of *two* entities, although they really relate to each other *per hiatum*.

Death and life relate to each other in an unmediated way as absolute opposites in the act of dying. Dying, as we have seen, is *not* mediation, but entirely belongs to the realm of life. If the living transcends beyond itself, it is *alive*, and the result is an increase in life. Ergo, death can only occur "from outside" as an intervening power and bring the act of dying to fulfillment. In this effectivity and for this effectivity it is its own being, existing independently in itself, and becoming effective when life is ready for it. But are we to assume that death, as a being only making itself known in destruction, has an existence beyond this, independently of dying?

Anyone who allows himself to be coaxed into such a belief has not understood the connection between this contact (in which the contacting elements set themselves apart from each other *per hiatum irrationalem*) and the law of the boundary. The *chorismos* between gestalt type and living being, death and living being results—not from the essence of life, but (according to our thesis) from *its underlying* relationship between the body and its boundary. Since (in case II) the boundary belongs to the body, we have here the so-called "distinctiveness of the boundary" by means of which the body is set apart from its surroundings and becomes self-sufficient in relation to them. Certainly this applies in the first place in a strictly spatial sense. Inevitably, however, to the degree that the boundary relation proves to be constitutive of the concrete phenomena of life, the set-apartness of the living body must lose its narrow spatial meaning and acquire a meaning applicable to every concrete phenomenon of life.

To set something apart is essentially to set it apart *from what* and to thus constitute a distance that does not remain vague but, [154] on the contrary, can only be a concrete thing. It is only in relation to this concrete thing that the living body remains set apart or maintains the distinctiveness of its boundary. Removing this concrete "from what" from its connection with living being, which is seemingly made possible by the hiatus inherent in setting apart, is to violate its meaning and ultimately leads to the most absurd problems.[4]—

Youth, maturity, and old age are the fated forms of life because they are essential to the developmental process. Fated forms are not forms *of* what exists, but *for* what exists; being subordinates itself to them and suffers them. The third kind of necessity in addition to the law of "what" (essentiality) and the law of process (consequentiality) is fate. The first kind determines everything that is something. The second determines everything that takes place in time. The third only determines living things.

The Individual Living Thing as a System

If the vitality of a body is based on its relationship to its own boundaries according to the schematic formula $B \leftarrow B \rightarrow M$—that is, it implicates these boundaries—the body has also drawn the moment of the boundary into the domain of being that is expressed in the contraposition of what is bounded and the surrounding sphere. The fact that a boundary brings about an oppositional relationship between the entities it separates and at the same time connects was already noted. Accordingly, the body implicating the boundary is out beyond the body that it is *and* into it. In order to satisfy this requirement, the thing must be in the position to move away from the thing that it is. The thing comes into this position as it becomes. The thing does not stand still; it passes over. As pure passage, however, the thing would give up its isolated, bounded existence and simply dissolve. This would contradict the fixating meaning of the boundary. In the required synthesis of passing over and standing still, of onward and halt, becoming takes on the specific character of "becoming something," of a process. The reader will remember the next steps from earlier.

An existing boundary means becoming. The opposing directions necessarily render becoming [155] bidirectional, a becoming that is out beyond the body and into the body. Set apart from itself, the body in its boundedness is one that is set into and set out of the body that it is; it stands outside of and within itself. As a body thus determined it has a *positional character* and, as a developing body engaged in process, it maintains this character of positionality until it dies.

In this respect, positionality in distinction to the determinations dynamically realizing the twofold directionality of the boundary (passing over—becoming—process—development) also demands a static realization. It is true that the more precise specification of dynamic realization can only take place with the help of its refraction in the boundary-conditioned essential moment of the static, just as, conversely, the more precise specification of the realization of the essential static characteristics requires their refraction in the boundary-conditioned dynamic. In both cases, however, it is only the matter of the modes of realization of the boundary *as* the in-between, of the realization of the "demand" made on the body to stay what it is *and* to pass over into what it is not and what it is.

Now it is our task to find an unconditional expression—that is, one that is not refracted by and dependent on the dynamic forms of realization—for the function of the boundary that consists in the contraposition of what is bounded and the surrounding sphere. It evidently belongs to the essence of the boundary to be able to completely fulfill its function in *one* direction: to close off against the "outside" just as much as open up into it—that is, to lead into the outside (lead the outside in). The essence of the boundary is not fulfilled only from the perspective of a neutral point situated above it, taking in the colliding spheres as if in one glance. A boundary is still a boundary for the area it bounds, even if it exercises its double function in regard to this area alone. A barrier, on the contrary, is a restraint that *only* exercises the function of closing off the area it confines.

In abstracto, every barrier can also be regarded as a boundary provided that it lies in a continuum of possible progression. *In concreto*, however, the difference between boundary and barrier quite often manifests with painful clarity [156]. Essential for the boundary from a point of view situated *behind* it is the unity of closing off and opening up, which from a point of view situated *above* it is the unity of standing still and passing over (from one area *into* another). The same state of affairs can be comprehended from both points of view. The notion that one has to have crossed a boundary in order to have determined it as such is false.

If, however, the boundary entity exists (in ontic terms) for a point of view that is enclosed by it, "behind it," if the full function of the boundary can be preserved exclusively in relation to the area bounded by it, the realization of the essence of boundedness by living things must also express this moment.

It is from this vantage point that the essential static characteristics of positionality can be seen in their purity. The requirement made of the bodily thing to remain what it is, to persist in the face of constant change, and to assert the monotony of what is given against total alteration only now becomes understandable and can be grasped in its ramifications. For the moment of separation belongs to the essence of the boundary just as much as does the moment of oppositional connection. Something that is bounded does not stop in the boundary but, in a sense, *before* it. Only in this complete restraint is the function of closing off expressed, a function that cannot simply be subordinated to that of opening up or leading over. This would rob the absolute change of direction when crossing a boundary, defined more precisely earlier, of its premise. In what sense does the direction change abruptly despite the steady transition in the essence of the boundary if not in reference to the equally essential moment of closing off, which *in*cludes the one area and *ex*cludes the other? The relativity of the point of view, from which either abutting area can be seen as the one that is enclosed (included), does not abolish the absoluteness of the opposition but confirms it.

To be bounded is to "stop," to "come to an end" or "to a halt" before one's boundary. These expressions do not apply in full force to inorganic bodies, but the circumstances as such are clear: the body extends *up to* such and such a place, up to its boundary. If it implicates its boundary—according to our thesis this holds for organic [157] bodies—it has to come to a halt before this boundary, thus does "not quite" extend all the way to it, although it, the boundary, in fact belongs to the body. It thus already comes to a halt before it reaches what it still is. It stops sooner than its being in fact implies.

How are we to conceive of this? It is an absurdly unreasonable demand on the imagination, but one that must nevertheless contain a kernel of truth. If we try to visualize it (which of course is not per se the way to arrive at the essence of the matter), we arrive at the intuition that the body is within itself. If we picture the boundary to be like skin, this skin encloses the entity to which it still belongs as part of this entity's bodily existence.

We have already come across this notion of "being within it," "set into it." When discussing the essential moment of "being into it," we discovered the necessity of ascribing to the living body that "ease in itself" that, as being lifted up, justifies determining the fundamental character of its being as positedness. We pointed earlier to the connection between this positedness or being positioned of the living body with the essential moment of its integration into a setting to which it relates. In the relatedness of the organism and surrounding field, which stand in opposition to each other, lies the indicator distinguishing the living from the nonliving body: positionality. A space-occupying thing is at one spot, in one setting. It is because the living thing is in itself that it also relates to the place of its being; it is positioned "into space" and thus against its setting. It has the property of a space-claiming entity that possesses a natural place.—We did not pursue this train of thought any further in order to focus on the characterization of the essential moment of "being in it" in its sense-determining influence on becoming.

The image we used does not get us anywhere, of course. The boundary is not an additional determination of an existing entity that can be thought of as a part of it, which is why the metaphor of skin, even if it is conceived as vanishingly thin, remains inappropriate. Instead we must free all the expressions used—the body comes to a halt before it is at an end; it does not quite extend all the way to where it stops—from their dependence on spatial intuition and understand them in a radical sense [158]. Then the situation would be thus: although the body extends into its boundary (for the boundary really belongs to it), it must come to a stop *before* the boundary (as a boundary) and leave it outside of itself—that is, leave it as unreal. Realizing the boundary then means to unrealize the boundary.

The hypothesis upon which this entire study is based demands the possibility of boundary realization. Thus the unrealization that is essentially required and co-posited should not exclude the possibility of realization. Since it is a matter here of *one* state of affairs, its individual moments should not render it itself impossible.

A living body fulfills this essential law by being in "it" or within "it"—the body that it is—without, of course, conceding "it" a greater size, independence from that which is therein, or any other spatial properties. Wherein and therein are one and the same being, thus determining a quasi-spatial relationship that connects the body *as a whole to* its spatiality.

There is only one way to make this "being in it" manifest in the body: the body relates to a central point within it that, while not spatially located, functions as the center of the bounded bodily domain, thus making the bodily domain into a *system*. The relationship extends to all the elements (parts) constructing the body and to the body as a whole. Insofar as the body is (posited) in the body that it is, this central relationship takes on a special character. The body is confronted by a point *within* the area it occupies, a point that is nevertheless of a nonspatial nature.

Language reflects this strange law by saying that the living body is a system having parts, or that the living being has a body with such and such parts. The mode of the body of coming to an end before the boundary that belongs to it, of really keeping this boundary outside of the area of its bounded actuality, is "being in it" or being in a relationship with it, with the body. Language has only one expression at its disposal for this kind of relationship: *to have*. The living body is thus a *self* or a being that is not only absorbed by the unity of all its parts but is equally posited in the point of unity (which belongs to every unity) as a point detached from the unity of the whole.

[159] "Self" and "to have" are to be understood at this point in the purely structural sense following their given derivation, without the weight of psychological definitions. A self is not yet a subject of consciousness; to have is not yet to know or to feel. Only so much is clear: that we have arrived at the crucial point in our investigation where the possibility of the development of consciousness emerges in the first place. If for the Cartesian two-world theorists and the subject idealists even just posing the question of how being can give rise to consciousness is childish hubris and an overestimation of our means of cognition and at its core the symptom of a complete ignorance of the nature of knowledge—at this point in our investigation we no longer have to be led astray by such objections. There is the one transition from extended being to interior being, from the world of being to the world of having, not only in the case of the human to the extent that he takes himself on philosophically and turns inward, but everywhere that he encounters life. The world also shows itself from the outside and from the inside to the eye gazing out, to the grasping hand. There are overarching laws of constitution that allow us to see what is within by looking at what is without.

Unity in the variety of parts has two directions: absorption in the parts as a variety and a gathering up into a central point of conjunction out of a pervasion of all the parts; it is a circulating process of alternation between the two opposite poles that turns back on itself. Where such a unity of a variety of parts exists, we can speak of a gestalt whole. It is only that this whole *does not have its own* position with respect to its parts. It is "in" them as the quintessence of all resulting effects of the parts. The parts are the whole, which they for this very reason do not compose piece by piece. The fact that the whole is more than the sum of its parts corresponds to its gestalt nature, discussed in more detail earlier.

Living bodies, however, are wholes in a way that goes beyond gestalt nature, as the central conjunction (which in other unities or gestalts appears simply as the *condition* of unity in variety) takes on a self-sufficient role *alongside* varied unity. They are systemic wholes. This "doubling" of the synthetic center is only the expression of the positedness in it of the [160] living body, given that the physical thing that at its core is substantively closed and bound requires a second, "deeper" core in order to be set into itself. This doubling is the underlying cause of the unique, systemic holism of organisms, which becomes particularly evident in the phenomena of regulation.

This holism guarantees the self-sufficiency of the whole thing with respect to its parts, a self-sufficiency that, however, is essentially unavailable to spatiotemporal representation. For that surplus of "being in it" that makes the body into a self, a subject of having, or a system is of a spacelike, but not of a spatial nature.

The Self-Regulation of the Individual Living Thing and the Harmonious Equipotentiality of Its Parts

Bodily things have the dual aspect of an inside that never becomes an outside (the substantive core) and of an outside that never becomes an inside (the mantle of property-carrying sides). This dual aspect is constitutive of their existence in appearance. If we analyze the physical thing, we of course find the constituted thing and not constitution. The thing can only be represented as the resulting effect of factors or parts that as a functional unity generates the gestalt image of a whole that is greater than the sum of its parts. This total can be referred to grammatically as that which "has" such and such properties, although the elements that can be distinguished in it *are* the total as a functional unity. There *is* no real subject that has properties here, but only functional unity and functional elements that separate into pure pseudo-self-sufficiency.

And yet the gestalt total appears as the real subject of properties in light of the way thingly solidity asserts itself intuitionally—namely, as being carried by a core and as having sides. If this appearance that in the hands of representation dissolves and truly dissipates into nothing is to be more than mere illusion, there is no other choice but to—as we have already done—rid it of the value of an additional determination of being and to think of it as an only seeable quality in which physical variety becomes perceptible. The thing has the dual aspect of substantive core (real subject) and property-carrying sides.

[161] According to the thesis set out, living things are distinguished from other things by the fact that the dual aspect they necessarily have as phenomenal things also functions as an additional determination of their being. Living things are colored, hard, pliable, heavy, long—as with all things, this can also be otherwise; essentially, however, they are dual-aspective. This means that they have real [*wirkliche*] properties because their being is such that it can *have* something: the core, understood here as the real subject, that "has," is actual. Living things are not only of the aspect of a core, appear "from the core," but are cored, core-holding. This core has the thing with all of its parts in their functional unity; it has the gestalt with its properties, because it is.

But what does this mean? Is the core demonstrable in space? Evidently not, for then it would itself have become a property of the being it carries and has. It would violate its essence to occupy this position, given that the meaning of its essence is to be the subject of having. The core, then, is not somewhere. Of course, this is also to say that it cannot be sometime, for this being in time would only be demonstrable under the condition that it can be determined by the clock—and such a determination necessarily takes place in measured space.

Since the function of the core, as I have shown, is essentially related to the boundedness of the body insofar as this boundedness is to coincide with the boundary itself, the problem of its location cannot be avoided. The core develops its function only as the *center* [*Mitte*]—whether we think of it as imaginary or ideal or whatever—of the space occupied by the body within its boundaries. Since, however, in relation to a spatial area the body's center can also only be conceived as a spatially definable spot, which, furthermore, since the core is to *be*, becomes a component of the area, a real and no longer imaginary or ideal, but rather precisely measurable point of its body, the center's function must exist in another sense as well.

The center merges all elements of an area into a unity; it is the throughpoint for all the relationships forming unity with respect to its elements. Every element as well as the "whole" existing in the functional unity of the elements are thus equally tied to the center [162]. The distance between the elements and the center is not yet of any relevance here. If the center is to be a core—that is, is to exercise a function in and toward the physical thing—this function must manifest (1) as spatially indeterminable, albeit not without a relationship to the spatial thing—in other words, as "into space" and (2) as a unifying function that is equally present for all elements (parts, factors) of the spatial thing and binds them all equally, thus ensuring the unity as such.

Something that is spatially actual can only be "into space" by *unfolding*; unfolding is the only mode in which something nonspatial nevertheless exists as extensive variety. Nonspatiality and spatiality cannot be together in ideal simultaneity. They come together only in the mode of passage from nonspatial into spatial variety, whereby what was previously nonspatial must have had the status of actual possibility. The *inexistence of the center* (of the real core, the subject of having) is thus only real as the actual possibility of the body or *its capacity* (potency). All the elements of the body are equally bound into a unity and ensured as a unity in this capacity. Insofar as the capacity ensures the unity *over against* the elements forming it in functional unity, it *represents* the unity in each of the elements connected through it. The fact that the unity is represented as capacity *in* each element of the living spatial thing and at the same time over against each element means that the elements are equipotential and as a total form a *harmoniously equipotential system*.—

It is well-known that Driesch created this concept in reference to regulation phenomena, in particular phenomena of restitution, whose importance for an exact biology his experiments and far-reaching analyses have demonstrated. In the course of these it turned out that in addition to regeneration in injured tissue and organs, restitutions of whole organs and organisms needed to be taken into consideration. For it is these that pose the greatest difficulties to a mechanical theory of development. Driesch went so far in his analysis of the harmoniously equipotential system that it no longer seemed possible to explain these phenomena with machine theory. One cannot assume the existence of a fixed system given in the seed or in the early stages of development [163] (in some circumstances, such as in the case of ascidians, in later stages) that would repeatedly organize the fragment into a whole after any kind of destruction. It is primarily this consideration that underlies the basic idea behind Driesch's first proof for the autonomy of the processes of life and of the existence of the intensive factor he calls "entelechy."

The current study synthetically confirms the results of Driesch's analysis of restitution phenomena insofar as they are of an analytical character. But at the same time it also shows (as does vitalism in the *content* of its concepts, albeit not in their *form*) that the potentiality of the elements and their connections of different levels of variety (cells, tissue, organs) in their potential quality belong to the body's layer of being that can be seen but not represented.

It is in experiments, in experience, that the phenomenon of restitution is observable. But its perception or its appearance as manifestation of a factor operating into space, as the demonstration of a capacity, is, while prompted by experience, not compelled by it. It is ontically founded and eternally valid for the sphere of the seeable, but not compelling for the exact biologist. This means that what is a factual finding, a pure phenomenon in the harmonious equipotentiality of the parts of the organism, cannot, in principle, hamper the progress of its exact physical analysis. Restitution does not define a boundary *on* the plane of measuring observation, but *above* it, in contrast to the plane of merely qualitatively identifiable what-structures and phenomenal forms of life that can only be understood in terms of essential laws.[5]

If, however, we take the opposition between vitalism and mechanism less from the methodological side and more from that of content, it becomes reduced to the opposition between assuming and rejecting a factor approaching the system of the bodily thing "from without" (this becomes particularly evident in the interpretation of autoregulation). Modern, scientifically trained thought struggles against this immaterial, adynamic principle of form with all its [164] might. It sees in its assumption a regression into the (itself falsely understood) speculative way of thinking that wanted to understand nature by way of occult qualities, effective ideas, final causes, potencies, nerve-spirits, or principles, without explaining it via measurement, observation, and calculation as a merely verifiable nexus of appearances. The natural scientists could not accept the introduction of an "entelechy" for the purpose of *completing* a—precisely empirical—explanation of life processes because it would then come up against an insurmountable barrier. In their field itself there is no such thing as an absolute obstacle that does not coincide with the limitations of the possibilities of scientific work.

But this is not the case when it comes to the phenomenon of autoregulation. We are not yet in the position of explaining it precisely. Conceiving of it as a manifestation of potency, however, is only permitted in a categorical analysis, if you will, in an ontological analysis in the layer of being specific to the seeable whatness of "life" and "vitality," and not in an empirical analysis.

No more does the philosophy of mathematics begin "behind" the Abelian functions—to modify a nice phrase of Husserl's—than the philosophy of the organism begins behind the phenomena of regulation. That is to say: the introduction of entelechy as a concept explaining empirical facts following Driesch's residual method is the side of vitalism empiricists rightly refuse to accept. The autonomy of the vital is an autonomy limited to the sphere of being. The exact scientist does not penetrate into this sphere; it lies entirely outside of his objectives. It is true that the phenomena of autoregulation bring the laws of this sphere particularly close and thus arguably induce a turn toward a philosophy of the organism. But this turn ought not to take on the character of a residual method for getting an empirically stalled explanatory process going again. Life is autonomous only in the special layer of phenomenality containing the irreducible what-structures, as is true for all of nature.—

In addition to phenomena of restitution, phenomena of adaptation also belong to the realm of regulation. Insofar as they follow necessarily from the principle of regulation, there is no need in [165] this context to treat them separately. They do, however, exhibit another essential law that can only be explained in reference to the integration of the organism into the environment. Not only is adaptation the organism adjusting itself; regulation is also mediated by the outer world.

Individual Living Things as Organized: The Dual Meaning of Organs

The harmonious equipotentiality of the parts of an organism presupposes, in addition to the undivided and indestructible presence of the unity as such, the qualitative differentiation of the organism in each of the parts conditioning it. We cannot speak of real harmony in the distribution of variety without also being struck by the qualitative differences in this variety.

In unicellular organisms, basic qualitative differences can already be seen in the different forms and specific functions of the elementary constituents of plasma, nucleus, flagella, cilia, cell membrane, etc., which must be considered the organs ("organelles") of unicellular organisms. In addition to these intracellular differences, multicellular organisms also exhibit a characteristic difference in the grade of the tissue and organ cell aggregates that are combined into special units. Unlike inorganic wholes, gestalts, the partial forms here are not simply forms of the whole, but forms of the parts.

This statement is not undisputed. It is true that there is universal agreement that the extent of differentiation exhibited by even primitive organisms could never be found in inorganic nature. But the mechanists refuse to recognize an essential difference between the mode of differentiation of inorganic and organic bodies and instead try to construe living organization merely as a high-grade formation of variety.

From the perspective of physics, no significant objection can be made to this. And yet it necessarily fails to do justice to the full meaning of the matter at hand. The relative self-sufficiency of the parts forming the organism only acquires its original character once these are regarded as *organs*—that is, as parts *not* immediately forming unity or as [166] *means* of creating unity. The difference between immediate and mediate unity formation is not representable, not physically comprehensible. The importance of each organ for the living body's survival differs, but these degrees of dispensability have nothing to do with the difference with which we are concerned here. The curious thing is that this dispensability or indispensability can be discussed as a pure question of facts, while the unity-forming function of organs as such is entirely independent of this question. Every organism has absolutely vital organs such as, for instance, the heart or certain parts of the central nervous system to which life seems to be tied, and this is strange in view of the undivided unity of the whole; furthermore, as regards the instrumental character of these organs as such, it is not necessary.

It is impossible for a pure gestalt to have organs, even if its different parts are combined into relatively self-sufficient unities. For every part is immediately gestalt-forming only if it does not maintain its self-sufficiency as in an aggregate, its contribution to the overarching connection ("and-connection") thus remaining visible and comprehensible alongside its isolated existence as a part. Organs, however, while also immediately forming the gestalt as does every part of a functional unity (and are thus also relatively independent of this gestalt unity, as it is of them), are *furthermore related to the unity* or belong to it in a mediate way. Organs not only compose the organism like floors, stairs, rooms, facades, roof, and foundations make up a house (which as such is more than the sum of its parts), but they also relate to the organism as a unity, mediate its unity in itself, thereby constituting the whole from which they as "parts" are detachable, for which they are "properly" dispensable. This is why the notion that one could remove all of a living being's organs while somehow being able to keep it alive is not entirely absurd. The carrier of the organs is regarded here as existing independently for itself beyond the limit of what is in fact possible, just as its actual independence from many of its organs is often surprisingly great.

The concept or fact of organs is of a priori necessity for the living thing. The notion of multilevel organization [167] does not refer only to external morphology. Even in the case of unicellular organisms, parts can be distinguished whose forms are not forms of the whole in the same way as they are forms of the parts. In all multicellular organisms, by contrast, the differentiation of the body in itself manifests in relation to independent, albeit vital, organs or organ-composing cell aggregates of the same kind by way of organization into levels, with the higher ones containing the lower.

A physical thing of positional character in itself or in its unity is not only functional in all its parts and actual with them, but—although only *potentia*—represented as unity (center, core) in every part. The living body is thus only potentially the representation of itself and in this respect undividedly present in each individual moment as the unity of all moments. This representation of itself in itself guarantees the harmoniously coordinated, mutually "considerate" equipotentiality of all constituents of the living body.

These constituents *immediately form* the body. *In actuality* [*aktuell*], it is nothing without them—that is, it is the functional unity of all of them and thus more than the sum of all of them, but no more than their actual total. If, then, the actualized [*aktuell*], actual, whole body comes to coincide with the total of its distinguishable structural elements, there is no more "room" for it to be represented as a whole, a unity, a total in the structural elements. How is this room for potency in the actual [*aktuell*], existing body created?

The state of affairs of representation comprises two entities: that which is represented, the object of representation, and the representor, the subject. In the case at hand, the physical body, as it exists, is to undergo this doubling in itself, to actually be object and subject of the representation in one. It must exhibit properties that do not permit any other conception than that of self-representation. We have seen how doubling as such is carried out in the form of the subject that "has" its physical body. Thus the living being has leaves, a stem, roots, or eyes, a torso, a tail, intestines, etc. What is had as such no longer only helps bring about the unity but can also be detached from the whole, exists outside of it, *as well as* being included in it.

[168] *Separated* from the subject of having, the body/its parts are an object that is had. Forming a whole with the subject of having, the body along with its parts, however, is itself a having "part" of this whole, of this physical thing carried by the character of the self. Only in one way can the combination of taking part in the whole with the position on the one hand of having, on the other of being had itself become manifest in the body: the body *organizes* [*gliedert*] its totality into "organs" that

1. in relation to its totality are simple parts,
2. in relation to it as self are members [*Glieder*] that it has (and that are dispensable or indispensable to it), and that
3. are means *through whose mediation* of its wholeness it is represented *as* wholeness in the parts. This is because the ontic form, the category, in which one and the same object exercises the function of having in the property of being had is that of becoming the *means* of having.

In itself as in all its parts the body is then mediatedly represented by organs. Their specification for certain operations takes place in an essentially harmonious way, in consideration *of* the unity of interaction *with* the others, which are of course not represented here, but are only "co"-present in the specification of each individual organ on the circuitous route of unity. The difference between genuine wholeness and a simple gestalt-type functional unity is very clear here. While unity is more than the sum of its parts and can be set apart and transposed from them, it is not also represented in them. Functional and morphological differentiation, on the other hand, takes place in consideration of the unity of the nexus, so that a loss of immediate or potential representation of the whole in every part is offset by the latter's actual [*aktuelle*] specialization as an organ and ultimately means nothing less than the maintenance of representation, than the making present of the whole (albeit in a mediated way).

The whole of the living body is itself immediately present in its parts in a potential way. This form of its representation is referred to as the "harmoniously equipotential system." The whole, however, is itself also mediately present in its parts in an actual [*aktuell*] way. This form of its representation exists in the harmonious divergence of specialized organs.

[169] And so what until now seemed to be purely empirical knowledge about the equipotentiality of the organism and its gradual receding behind real specialization in the course of normal development turns out to be an insight into an essential state of affairs, a condition of the possibility of life. As the organism ages—that is, its original being changes by quantitatively and qualitatively increasing in variety—by unfolding, in other words, its ability to adjust and regulate decreases and it becomes increasingly fixed and specialized in all its parts. It is on this that the concrete change in stages of aging rests, which were characterized previously only in their necessary succession, as a developmental curve.

This unfolding or development [*Entfaltung*] is a priori limited in itself. Opening up [*Auseinanderfaltung*] comes to an end, but not because the dispositions of the living organism are limited or because the resources of matter and energy that the body makes available to the process of life are insufficient, leading to attrition. It is rather that it belongs to the essence of a living being to realize potencies that in their realization fall under the law of the whole represented or present in it. That which was potential has now become actual [*Aktualität*] and represents the whole to the degree that it has relinquished its centrality for the sake of the greatest possible specification. Since the whole is bounded, specification must be bounded as well. To unfold is to forgo possibility and to gain it at the same time. But the loser and the winner are not one and the same; between them lies time: age.

If it is true that the thing only lives because of its boundaries, then it is just as true that it dies because of them. It is in this—restricted—sense that the statement holds that life perishes of itself.

In the organ the living being has its means: to live. In its body the whole becomes mediated into a whole. The positedness in itself of the organic body is in actuality *mediated immediacy*: the whole is present in all its parts by virtue of their agreement with the whole given in divergent specialization; the parts *serve* the whole. Or, to put it concisely, the real body in all of its effectively attained phases is its own *end*.

To cite Uexküll, organization is [170] the combination of varied elements according to a unitary plan for common effect. If this definition is understood in a strictly descriptive way, then in reality the unitary plan is not there first, followed by the combination of elements, but in ideal simultaneity, variety and unity become actual in one move. Organization is the mode of being of the living body, which must differentiate and which in and with this differentiation *brings out* that inner teleology *according to which* it, at the same time, appears to function and to have been formed. This insight should discourage us from invoking the efficacy of ideas transcending the body or the architectural imagination of God when trying to explain the wonderful expediencies and harmonies found in the ever-newly designed construction plans and schemes of organisms. Like life, organization explains itself.

It is true that the fullness of the construction plans cannot be developed from their underlying laws (the significance of this will become clearer as we continue). In the tendency of organic realization there is a moment of absolute arbitrariness, whose essentially necessary consequence is the irrationality of the basic forms of organization. Without this characteristic trait of playful capriciousness, life would no longer be life. It would be absurd, however, to conclude from this that a plan's *conformity* to a plan would have to come about in an equally irrational way by means of the intervention of an ordering power transcending life. The living physical thing carries in itself the order-creating conditions that come into play, themselves hampering and furthering, once the game has begun.

The living body's positedness in the body that it is or its doubling in the body is (in contrast to the immediate existence of the nonliving body) a *passing through the body that it is*. The living body mediates its own existence. And yet this mediatedness is supposed to make up the basic character of an actual being—that is, be expressed in it. The body must thus confront this being and at the same time be this confrontation itself.

This essential requirement is fulfilled by the organization of the body into organs, whose totality it is *and* at the same time also confronts, individually and overall, so that the body lives out its actual existence confronting or passing through the body that it is [171]. It is all of its organs, and it has all of its organs, so that *in* them the different elements come together *according to* [*nach*] a unitary plan *for* [*zu*] common effect. The whole of the organism is not only logically but also ontologically able to set itself apart from the body that it is as a physical body; indeed, it even constitutes itself in and with this setting apart that is captured by the words "according to" [*nach*] and "for" [*zu*].[6] Only as a unity of means and end is the living body whole or an autonomous system.

What is significant here is the essential compulsion for the body parts to take on a life of their own as organs—that is, as means of life. It is popular to view the physical lived body of the living being as the pure field of expression of the flow of life. What is forgotten, however, is the implication of the opposing tendencies that have their origin in the fundamental laws of life itself. Organization, whether centralized or decentralized, outgrows life, which in turn only becomes physical in organization. In passing through the body that it is, the living body "loses" its undivided centrality (which, however, it only "possessed" for abstract consideration, not really [*reell*]); it is this centrality only mediated by its organs without which it is "no longer" able to live.

The Temporality of Living Being

Does [the German] language hit on something essential to reality [*Wirklichkeit*] when it describes this characteristic setting apart of the living being from its organs, which becomes manifest in the double setting apart of the body from itself (and thus in a dual mediation of the whole to the organs and from the organs to the whole), as the coming together of different elements *for* [*zu*] common effect *according to* [*nach*] a unitary plan? Would the difference between the two prepositions then no longer be antithetical to the insight that they are seeking to grasp one and the same state of affairs from different sides? And could we go even further and claim that the reference to the *future* in the first and the reference to the *past* in the second contain more than grammatical value?

The error of attributing the unitary plan of the organism to already given guiding ideas brought to the living being from without would then be, if not more forgivable, at least more understandable. We have come to the point of deciding [172] whether the temporal difference hinted at in the effective community of organs and the plan that is manifest in them point to a special *position of the living being in relation to time*. This decision must be made without reference to the determinations that traced the boundary relation of the living body in a positive sense as being out beyond the body, thus emphasizing the dynamic sides of the organic, and rather be limited to the static meaning of the boundary.

As a body with the essential feature of being in the body that it is, the body is a self that can have. How does this internal nature of the "core" become real? By appearing as potency, capacity, actual possibility. A being that can only be determined in the mode of capacity and ability cannot be approached as a complete, genuine being in the ordinary sense. For it is not the matter here of an attached ability, of the ability or non-ability of something that already exists, but of the quality of ability as such. Being as the pure quality of ability is not-yet-being, non-being that has the prerequisites within it for transitioning into being. Non-being (or nothing) is not taken radically enough in this definition, however: in order to have prerequisites, it would itself have to be something; in any case it would have to be (in the mode of actuality [*Aktualität*]). The not-yet thus threatens to become an ontic contradiction in itself. And merely as subjective category and mode of perception it does not fulfill the task of potency, of actual capacity, and is thus equally untenable.

A conceivable solution would be to have the actual [*aktuelle*] and potential being of the organism cancel each other out and to only retain the passage from the one to the other, or becoming, as the mode of its existence.

Modes of Being   161

This, however, would rob the living body of its presence and turn it into pure flow, the nature of which contradicts the nature of genuine boundedness. In a true synthesis, furthermore, the determinations of potentiality and actuality [*Aktualität*] ought not to be offset against each other but must be retained with the full weight of their meaning. Being in the mode of potency precisely has the specific weight and fullness that is not expressed in potentiality as pure not-yet-being. We must find a way to understand the quality of ability *as* a quality of being, as *existing* possibility.

[173] Certainly empiricism refers to the living body as "having" potencies, as being able to compensate for damages, or as possessing morphogenetic capacity, so that it seems reasonable to regard potency as an attached determination of the (actual [*aktuellen*]) being of the body. But this way of speaking, as we have seen, is imprecise. By (rightly) conceiving of the whole of the body as the subject of having its properties and thus recognizing it as set against itself, it obscures the equally existing—physically existing, in fact—overcoming of this contraposition, in which the living body as holistic system comes to exist in the first place. Hence the potencies exist because the living being has them, and the living being has them because they form the total existence of its real being. We have no choice but to understand living being as existing possibility and to define its relationship to the existing actuality of the present, tangible body more closely.

That which *is* actually present is definitely now. That which *can be* present is definitely not yet and not until then (when its mode changes). To be actually present is definitely distinguished from being not yet present by a different relationship to the modes of time. They relate to each other like the mode of the present relates to the mode of the future. An actual possibility, an existing ability thus has a double relationship to the modes of the present and the future: a not-yet standing in the now. The fulfillment of the relationship to the mode of the future must be of the same kind here as the fulfillment of the relationship to the mode of the present. It is thus equally a now standing in the not-yet.

If we apply this definition to a physical thing—which the living body is, but also the pencil on my desk or the house in which I am writing—it immediately becomes clear that we have not yet grasped the character of potency. The pencil too is a red object lying on the table standing in the not-yet of being picked up by my hand. The house too is a space standing in the not-yet of its renovation, now coolly shielding me from the pressure of the summer heat. What is the difference between the echinoid embryo that is a pluteus larva *in potentia* [174] and the three-story house that is a four-story house *in potentia*?

This difference lies in the direction of dependency between the mode of not-yet and the mode of now. The echinoid blastula "already" carries the pluteus "in itself." The three-story house does not carry the four-story house in itself. It only *offers* the possibility of becoming a four-story house, or it has the possibility of this being done to it. The relationship of its being to the mode of not-yet is not of the same kind as the relationship of its being to the mode of now, for the latter is fulfilled while the former is not; it remains to be fulfilled. *Nota bene*: The fulfillment of the *relationships* to the mode of the present and to the mode of the future is independent of the fulfillment of the modes themselves. Just because the existing possibility, the potency, exists does not yet make it actual actuality [*aktuelle Wirklichkeit*]; it still has to become it. But its relationship to the fact that it must "become" is fulfilled in the same sense as its relationship to the fact that it "already" is.

Of course, the pencil has the possibility of being picked up by my hand, of being used to write, of being sharpened. But these possibilities do not belong to it in the same sense as do its redness, its woodenness, and so forth. The pencil's physical, thingly existence does not exhaust itself in the possibilities of using it. The possibility of doing something with it does not belong as possibility to its own stock of being, but precisely as actuality, as this or that property of its actuality. The pencil can be used to write—that is, it has a slender shape that fits into a hand; it has a graphite stick that leaves traces on paper, etc. Potency as existing ability is not a possibility of doing something *with* the physical thing in question, but a possibility contained *within* it. As a possibility, this potency belongs to the thing's own stock of being and is this possibility. In the same way, the living body is not yet what it is now.

If it is to fulfill this law of equally signifying a body that is not yet standing in the now and a body existing now standing in the not-yet, then everything depends upon the direction prevailing between these determinations. The character of potency cannot be grasped until we are able to conceive of it as a now *dependent* upon the not-yet. Capacity is a type of being whose relationship to the mode of the present is dependent upon its relationship [175] to the mode of the future. As actually present, this type of being has a fulfilled relationship to the mode of the present. Its specific nature as possibility gives it a fulfilled relationship to the mode of the future. Since the mode of now cannot be dependent on the mode of not-yet without reversing the direction of the successive order prevailing between them; since, furthermore, the relationships between the mode of now and the mode of not-yet apply in quite the same way to the being of "potency"—that is, should be fulfilled in it—all that remains is to regard a mutual dependency, not of the modes and not of the relationships to the modes, but of the fulfillment of the relationships as that which characterizes real potency.

As form, time is the unity of succession in a nonreversible direction. Time, then, one is inclined to assume, should impose the direction of dependence at a particular time on what exists by being a unidirectionally focused, nonreversible succession. Everything flows in time; what it was yesterday it no longer is today. What it has been, however, determines its today and tomorrow in an immediate or mediate way; its past helps shape its current present, brings it about. How can what is not yet institute a relationship of dependency for what already is? How can something depend on non-being, on nothing?

This conventional line of argument, however, misses the point of the precisely posed question. It is not a matter here of the relationship between two existing elements but of a mode of being: capacity, potency, as a type of being. This type of being is characterized by its relationship to two modes of time that must be fulfilled for it to obtain. If, however, the relationship to the mode of the present is fulfilled in the *same strict* sense as the relationship to the mode of the future; if, in other words and as set out earlier, the living body is just as much now as it is not yet and there is no difference here, the physical existence of the body would simply be abolished. Its actuality [*Aktualität*] would then be consistently and essentially contested by its potentiality.

It is precisely this that we want to avoid. Actuality [*Aktualität*] and potentiality should harmonize with each other in the actual [*wirklichen*] life of the organism, should by permeating each other create the unity of instantaneous existence, without thereby canceling each other out. Again the notion tentatively raises its head that things are [176] the way common parlance portrays them to be when it describes the body as having potencies, suggesting that possibilities exist on the ground of a completed, actual [*aktuellen*] being (in the mode of now), *conditions* for positing this or that in actual actuality [*aktuelle Wirklichkeit*] as the opportunity arises. But falling back on an idea we have already rejected is not helpful here. At stake is precisely the exact understanding of this type of *conditional* being, which *qua* being belongs to actual [*aktuellen*] reality, *qua* possibility does not.

Existing possibility, real potency definitely *is*, and is thus in the mode of now. The relationship to the mode of now is fulfilled. "Possibility" means the ability to be and is thus in the mode of not-yet. The relationship to the mode of not-yet is also fulfilled. Under what condition does the fulfillment of the relationships to the two modes of time constitute a type of being if both relationships have to be fulfilled?

Being possible does not simply coincide with non-being, nor is it defined as equivalent to non-being. Being possible is non-being that, as stated earlier, has the prerequisites for transitioning into being within it. This is an imprecise way of speaking, for what is essentially without being cannot have prerequisites. Being possible, then, refers only to a particular direction from non-being to being, captured in the expression "not yet": the direction from the future into the present. That which has the prerequisites for passing over into being anticipates something in becoming that guarantees its unity. Possibility thus carries a unity of direction oriented *against* the determinate direction of what exists in the time of past—present—future. It is thus ultimately nothing other than a relationship of anticipation [*Vorwegverhältnis*], in which the direction of dependence moves from the future to the past, that is established in the ability of *being*.

This means that a real potency is given when the fulfillment of the relationship to the mode of the future is ahead of or conditions the fulfillment of the relationship to the mode of the present. A real conditioning relation enters into time here, whereby that which conditions is not that which temporally precedes the other, as this would reverse the order of dependency. The conditioning relation between the fulfillments themselves is thus "timeless"; it runs counter to the successive direction of what exists in passing time and amounts to connecting the future back to the present. This proteron hysteron [177] is the schema of *material, semantic* dependency, or *grounding*.

Modes of Being 165

In its potencies, the being of the living body is ahead of itself. Real potency is mediated being that is not grounded in itself as something that is present but rather that is in the future. A real possibility is given if the fulfillment of the relationship to the mode of the future conditions the fulfillment of the relationship to the mode of the present: potential being is subject to this condition of being grounded in the future.

If the living body is set into the body that it is (spacelike) and with this character of positionality constitutes a space-claiming body, it is potential in its actual [*aktuellen*] existence; it is ahead of itself. The space-related essential properties of positionality thus allow us to determine the essential laws of the living body's relation to time. This means that positionality itself includes the relationship to time.

As a living body, the organism is not only into space, but equally into time. It is not simply temporally determinable like every thing, in time, at a particular time or at any time; rather, its own essence is timelike. This is the true foundation of the timelike determination of immanent teleology that is manifested by the unity of the members in the whole of the organic body and that formed the starting point of this investigation. The members come together *for* [*zu*] common effect *according to* [*nach*] a unitary plan because the whole of the body, which is immediate to itself in its potencies and mediately present to itself in its members, is ahead of itself. Indeed, even in the organism taken at rest and outside of time, the whole is its own end in all its parts, and the already given nature of this plan is essential for the construction and function of the mediating organs.

Mere gestalt unity is immediately given with its parts, but its surplus over the and-connection of the parts, expression of the interplay between all of the elements, is not constituted for itself alone. A unity as a whole is given with and mediated by its parts; and yet in this unity with them it still confronts them. The expression of this self-mediation lies in the space- and timelike character of positionality. Being-in-it appears as organization, being-ahead-of-it as full present being, as true [178] persistence, genuine actuality [*Aktualität*]. Thus the final difficulty in the concept of real potency and of its relationship to the actuality [*Aktualität*] of the present body in the now is solved. Potency is only the way in which nowness is mediated to the present.

All things exist in time, last for a shorter or longer amount of time, change, and disappear. If they are left to themselves and are not given a specific end, they move from change to change following the law of cause and effect. The direction of determination is the sequential direction from the past into the future. If things are given an end, this only seemingly reverses the direction of determination by orienting them toward an end in the future, thus binding them from the future. In actuality nothing changes. The end is posited at a particular time and causes the development of all further steps into the future. Only the content of the cause or its purposive sense sets in advance the schema according to which the effect is meant to ensue. In its causality, as that which determines actual being, the effective end, counter to its ideal meaning, also belongs to the past.

Novalis famously wrote that nature is pure past. Such a nature, however, would be devoid of all life. It would be pure paleontology, a battleground littered with corpses. Novalis's statement applies only to nonliving things in space and time, and even there not in its full weight. For the being of inanimate things is the pure transition from no more into not yet. Their being in the mode of now is only a limitative one. That which they can be observed to be goes by and becomes, but is no more related to now than it is to back then or to later. Nonliving being elapses with time because it lacks a positedness in itself, a centeredness in which it could persist. To this kind of being, the modal accentuation of time remains unessential and closed. It is true that the schema of causality can be applied to inanimate being, revealing it to have been determined by the past. Characteristically, however, everything that can be presented in the form of causal chains can be transformed into conditioning connections or be understood by means of the computational equivalence of space and time. The specific directional quality of time is of no significance for knowledge in the physical sense. Pure measured time finds its adequate expression in the clock. An excellent example [179] for this way of treating the time "in" which everything exists can be found in the equations of the theory of relativity.

Time in its modes is only essential to living being, which constitutes itself by means of time insofar as it is being that is ahead of itself. Certainly one can say that every thing that moves needs time. That is an analytical judgment. But rest does not need time. The observer might find that a thing remained at its place for "such and such a time," and since its rest as standstill can only be measured in relation to other movements, this cognitively relativizes the qualitative difference between rest and movement. The meaning of rest, however, does not include a reference to time, even if the resting thing is subject to the form of time just as much as it is to the form of space. With the living thing it is different. Its positional character itself implicates the form of time in the being-ahead of being; it is time. Like every thing, the living thing in regard to its objectivity belongs to a particular point in time; time in this function is the condition of objectivity and does not affect its being as such. Living things, however, are distinguished from other things in that they "contain" a reference to the future in their being as such.

The organic body, as set into the body that it is, is ahead of itself. It *is* inasmuch as it stands in a relation of anticipation to the body that it is (to itself). Or its being is grounded in a timelike way, determined by the direction "from the future." The being of the organic itself stands in an essential relation against the flow of time, which approaches it and elapses behind it.

But what does this in fact mean if we want to avoid using metaphors that portray being as a solid body and time as a flowing liquid? If the living body has this unique character of anticipation, then it follows that being standing in the now is being that is determined from the future or being that has come. That which is factually present is thus the fulfillment of an already given direction into the future, of an expectation (if the reader will permit the psychological expression), of a tendency. *Not* in the sense that the tendency is factually in advance and already given, as this would imply that what is only meant to ground is present. Anticipation is the mode of living being, anticipation not of [180] something determined that is still to come, to become, to enter into being, but rather the anticipation of itself as something determined. (Putting this in terms that are very common today, albeit generally used in reference to the structure of consciousness, one would have to say that only the fulfillment is there without an actually foregone or corresponding intention. Provided that it is alive, the body as present being, in these terms, stands in the light of a fulfilled intentionality.)

"Being ahead of itself" and living being are one and the same thing. Living being is thus just as much *according to itself* or the fulfillment of itself. This essential trait secures for the living thing something that is not given to any inanimate thing: *presence*. Connected back from the future the living body is ahead of itself—that is, is its own end; it stands *over against* its constant passing over from the not-yet into the no longer, or persists. The abstract now between future and past is no longer suitable as the schema of the living body's existence, but only concrete presence, whose differential is the instant, or *blink of an eye*, the unity of future and past. This is why the living body, which, connected back, is according to or after itself [*das ihm selber Nachseiende*], still *has* a past. It does not simply pass away, losing what it was in what it is, or like a mountain conserving what it was as what it is, but rather conserves it in its having-been as the stock of its own being. As that which is according to or comes after the body that it is, the living body is pure past. Thus it seems appropriate to speak of "memory," as Ewald Hering did, as a general function of living matter.

Living being stands in the mode of the present because it is being that is ahead of itself (is in accordance with/after itself). Its presence is actuality [*Aktualität*] that ought no longer to be thought in irreconcilable opposition to potentiality, but rather has potentiality as a prerequisite: fulfilled potentiality. This being—mediated in itself—(in the figure of the eternal cycle or the quiet flame) signifies the continual transformation from one mode of time into the other *and* the unity of this transformation—that is, presence.

The Positional Union of Space and Time and the Natural Place

If advances in physics, the use of non-Euclidean geometries in its measurements, and especially the equations put forward by the theory of relativity were able to shake people's faith in absolute space and absolute time [181], this always occurred in collision with the living evidence of dispassionate intuition. Above, below, in front, behind, left, right still maintain their meaning as specific directional possibilities, as do earlier and later, now and at the same time. Of course, this is to posit their meaning in relation to an intuiting subject, but in the relation this meaning remains absolute. As such, the quality of above, the quality of later are only meaningful in relation to an intuiting, experiencing being and are not interchangeable with the quality of below, the quality of earlier. Only the choice of a certain direction in relation to the position of above, right, behind, earlier, or later is arbitrary—that is, dependent on the location of the observer. The fact that all measurements depend upon the location and speed of the observer, that the sought distances and times in their spacing are functions of the place and time of the measuring instruments being used, that the measurement of the actual processes in bodies does not take place with the forms of intuition, but like every measurement constitutes a detachment from what can only be characterized qualitatively, transforming it into a continuum of quantities—all this does not run counter to the structural laws of space as it is experienced, of time as it is experienced: it has no bearing on them at all.

Above, right, now, later should only be understood in an experiential way, but by no means *as* lived experiences. And because they cannot be demonstrated mathematically, their being cannot be plumbed by mere representations. The possibility of distinguishing between front and back can be denied to radially bound beings such as starfish, and if the human were Janus-headed—or, better yet, had two front sides—it would be conceivable that, barring a pointed focus of attention, he would not be able to differentiate between front and back. But they are still specifically different possibilities of inscription on the continuum of divergence and juxtaposition in lived experience. The same holds true for time. As an irreversible, unidirectional succession available to lived experience, it represents a structural form of being that in its absoluteness is not affected as such by the fact that it can only be understood as "time" in experience. Dependence on experiential and potentially intuitional mediation and the possibility of excluding certain lived experiences are no more an indication of the existence or nonexistence of what is experienced as is its measurability or nonmeasurability.

[182] If, given the previous, we regard a nonliving physical body in its relationship to space and time as forms of being that can only be grasped experientially—in other words, in its intuitional relationship to alongside, over, behind, now, later, back then, etc.—we see that it is essentially indifferent to its location and duration despite the fact that its being is spatially and temporally "determined." To be spatially and temporally determined here always means to be determined in relation to other spatiotemporally positioned things. The body itself can be arbitrarily shifted around in space and time. The idea that the body suffers spatial deformation at high speeds or that space and time contract in relation to it does indeed conflict with this independence of its being from its where and when. But even if, as physics shows (in an apparent overcoming of this structure of independence), the body is fully absorbed by its locations and "is" nothing other than what is indicated

by the measurements of clocks, galvanometers, or scales, it nevertheless appears outside of its relation to the place it occupies in space and time, those empty forms to be filled by dimensions and paths of movement.

The living body is not indifferent to space and time in this way: it grows and ages. The fact that it is located somewhere and will change sometime can, however, only mean the same relative determination as is characteristic of all bodies. But while all bodies are absorbed by their location as measured by the coordinates of space and time (and, on the level of appearance, are independent of empty space and empty time), living things have a relationship to their place in space and time. The growing body has an absolute spatial measure as its boundaries expand, an absolute temporal measure as it ages.

It is true that this measure does not supply a universally valid unit of measurement to be used at will. The times and spaces of life as captured by a unitary system of measurement are completely diverse and, if made into each other's measure, would each render completely different numbers. But this has nothing to do with the absoluteness. The organic body is its own measure; space and time are measured in relation to it. This is why we can say, for instance, that a three-year-old rat is as old as a sixty-year-old human. And why we cannot say (as has been done in popular explanations of the theory of relativity [183]) that an organism moving at the speed of light in the opposite direction as the rotation of the earth is becoming younger.

Crystals too grow and age, and there is much in nonliving nature that over the course of time grows in size or loses its resilience, wears out. We speak of old rock formations, landscapes, etc. But growing and aging here are of a purely extensive nature. They refer to the gradual change undergone by the original body as a result of accretion or the interference of other substances, making it into another. The original body passes away with every step of these external influences, even where the composition and changes in its form are brought about by processes of rearrangement and development of the elements making it up. The processes that are referred to as aging or growing proceed here from the thing just as much as they form it. There is no nonlinguistic reason here for contrasting a carrier of the processes and the processes themselves.

Only the positional character provides such a reason. Being beyond the body that it is, being into the body that it is—constitutive characteristics of a body "within" its boundaries—makes it into a being that is into space,

Modes of Being   171

into time. The boundary's original indifference to spatiality and temporality (which allows it to delimit the body spatially or temporally) gives the body, which has the boundary not only as condition of the possibility of its boundedness, but as actual constituent, the properties of spacelikeness and timelikeness.

The discussion so far has shown how these properties come about: positionality means to be posited, to be mediated in itself (to be lifted up and set down, whereby the difference between the phases here is conceived as having been annulled). The spatiotemporal body is thus one that is mediated in itself—that is, the form of space and the form of time shift from the position of conditioning external forms into that of conditioned "inner" characteristics of being. The organic body, as one that is "within" itself, ahead of itself, exhibits the same distinctively spacelike and timelike traits as that which is beyond it, that which is becoming, developing. The fact that we have been able to develop the transition from the dynamic range of properties into the static and vice versa from the two basic functions of being a boundary—the first of which leads into the dynamic, the second into the static [184] range—serves as proof of the union of this space-timelikeness.

An organic body is not simply, like every physical body, a four-dimensional formation, but in its essential property as positionally claiming space and time, it is in itself an absolute union of space and time. This is the meaning of the doctrine going back to Aristotle of the natural place, the essential place of things, which can only be confirmed for those that are alive.

[185] FIVE

The Organizational Modes of Living Being: Plants and Animals

The Circle of Life

Organization is the self-mediation of the unity of the living body by its parts. Unity here is thus wholeness; the single thing an individual. A whole is not only more than the sum of its parts—as is every gestalt as functional unity—but is as one in each of its parts, as one beyond the unity of their variety. If, then, a closed system of elements forms a whole, the system as such is the connective condition for the variety of all the elements and at the same time a unity alongside (and by virtue of) this manifold unity. Unity in and beyond the manifold does not constitute a simple contradiction only if it is conceived as unity mediated by the variety of its elements. The realization of the contradictory determinations hence also takes place on this condition.

The organ represents the whole *and* is a part of the whole; it mediates the whole to the whole or is a means for it upon which the whole is dependent to the point that it cannot live without it. In the organ life creates its own inhibition, without which it could never be life. The necessary self-inhibition of life is not compelled here, as if life's unbound, flowing essence suffered congestion and refraction as a result of its unavoidable ties to physical things. Inhibition rather springs from life itself, as the previous has shown, from its own nature of primary ontic boundedness. The organ inhibits life because it mediates the immediacy of the process (by effecting a separation within it). At the same time, however, this inhibition is a furtherance, a means of life—a means for life, serving life.

[186] If this definition is taken in its fundamental meaning—if, in other words, we do not prevaricate empirically by contrasting organs with, for instance, interstitial tissue (which would be to no avail anyway, for by "organ" we mean here a relatively delimitable, functionally and morphologically specialized subunit of the body as such)—this leads to the recognition that the organism as the unity of all organs is the unity of the means for life. The organism, however, only represents this unity over against its organs because it is mediated *to* [*vermittelt*] unity in these organs. Held together by means [*Mittel*], on the basis of means, the organism is in truth its own means.

Clearly, however, the correct way to look at this is to say that the organism is its own end and possesses and uses its organs as means to this end, without simply becoming substantively absorbed in them as owner and user. It is, after all, the contraposition of a subject to its body, a body that it has with its parts, that necessitates the organization of the living body. The subject here is precisely at the same time its object; the carrier of life coincides with its goal. And if it then turns out that the carrier of life (that is, the unity of the organism, standing, as core and center, over against the variety of the parts) is itself only a mediated unity based on a reference back to its parts, this reveals the organic nature of the parts—but at the same time gives up the separate being of the unity of the organism, the sovereignty *over* its organs. This, however, cannot be. It is impossible for the living body to be, in the same sense, its own end and its own means.

The fact is that we do not think of the essence of organization as fundamental, but rather as an apparatus coming *alongside* the unified-undivided process of life, which "life," "the organism," draws on in order to live. Or we fall into the other extreme and identify the organism with its apparatus, wholeness with the unity of the parts. Both of these views are incorrect in their one-sidedness and only become correct once they have been successfully joined. This synthesis will only succeed on the condition that the twofold nature of the organ is maintained in its entirety: that is, it is both means *and* mediation of the living body.

[187] Wholeness is mediated unity. Mediated by what? By the parts that form the unity in an immediate way. Consequently, in a whole, the unity is in the variety of the parts and outside of ("next to") it. It is not only the connective condition but at the same time the center existing for itself, the central *one*, the core that is enclosed by all variety.

This is the point of unification for all parts, which at the same time must be able to exist without them—without this independence, however, going so far as to break up the unity of the whole into a functional unity on the one hand and a unitary core on the other. This is avoided by the parts being the unity just as much as the unity has the unity that it is. Then we have the case described earlier: that which is had as such is no longer only the part co-conditioning the unity of the whole, but also the part that can be detached from the whole (that is, the unity exists for itself) *and* is included in it.

We are speaking here only of the body: the body is the whole that has the body *and* that is had by it. For the body is itself the living thing, the thing with the property of positionality, the thing in itself, whose structure includes the presence of the unity in each of its parts. It is doubled in itself, but unified in this doubling: unity for itself (core, the subject of having), unity in the manifold parts (functional unity, gestalt, overall function that is greater than the sum of its parts, object of having), and unity in every part (harmoniously equipotential system).

The unification of the first two determinations ought to lie in this third determination, so that we could say that the living body *as* unity in every part is unity for itself and unity in variety. We cannot say this, however, as long as "unity in every part" means harmonious equipotentiality. As harmoniously equipotential, the organism is only actual potentiality for unity for itself and unity in variety. It is not yet this thing with hide and hair in actual actuality [*wirklichen Aktualität*].

The necessary and sufficient interpretation of this emerges of its own accord, however, if we meet the conceptual requirement of determining "unity in the part" as the unity of: unity for itself and unity in variety. This can only [188] succeed if we change the meaning of the concept "unity in the part." It ought no longer designate an immediate presence of the unity in the part; indeed, it cannot if the two other modes of unity are to have their inner connection in this presence. An immediate relationship would only bind the two other modes of unity together in an external way without synthesizing them. Instead, *the unity in the part must itself be the unifying mediation for the two others*. It is through this unity that the unity must come to be, that the unity must exist. Otherwise, instead of *one* stroke of unity with the three determinations as non-self-sufficient moments of itself, we would have three self-sufficient modes of unity, requiring a fourth form to synthetically bind them—a process that could be continued ad infinitum without ever leading to success.

Unity in the part can thus only mean mediate presence. By virtue of its mediacy it *synthetically* binds unity for itself and unity in variety—that is, into an overarching unity. Unity in the part is nothing other than the binding mode of the mutual absorption of unity for itself and unity in manifold; it is the pure passing-through of their undivided unity *and* separateness: their mediation.

The mediate presence of unity in every part is the mediation of the unity for itself and the unity in variety into the unity of the whole. The mediate presence of unity in every part, as the previous has shown, is the organic character of the part, its nature as member that in its specialization immediately denies unity but, by taking "the circuitous route of unity," manifests it. The organ thus mediates the unity (of the whole) for itself and the unity (of the whole) in manifold into the unity of the whole; it is the pure passing-through of their undivided unity and separateness.

But an organ also (and for empirical intuition certainly primarily) refers to a means [*Mittel*], an aid [*Hilfsmittel*], an evolved tool. If, radically speaking, the organism is organized into a whole host of organs, this means that it is organized into a whole host of tools, and, as, unity in variety, it is nothing other than a unity of tools. As long as this unity stands over against the unity for itself, the core, the center [*Mitte*], the subject of having, we can maintain the normal view according to which the subject of the living whole independently has and uses its means. In this view, the body synthetically combines [189] the property of being the subject of having with the property of being the object of having (its body) by becoming the means of having.

But it is the whole body that is thus determined, since in its organs it is the combination of unity for itself and unity in variety. The means of having that the body has is the unity of having and being had, of subject and object in the living body, their mediation to its wholeness. The body is thus its own means, just as it is itself mediated.

If, then, we said earlier that the whole is present in all of its parts by virtue of their accordance, given in divergent specialty, in a whole; the parts, in other words, serve the whole or are assigned to it as their end, making the actual body in fact into its own end—then this determination is contradicted by the equally conclusive determination that as mediated unity, the whole of the actual body is its own *means*.

One is tempted, of course, to balance these definitions out by subordinating the latter to the former: something that is its own end can never become the means for another end, but something that is its own means presumably could. As its own end, the organism would of course also be its own means. By means of its organization it would achieve its life purpose. With the help of its body, it would be able to master the biological tasks confronting it, etc.

Subterfuges such as these, however, revert to old antitheses and fail to see what is at stake here. Something that is its own means is not the same thing as something that is the means of the end that is present in the unity of a system and that has all the parts of the system as its means. The words "its own" would then refer to an entity that contains the means but does not coincide with it, and "its own means" would merely be the expression for the contraposition of the organs and the implementing body-subject. In actuality, however, what is referred to here is the overcoming of this opposition, the mediation of the unity of the antitheses into the unity of the whole containing them both.

Under what conditions is something that is its own end its own means? In concrete terms: how is it possible for the physical organism to be its own means [190] without sacrificing its immanent teleological self-sufficiency?

The solution lies in the concept of the organ. The organ is a tool. For what purpose? For life. For eating, fighting, running, attracting insects, for reproduction or metabolic conversions—the most specific and fundamental life processes are tied to organs, are mediated by organs. Life as the fundamental property of bodies whose bounds are boundaries expresses itself in a variety of processes, none of which however are themselves life, but only manifest it—as they serve it. It is by the very fact that in its essence, life, vitality brings about the organization of the physical "mass" that it raises itself to the position of end, is out beyond this end as the epitome of individual cooperative processes and becomes that which everything serves.

As little then as life is more than its means, the apparatuses of its existence, the more absolutely it succumbs to them. For life does not float like a delicate breath above the body, is not drawn through its pores; it is rather completely tied to the body and, by virtue of its ontic structure, is a property of the body, nothing more. Even if the interplay of the organs and the organ functions is merely the apparatus of life, its very existence is tied to this interplay. It is true that this or that organ is expendable, and thanks to harmonious equipotentiality one function can be taken over or compensated for by another. But without any organs whatsoever there is no life, for life means being in mediation with life.

Only life is its own means and its own end. It is being mediated in itself, raised above itself and thus end; but at the same time it succumbs to itself in its means, which are set off against life because life is raised above itself as organization: the mediated immediacy of the whole. The physical organism can become its own means without sacrificing the self-sufficiency of its inner teleology if it, the living being, is a means of life—*that is, physically carries out in itself the distinction between it, the living being, and life.*

This is not about a conceptual distinction, which has always been possible. [191] It always amounted to setting apart the property of being alive from the living body as it is physically present in space and time with all of its other features and looking at it separately—ontically speaking, to holding up a non-self-sufficient moment next to a self-sufficient thingly being. This changes nothing in the state of affairs of living reality [*Wirklichkeit*]. If this state of affairs contains a conflict of being—and this is what the previous has shown—it has to be settled in being and not only in the reflections of writers.

Life is to be set apart from the living and to be united with it through this set-apartness—how is this possible? Above all, how can this be carried out in the physical organism itself? The restrictive condition of solving this task with only the means of the physical organism eliminates from the beginning concepts of life that treat the living body as an extract, as it were—for instance, as the subject of the conduct of life, the goal of all activity, the scanty encasement of existence. Life in this sense as an existence, measured in space and time, and at the same time as a wealth of possibilities into which the living body is positioned, life, in other words, as existential sphere cannot enter into our discussion at this point. That would be getting ahead of ourselves.

And yet the solution to the task at hand does go in this *direction*. In the unity of the organic body, it is only the organs—engaged in functional unity as carriers of the unity that they mediate into a whole—that can be set apart from each other/from the whole. They alone stand opposed to each other in the unity of the whole; it is upon their opposition alone that the whole at the same time rests. There is nothing else there that could take on this function, which takes place on the condition that the physical setting apart causes the organized body as a whole, as it lives and breathes, to become *a means of life*.

The organism, as its own organ, the means of its life, presupposes that the organism, as life, is out beyond the organism that it is or, to provide an equivalent image, presupposes that the organism sets itself apart from the organism that it is—a setting apart, however, a being beyond, that leads it back to itself, in which it mediates itself to the unity of the whole.

The physical carrier of this mediation is the organ/the functional unity of the organs. In its organs the living [192] body comes out of the body that it is and back to it, provided *the organs are open and form a function-circle with whatever it is they are open to*. The organs are open to the *positional field*. The result is the *circle of life*, whose one half is formed by the organism, the other by the positional field.

The living body's openness to the medium of its positional field does not contradict its fundamental properties of being delimited and closed off, but neither is it immediately in line with their essentially necessary counterparts of openness and being out beyond borders. (All the fundamental features of being alive can be developed from the interplay between the properties of a body arising from the nature of the boundary. Such a development must take place layer for layer, as it is not given to thought to be able to analyze and take stock of the essential relationships all at once. At the same time, progressing from layer to layer entails tracing the conditions for the compatibility between the essential traits of being alive and the essential traits of physical thinghood. It is not irrelevant, then, in what layer the observation is taking place. If I refer here to the openness of the organism by way of its organs, this openness *rests on organization* as an essential property of the living body as such and only implies a consequence of this organization's premises, not a conflict with them.)

Not only does the organism's organ-induced openness to the medium not contradict the elementary determinations of its nature, it is rather only thereby that it fully enters into contact with its setting. This contact, as the following will show, is warranted by the organism's physicality by way of assimilation/dissimilation and adaptedness/adaptation. The organs thus have nothing to do with the existence of the reciprocal forms of equilibrium found in the metabolic and energetic cycles and the morphological and functional coordination between organism and setting as such—in other words, these are not brought about by the organs. They are, however, mediated by the organs, which, precisely, can only take on their role as mediators because the reciprocal equilibrium between living being and medium already exists as such. This equilibrium is the simple and essential [193] consequence of the conflict between the vitality of a body and its integration into a spatio-temporal, reciprocal community with other bodies. If the living being's contact to the medium depended only on its organs, these would have to be more than mediators; they would have to be producers of the metabolic, energetic, and adaptation cycles.

In actuality the organs mediate the contact without creating it. Only in this sense of mediating something that already immediately exists do the organs remain within their essential boundaries and "open" the organism up to the medium, integrate it into the positional field, and thereby rob it of its self-sufficiency. For now the organism must become part of a comprehensive whole. While the scope and nature of this whole are in the organism's power insofar as its organs fit into this whole in a precisely measured way, so that the whole cannot put forward anything to which the organism is unable to respond, just as the whole only supplements the purposive system of the organism's body and genuinely coincides with this body, the organism is still a part, in need of supplementation. Its *autarky* has been lost. It remains autonomous, as nothing gets close to it and nothing gains influence on or in it that it does not subject to the law of the bounded/bordered system. Thus autarky belongs to the whole circle of life, which provides the autonomy of the living with the means of subsistence, nutrients, light, warmth, water, gases, and other living beings, which make life possible in the first place.

We find confirmed here what was said earlier about the body parts' essential compulsion to take on a life of their own as organs. The physical lived body of the organism implies (and is already the manifestation of this implication) a tendency running counter to the original direction of life while at the same time originating from its fundamental law. Organization outgrows organizing life, which only becomes physical in organization. As it mediates itself into unity, the living body "forsakes" its immediate centrality; it "still" is this centrality only with the help of its organs. It forsakes its absolute self-potency because it can no longer live without organs. It loses its self-sufficiency because its organs, as they mediate it into a unity with itself, only enable this unity by means of contact with that which it is not: with the field of its position.

[194] The organism as a whole is thus only half of its life. It has become absolutely needy, demanding supplementation without which it would perish. As a self-sufficient entity the organism is part of the life circle of an overall function between it and the medium that channels life itself through it.

The organs have now reversed their relationship to the organism: at first it was the organism that by virtue of its positionality revealed the ontic characteristic of "passing through," thereby establishing the necessity of its own mediation of its immediate unity to the unity of the whole, and thereby in turn showing itself to be the necessarily organizing body. Now the organs must exercise the power that the living body was essentially compelled to shed and delegate to them, thus turning against the unity (immediately) existing for them. Their essential ambiguity undoes them *as well as* the immediate unity: they open up the organism, chain it to the medium, and, by mediating, take away the self-potency of its own life not only from the organism as the immediately central unity of the whole, but also from the organism in its totality and, in so doing, of course also from themselves. They turn the whole into a means of life, into a link in a circle, which is now truly sufficient to itself.

Empirically speaking, this seems to be stating the obvious: the living body is, simply, a physical thing like all other things in space and time; it is in contact with them, so of course it is open to their influence and thus, like everything in nature, automatically becomes dependent on them.—This is not, however, how things stand. The following will show how the living body as a physical thing takes into account the reciprocal community with other physical things in which it finds itself and how the disintegration within it preserves the closed nature of its own system. In order for this to be the case, however, the body must have the means, which naturally do not bring about the *contact itself*, but are supported by it: the organism of its own accord makes contact, a contact that falls to it not simply by virtue of its material thingness, but as a result of the fact that it is organized, alive. And again we must point out here how wrong it would be to blame this essential conflict between the organs and the organism on the physical "bond" [195] life enters into. Vitality creates this conflict itself.—

Biologists have referred to the struggle of the parts in the organism.[1] Valid from the standpoint of a whole gaining an advantage as a result of its elements competing for the best possible living conditions, this is only one side of the coin. If the parts are struggling *in* the organism, they are in truth struggling *for* it. The organism here is not simply the "unmoved mover" disconnected from the struggle for it, but rather the unity of the struggle as it is mediated by the opposition of the organs to the central unity. And because this mediation is brought about by the organs and by every organ, the unity is involved in the opposition—indeed, it is nothing less than this opposition, this disintegration into the abundance of individual organs.

Are we to conclude that unity coincides with its loss? This cannot be, for it contradicts the simple analytical truth that cooperation (accordance of the organs in the whole) "also" always means opposition (isolation, specificity, difference between the organs, their separate march to a unified beat). But here the plan of the whole still hovers above variety, the end "in front of" the means. And thus the organism *as* means of life is finished. Arguments such as these, which in a limited sense are certainly correct, cannot save its unity. *The organism is a unity only via entities other than itself, a body mediated in itself, member of a whole that is out beyond it.*

We have now demonstrated the necessity of the claim made previously—namely, that in the unity of the organized body, it is only the organs engaged in functional unity as carriers of the unity that they mediate into a whole that can be set apart from each other/from the whole. Life should be set apart from the living and be united with it through this setting apart if the organism really [*wirklich*] is its own end and means in one. The circle of life into which the organism is integrated is the possibility and truth of the fact that it may be called its own end and own means [196]. The unity of the life circle and the loss of this unity in the revolt of the organs against the immediate unity of the organism define one and the same state of affairs.

Assimilation—Dissimilation

Thus physical being, if it is bounded, is connected at the place of its individual existence with other physical being. Manifestly, space and time conduct action and reaction from one formation to another, and every thing persisting in its contours puts up resistance to the degree of its solidity. The thing's specific inertia causes it to act back on itself. Its isolation at the same time warrants its incorporation into the overall context of cause and effect.

The fact of this reciprocal correlation takes on a special character for the living body because—according to the premises set out earlier—its boundary belongs to itself. The living body is the carrier of the boundary between itself and the other—that is, the medium that borders on the body and conducts all the influences of other bodies and processes toward it and its reactions away from it. As this in-between, the body separates its own domain from that of the other and relates them to each other in an absolutely opposing way. The merely relative opposition between physical action and reaction, whose mutual transformability is warranted by the space-time continuum, becomes an absolute opposition between the living body's own zone and the alien zone of the medium bordering on it.

As carrier of the boundary that both runs in between and bridges the in-between, the living body separates the alien zone from its own zone in order to join the two zones together. That is, the body's own zone, irrespective of its positioning over against the alien zone, *disintegrates in itself* [*zerfällt in ihr selbst*] in order to bring about a connection with the alien zone (see Fig. 5.1).

If a living body is part of the general circulation of matter and energy, it has to "fall out with itself" [*mit sich selbst zerfallen*] in order to conduct this circulation through itself, has to empower itself both to absorb as well as to discharge matter and energy. In empirical terms: if destruction is not fol-

Organizational Modes of Living Being

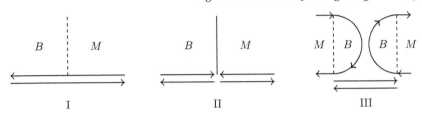

[197] Figure 5.1. *B* stands for the body; *M* for the medium bordering on it. The direction of the arrows and dotted line between *B* and *M* in diagram I represents the reciprocal boundary between body and medium that generally holds for spatiotemporal formations. In diagram II, the direction of the arrows and the solid line of separation indicate the absolute opposition of the living body to the medium bordering on it. The break that the boundary really posits between the two zones would seem to make the physical influence of *B* on *M* or *M* on *B* impossible. A physical life-form would have to find a way to prevent this impossibility, which also contradicts the (equally to be realized) bridge function of the boundary. The body in its vitality obeys the laws of physics. Diagram III shows how the *B* zone can continuously interrelate with the *M* zone through self-disintegration (self-construction) without relativizing its absolute opposition to it.

lowed by an equivalent construction, the physical prerequisites of life would cease to exist and the body would perish as a result of self-decomposition [197].

The decomposing phase is balanced by one of construction: "This denotes the essential trait characterizing life: the *duality of autonomous self-modification*. If living matter did not recreate and supplement itself, it would become consumed by destruction and decay and fall prey to death. When destruction and recreation evenly balance each other out, living matter appears from the outside to be stationary. In this case one can say that the duality of life expresses a striving for the preservation of a particular condition by recreation removing the disturbance found in destruction. If recreation predominates, living matter propagates and grows of its own accord. The dominance of destruction leads to exhaustion, reduction, and atrophy, and finally, in a longer or shorter process of dying off (necrosis), to death. There is no life, however, without active self-replacement, regardless of whether this manifests outwardly as a predominance of recreation over destruction, i.e., as obvious growth, or not."[2]

[198] This empirically comprehensible dialectic of life has given rise to a particular fondness for speculations of a general nature. Felix Auerbach's

concept of vital ectropism stands out here as one that appealed to this set of facts. In contrast to the entropistic process of dissimilation, the notion of ectropism was the attempt to show the assimilative generation of less likely conditions from more likely ones to be a fundamental physical characteristic of life. All physical being is subject to the second law of thermodynamics, to the principle of entropy, according to which all conversions of energy give off heat, so that the overall energy of the universe is tending toward a minimum (hypothermia). Life, Auerbach argues, however, increases and binds energy, is ectropic, and it is only this principle of ectropy that allows us to understand development.

Even if the low-heat effect of certain metabolic or developmental processes in animals enters into such theories, the qualitative nature of this, as it were, intuitive ectropism ought never to be overlooked and ectropism made into a principle analogous to that of entropy. If ectropy only refers to the developmental processes increasing matter, energy, and form, as intuition shows them, then we have no objection. These processes would then still conform to the principle of entropy, which has no bearing on the intuitional character of the physical.

Assimilation and dissimilation in metabolic, energetic, and metamorphic cycles determine the phases of a circular process containing all the functions of the self-preservation of the living individual. Self-preservation is necessarily tied to self-abandonment and self-destruction, as it is only by means of inner contraposition and continuous splitting that the zone of the living body comes into sustained contact with the alien zone of surrounding nature. Here we find what is called the "unstable balance of living matter." And here too one must guard against offhandedly depicting an essential trait distinguishing life in physicochemical terms as an unstable mixture, etc. It may be consistent with such a mixture, but never identical to it.

The self-preservation of the living individual is based on the antagonism between an assimilatory and dissimilatory [199] process—does not, in other words, coincide with either of the two themselves. This means that there is no contradiction at all between the irreversibility of the course of life from the seed to death and the circularity of a cross section of life. Self-preservation is not primarily directed against death, but rather against the immanent self-abolition of life. It is not a defensive act for the sake of an already completed living being, threatened only from without because of its instability, but rather the way in which the living being becomes possible as a present thing. Neither the assimilatory nor the dissimilatory phase plays a privileged role in the conditioning character of their unity in the antagonism. Preservation *and* destruction of the self constitute the living thing as one that acts and reacts in the present and in the cycle of the overall energetic processes of nature.

What biology defines as nutrition and elimination in the narrow sense and relates to other oppositionally connected processes such as differentiation and fusion or movement and rest falls under this essential law just as does the specific coupling of stimulation and reaction. The fact that these processes take place antagonistically within the body does not exclude their task of keeping the body connected to the outer field; indeed, it includes it. This opens up possibilities for understanding the presentation of bodies and events located outside of the organism in the sense and nerve organs within the organism itself—for understanding, in other words, external perception. For here too the only assimilative and dissimulative processes found in the lived body are ones that of their own accord connect it with the outer world.

By virtue of its positionality alone, according to which the living thing is (set) into the living thing that it is/out beyond the living thing that it is, it disintegrates in itself into two processes running in opposite directions and, through them, becomes integrated into the world of physical things as a self-sufficient unity. The medium of its integration is the positional field, which in turn is related oppositionally to the living thing, or the presence surrounding life. No relational act or even a conception and interpretation by means of complicated mediations in sensations, perceptions, ideas, or judgments is required for this. Even in those cases where [200] such contents of consciousness, where consciousness itself in even the most primitive form is absent, there is "for" the organism an opposing field of its life in which it *exists*: a field with which *and against* which it lives.

Only the inner disintegration into processes running in opposite directions from each other allows the living physical thing, which is its own system absolutely enclosed in its boundaries, to "open" up to the influences from without and to the outside. Its absolute isolation (in the sense of the boundary) is balanced out in the sense of the boundary, not abolished. This balance preserves the contraposition of the living bodily system as a whole against the outer field so that influences on the organism do not automatically compel reactions from, or arbitrarily change, it. Given with the sharpness of the organism's boundary is the imperative precondition that it have the leeway to either allow influences to affect it or to dampen them instead. If it were of a piece with the continuous context of cause and effect like an inorganic body, and thus in contact with all spatiotemporal "coexistents," we could no longer speak of autonomy as a fundamental characteristic of life. The vital system must have leeway for vital processes to be self-inducing. It reacts to influences, but this reaction is not only the continuation of the effect of given influences on the organism, it is also an *answer*; that is, it is an effect directed in the opposite direction from its stimulating cause.

Adaptedness and Adaptation

To the essence of an organism existing individually at a particular time and in a particular place belongs a positional field relative and in opposition to its body—a habitat, or living space [*Lebensraum*]. This space comprises everything in terms of strength and quality that could possibly have an effect on the organism and that could in turn be affected by it. Living space or the sphere of life, however, does not have a purely spatial character. As was noted earlier in reference to potential effectivity, this sphere contains a temporal moment insofar as it coincides conceptually with the mode of standing in the present. Thus it is best to avoid the expression "living space," with its predominance of spatiality, and to instead [201] use the term "positional field" or "sphere" (= field of presence).

As the [German] term *Gegenwart* (presence) suggests, that which is present is not only on hand or simply there, but as it continues and lasts is also "against" [*gegen*]. This necessarily characterizes a life that, as a closed system, is enclosed in its positionality by a surrounding field set in opposition to it. That which occurs in this surrounding field is on hand, is there, encounters the organism, and forms its presence. The organism exists in this *opposing* field [*Gegenfeld*]: it is with and against it. Natural scientists, trained in physics, tend to overlook this duality of the relationship between organism and medium. And yet it is of significance for the difference between two properties characteristic of life, which, as Driesch has shown, have not been carefully distinguished because they are subject to the same basic law.[3] These are adaptedness and adaptation.

If we assume that the organism behaves toward its medium like any body to its surroundings, the relations between it and the medium would have to be reversibly oppositional. The organism would be in a relationship of reciprocity (of varying degrees and kinds, of course) with everything and everything with it. A maximum of viability would coincide with a maximum of conformity to the surroundings: the organism would fit into its surroundings and the changes in these surroundings in a precisely measured way, inserted like a metal core into a mold—statically as a system of forms and functional harmony, dynamically as development, regulation, and functional change. The two would relate to each other in their conditions and processes as a positive image to a negative one. Since experience, however, shows dissonance between organism and environment to be everywhere, to a degree that, at minimum, balances out their attunement, disruptive factors were attributed to external causes, or—in a certain opposition to Darwin's view—their significance was downplayed. One sought to comprehend the abundance of species in an aesthetic and optimistic way as a kind of artistic manifestation of nature's primary stylistic ideas guiding how life fits itself into the world.

[202] Or, on the contrary, the moment of primary coordination between the world and life was entirely disregarded, and the organism was seen instead as a physical thing simply exposed to being jolted by other things in space and time. This established the notion of "look out for yourself" as a law of nature governing the living thing, now reduced to a mere plaything of all forces. Life in this view was nothing other than an incessant striving for adaptation in order to decrease the disruptive influence exerted by other organisms and by all of reality.[4] The struggle for life as the struggle for adaptation by means of adaptation had to provide the motive for the formation of species abundance via the concurrent selection of those most able to fight.

Both the theory of exclusive adaptedness (conformity) and that of exclusive adaptation overlook the fact that life is essentially both and needs both to live. This is because both theories see the relationship between the carrier of life and the medium as reversibly oppositional in terms of a simple physical relation between things in space and time and not, as is in fact the case, as an irreversible oppositional relation. In one case it is the outer world or some other cause that brought about the disrupted conformity; in the other we find no conformity at all. In the former we lose the dynamic invariable (Buytendijk), the range of arbitrariness, and the unpredictability of the organism; in the second we lose the factually existing equilibrium with the environment and the equal status of the manifold species in the plant and animal worlds in favor of an anthroprocline process of development.

Initially, the living being is in harmony with the medium as its positional field. This relationship precedes all particular relations between living being and medium. We could almost compare it to the forms of intuition of space and time, which in Kant's theory of knowledge preform the material of knowledge, although the positional field is not the form of the living being in the same way as for Kant space and time are forms of the subject of cognition. The opposing field is, with respect to form, posited *with* the body as it exists within its boundaries. The opposing field encloses the living body, but not in the sense of merely surrounding the place occupied by the body, in which case it would be located outside of this natural place, but it rather includes the natural place within itself. The [203] location of the positional field can be understood from the essence of positionality, according to which the living body is just as much itself (within its boundaries like every bounded body) as well as out beyond it/into it. This means that the living body itself is part of the contents of the positional field, even if, as its central point, it is lifted out of it. The organism is the positional field's *excentric central point*. In this way it comprises the transition between itself and the medium, without thereby giving up the determination of its actual boundedness and closedness.

If the positional field were identical with its natural place, regardless of how expandable in space and time it may be, it would be nothing other than the system of the organism. The positional field would actually belong to the organism, just as Kant's forms of intuition of space and time belong to the subject and only the subject. The organism would then move in its positional field like the monad in the world, like a solipsist; in a surrounding field, but not in the actual world independent of it: in absolute immanence.

If, on the other hand, the positional field were only the opposing sphere to the natural place of the organism without at the same time including it within itself—only, as it were, the "over there," from which it can be affected by countereffects and that it in turn affects with its own countereffects—then any adaptation could only be the result of coincidences and trial and error. The organism in its positional field would find itself as if in a zone of total alienness, unpredictability, and independence: isolated and at the same time abandoned to an absolute transcendence.

If the positional field is characterized as immanent, there is only adaptedness, not adaptation. If the positional field is characterized as transcendent, there is, at best, adaptation, but not primary adaptedness.

The essential law of positionality excludes these two extreme cases from the beginning. The living body as boundary-containing body comprises the transition between the body that it is and the medium enclosing it. This characterizes it as center and periphery in one. In the living body's relationship to the positional field, the two extreme cases described earlier are conjoined into one reality: the body as element of the periphery belongs to the field, but as center finds itself opposite the field. The field neither belongs entirely to the body as simply an expanded zone and mirror image [204] of its organization nor is completely separate from and absolutely opposed to it. The hiatus between the organism and its setting is not destroyed, but rather bridged.—

The counterpart to this, clearly, is the law of assimilation and dissimilation. This law states that the living being bridges the essential chasm between itself and other things by, as it were, moving the boundary antagonism inward, into its own central fullness of being. With the antagonism of the circular processes of construction and destruction, it opens its boundaries to the influx and discharge of matter and energy and thus to its integration into the context of things. The living thing becomes bordered even in metabolic, energetic, and metamorphic cycles. In this case, on the contrary, we are concerned with the position of the living thing as a whole in relation to the reality surrounding and carrying it, not with its material and energetic integration into this reality.

Under what conditions does the living body fulfill the requirement of being positioned both in accordance with and in opposition to the positional field? Under what conditions can it avoid a merely partial compliance with this law of genuine synthesis by taking on a dual mode, two coexistent properties? Only by joining together both functions into one and the same property: that of *adaptability* or *adaptedness* as it is ensured by the concrete gestalt and its regulability. We have already discussed the regulatory prerequisites for adaptation. We can now turn to the specifics of adaptive behavior.

When an organism is positioned in accordance with the positional field, it is a content of that field—that is, its relationships to it are reversibly oppositional (physical and chemical), and it is entirely integrated into the continuous chain of cause and effect. When it is positioned in opposition to the positional field, the organism corresponds to this field, and its relationships to everything that conceptually coincides with the field are irreversibly oppositional: these are relationships of stimulus and reaction, of compatibility, of becoming attuned to one another, of individual correspondences and specific harmonies.

[205] According to the law of positionality, it is the essence of the organism in every moment of life to join together both positions and to occupy them not simultaneously and side by side, but in one. We can see how inaccurate it is to define a stimulus simply as a triggering process followed by the pure effects of excitation and then reaction. Such a definition would never allow us to understand the individuality of the correlation between stimulus and reaction and the meaningful correspondence between them, and Driesch was completely right to base a proof of the autonomy of vital processes on this phenomenon (specifically in the area of organic action).

The position in accordance with and the position in opposition to the positional field are able to unify synthetically because, within certain limits, the organism *harmonizes* in substance and form with the medium *without thereby entering into an absolute bond with it*. The organism has to fit into the medium and at the same time have enough leeway in it to not only exist within the fixed forms of harmony but also to survive dangers with them.

The organism in accordance with the positional field is a body and thus exposed to the influence of other bodies. Like all centers of physical things it enters into open exchange with other centers of energy, is an element of reality alongside other elements. The organism in opposition to the positional field is a closed life system, embedded in a surrounding field relative to its essence, immanently secure. Both of these moments—openly belonging to actual nature and being secure in it by being closed off from it—can only be realized in the organism by way of an unstable balance between it and the medium.

Adaptedness is one of the prerequisites of organic life in the world, but it cannot go so far as to rob the organism from the beginning of every possibility of improving or modifying itself. The primary harmony of its organization and organ functions corresponds to a primary harmony of its overall correspondence to the medium, but, like the former, the latter must also be regulable. The closed system of the living body corresponds to the closed system of the positional field, but just as the body possesses its own boundaries by at the same time being out beyond them and opening them, the closed system must too be an open system, allowing for and demanding constant [206] correction. This means that adaptedness is not a self-sufficient prerequisite of organic life but one that already co-conditions adaptation.

A certain simplification has led the names Darwin and Lamarck to become associated with two theories that characterize the organism's process of adaptation as either active or passive. Following Lamarck, active adaptation is understood to mean that the organism striving to adjust to altered conditions results in the emergence and development of organs and functions corresponding to its surroundings. Thus the atrophied eyes of moles or the loss of extremities in snakes are attributed to non-usage, and the development of webbing between the toes of frogs and water birds is explained by their sustained efforts to swim on the water's surface. More recently, the concept of functional adaptation was developed in order to describe the adaptation of a function to a function in morphological and physiological terms and to distinguish it from primary function-forming adaptation. The uncontrolled psychological speculations surrounding the latter (psycho-Lamarckism) have not made it any more accessible to knowledge.

The doctrine of mechanical natural selection, associated with the name of Darwin, can be described as a theory of passive adaptation. Since here the outer world has priority in the struggle for life, by means of which it brings about the selection of the fittest, it would seem redundant to postulate that the living being also strives to adapt. It is changes in the medium that compel adaptation, and only in a second step does a reflexive reaction to altered surroundings initiate the adaptation process in order to restore the balance that existed under earlier conditions.

The cultivation and exaggeration of the contradictions between the two theories was of course the work of Lamarckists and Darwinists, who wanted to define the entire relationship between organism and medium as one of adaptation, leading them to forget the prior givenness of a primary harmony between life and the sphere of life, which is not itself a result of processes of adaptation. Once a tremendous [207] increase in the body of facts undermined the theory of the invariability of species, the obvious conclusion seemed to be that given the fluid transitions between the individual organic forms, their equilibrium with the fluid environment could only be established by means of ever new adaptations. It was not until the emergence of comparative physiology, which gave rise to biology as the science of species-specific life plan forms, that the side of primary harmonies was recognized—not, it is true, without tending toward the other extreme of espousing the absolute adaptedness of life systems, giving way, as it were (as in the work of Uexküll), to a biological monadology.

Insofar as the relationship of adaptation (adaptedness) is unstable—*that is, while its success with respect to "form" is given beforehand, with respect to "content" there can either be success or failure*—the organism remains endangered, regardless of how secure it is. The positional field or milieu is, in its essence, the scene of struggle and the sphere of protection. It is in this way that the law of the living being's own body is fulfilled in it, as a body that, by virtue of its positionality, is its own transition into the medium, is the excentric center. The body must exist in the field that contains its natural place, positioned with it and against it—in peace and at war, in life and in death. That is why life means to be in danger, why existence means risk.

The life of the organism plays out in relation to its surrounding field in a way that is anticipatory in its form, that seeks contact to the medium in the concrete living act, that is adapted and adapting. The form, of course, cannot be separated from the concrete individual content. One cannot say that the living being is adapted up until this point; this is the framework within which it can move with absolute security—and there begins *terra incognita* and thus the area for which adaptation is responsible. On the contrary, as the notion of an "open system" implies, adaptedness *and* adaptation are realized in every living act. Otherwise, the guarantee of certain success would protect the organism from dangers, exertions, and accidents without any effort of its own. It would run into danger only if it ventured beyond the borders of adaptation.

But that is not the case, of course. Adaptedness and adaptation can only be separated from each other in hindsight, when considering a completed

[208] act, an executed organic process, similar to the way in which only subsequent epistemological analysis of a completed cognitive act can set the a priori element apart from the a posteriori element, anticipation—that which is guaranteed to succeed—from that which is in fact found. For the observing biologist, then, a living unit becomes separated into two states of affairs.

The relationship of the organism to the positional field is determined by the nature of positionality: the structure of this relationship is strictly commensurate with the relationship of the body to itself, upon which positionality rests. As this body that is both into the body that it is and beyond it, it is ahead of itself and is thus present. As such it is enclosed in a field or confronted with a present. Presence [*Gegenwart*] is only possible where something stands opposed [*gegen*] to the anticipated direction in its not-yet. Something is present as long as it is grounded in the not-yet, in the future. This means that the confrontation of the organism with its surrounding field presupposes the organism's anticipation—that is, the risk of the living act. It is this essential characteristic of risk that allows for the adaptedness that becomes fulfilled in the living act or for adaptation that goes beyond this adaptedness. Thus every act that relates to a surrounding field and confronts the organism as a whole with its setting involves going *along with* the contents of the field and going *toward* the contents of the field on the basis of an anticipation continually executed by the being of the organic body.

The organism is only within a setting if a background, that which is not present, is ahead of it. The organism lives in this structure of anticipation; there is no need for it to perform any particular anticipatory act. The positional field, to which the organism belongs as content and central point, in accordance with and in opposition to which it is positioned, constitutes itself by virtue of this being-ahead. Every relationship between the organism and the medium thus proceeds with the organism and to the organism, or, to put it succinctly, through the medium (the organism) to the organism.

Thus the riddle of the inner potential for adaptation or adaptedness is solved in a way intuitively grasped by Goethe when he said that the eye was formed by light for light. He was not referring to the simple fact of functional adaptation. The question is rather how to think about the *first step* of every specific [209] act of adaptation, about how adaptedness, harmony as such, comes about. The famous couching of this insight in Plotinus's notion of the eye's luminous nature, of the divine nature of the human who recognizes God, reminds us that what we have here is a root of the age-old doctrine of like knowing like.

Furthermore, the perspective of adaptation and adaptedness permits us to take a fresh look at the whole problem of an antecedent necessary for knowledge, whether this is thought of in terms of the innateness of ideas or of familiarity with them from an earlier life allowing us to recognize them in this one, or the inborn nature of the a priori forms of the subject. Now that the dichotomy between objects conforming to our ideas and our ideas conforming to the objects has been overcome in the adaptation relationship, we should succeed more readily than Spencer did in overcoming the extremes of apriorism and aposteriorism in equal measure. In this relationship—if I may be permitted the comparison—the object (to which the organism must adapt) is at the same time mode, form, and means of adaptation.[5]

It is symptomatic of this problem that it led to psycho-Lamarckism and psychovitalism. The laws of positionality had not been penetrated, and the relationship of the organism to its medium was seen entirely with the eyes of an empirical natural scientist, ultimately compelling a radical reversal of perspective. Let us assume for argument's sake that the issue at hand is the formation of light-sensitive organs. How are we to think about this specific process of adaptation? Before the organ has formed, there is no such thing as light for the organism—and thus also no need for light. Where, then, does the tendency to form a light-sensitive organ come from and, on the side of the organism, from where its specific capacity for such formation? The psychological solution to this problem cites cellular instincts that guide the cells along their way, so to speak—clearly a hypothesis invented ad-hoc [210] in an attempt to solve one riddle by posing another one. The problem of the formation of these instincts or judgments or perceptions immediately poses itself, and it is just as impossible to grasp the possibility of such psychological properties as it is implausible—and useless—to endow the elements of the body with them.

Even if we try to get by without such micropsyches and instead ascribe adaptive drives and tendencies to the entire organism, we have only displaced the problem. There is no way around the anticipation inherent in the adaptive relationship. The peculiar selection of a particular component of the medium to function as the adequate stimulus for an organ yet to be formed, the predisposition of certain cell parts, cells, cell groups to become specific stimulus receivers or reactants must simply remain the secret in the foundations of adaptation theory. (We should certainly be wary of looking for reasons why this certain life form is predisposed in this particular direction and that life form in that direction—why, for instance, this one developed visual organs, that one olfactory organs. The moment of arbitrariness in morphogenesis—that is, in the plan formation of a type, is part of the essence of life and in and of itself eludes rational insight. But this is not our question here; rather we are interested in understanding the possibility of adaptation as such—by means of positionality.)

At the same time, attributing adaptation (adaptedness) to positionality corrects the theory that like belongs to like, repeated also by Goethe and still hinted at in the first version of Johannes Müller's doctrine of the specific sensory energies. In reference to the case of the eye, mentioned earlier, Goethe gives this idea the twist that light engenders an organ that is like it so that the inner light can encounter the outer. This makes it seem as if successful adaptation is based on two components that are essentially alike finding each other, one of them belonging to the organism, the other to the medium. Just as the circuit of an electrical conductor is only closed if two free wire ends are touching each other, just as a tunnel only comes about if the two boring teams meet in the middle of the mountain, so all adaptation, in this view, rests on an encounter between two elements that share the same essence.

[211] This interpretation surrenders all the advantages of attributing adaptation to positionality. It is precisely without assuming that the organism has specific, finished predispositions that a positionality-based theory of adaptation makes possible the formation of organs and functions corresponding to the world that fit being, that fit the medium, without "knowing anything" about it beforehand or having been specifically "predestined" for it.

The eye was formed out of indifferent auxiliary organs by light for light: Goethe's phrase gets to the heart of the matter. But it too remains opaque if we are unable to define the principle of this by—for (or, as we said earlier, along with—toward; through—to) as the principle of positionality and thereby attribute it to the fundamental essentiality of the animate body: its relationship to its boundaries. The existence of the positional field is an anticipatory empowerment of the world by the organism, which in itself denotes a being-ahead of the organism that it is, and it is precisely this that renders unnecessary the assumption of special guiding instincts, judgments, or similar adaptive psychological or physical factors. The anticipatory element is based on living being's structure of being ahead.

Reproduction, Heredity, Selection

Life is being that is set off against the sphere of its non-being, that relates to it as to its opposite and contrary. This relationship is not simply one between two autonomous systems closed off against each other, that have no need for each other, that are only posited together in an irreducible way like shadow is given with light, but rather a relationship in which one counterpart is the condition for the autonomy of the other. If this is not to lead to the loss of each part's autonomous character, they both have to be structured oppositionally within themselves. This allows the boundaries of the two zones of being to open up to mutual influence without the boundaries being destroyed. Autonomy does not turn into heteronomy but is preserved by virtue of heteronomy.

Adaptation and the metabolic and energetic cycles show how the organism realizes the law of the oppositional structure of its own being. It is in the inner antagonism of assimilative and dissimilative processes that it opens its boundaries in order to exchange matter and energy with [212] surrounding nature, into which it is integrated as a physical thing. In the—as it were external—cycle of adaptedness and adaptation affecting it as a gestalt, the organism realizes the synthesis of positionality and physical thingness—that is, of the properties that are specific to it as a living thing and the properties that connect it with all things as a physical thing in space and time. It realizes this cycle in the antagonism of being positioned both in accordance with and in opposition to the positional field, as the contradictory unity existing in and with the medium *and* against the medium, vulnerable and secure, in a struggle with the medium and in equilibrium with it.

The circulation of matter and energy and the adaptation of its organization bring the living being into contact with actual actuality [*aktuellen Wirklichkeit*]. Circulation and adaptation merely determine this contact in an ideal cross section of time, in the mode of the present, for it is their "goal" to connect the living being with the effective factors of the medium and to maintain this connection, to balance out the living being's physicality with the laws of vitality. *The living individual, however, finds itself in an irreversible developmental process of relentless aging*, and as this downward grade it carries a reference to the future in its being. We ought not, however, to confuse the characteristic of being ahead of itself essentially immanent in living being with this reference to the future. It is not living being that relates to what is to come, but the individual, the developing whole, that stands in relation to the sphere of its not-yet-being as its opposite and contrary. Living being is grounded in the future; it does not relate to the future. In its groundedness in the future, the individual, by contrast, undergoes becoming and development, becomes something, and goes toward its death. Hence the positional field, in accordance with its character of being present, is open toward the future, while the individual is in *relation to* the future.

The organism is supposed to be in contact with this just as with all the properties of the positional field, which leads to a fundamental difficulty arising from its absolute boundedness by death. The organism's contact with the field steadily wanes as it ages and finally expires in death. [213] Because aging, as it is caused by the gradual decrease of unevolved potential due precisely to its evolvement in the developmental process, increasingly limits regulation in its radius, the circulation of matter and energy and the adaptation of organization are also affected by the irreversibility of the overall developmental process. This limitation poses a crucial threat to the process of circulation and adaptation, which are both based on the capacity to regulate. *Something has to counteract the irreversible decline of aging* in order to protect from inner decay the relations of nutrition, movement, and adaptation that bring the organism into contact with the world and maintain this contact: *regeneration*.

It is not because nature brings forth eternal and indestructible life that there is regeneration and reproduction, not because the individual cannot hold the abundance of life and, like a narrow bed, only allows the current to pass through it that individuals descend from each other, but it is precisely the boundedness of life and its decline into old age that induce renewal. Regeneration only balances out aging; it does not cancel it out or overcome it. All artificial regenerations that extend the life of an individual or species are not able to compensate for fate itself, for the decline of the lifeline as a whole.

If the process of regeneration were able to confront aging in a truly antagonistic way, life would have to take on a stationary character. It would be pseudo-life, no longer be ahead of itself, no longer have a presence, and so perish in itself.

Compensatory renewal must allow for the developmental process of the individual to be maintained as a whole. If it is aimed immediately at the individual carrier of life, it leads to death. Thus the act of renewal in unicellular organisms under certain conditions coincides with their death when the cell divides: this is the type of reproduction from which Weismann deduced the concept of the "potential immortality" of unicellular organisms. In the case of multicellular organisms, compensatory renewal succeeds by means of the formation of germ cells, which in relation to the whole body contain the maximum of unevolved potential. The act of fertilization creates the basis for a new individual, which, descending from the living mass of the parents, must become similar to them. Thus the individual regenerates itself in *another* individual, to whom [214] it passes on its properties, subject to their physicochemical representation in the cell mass and their influenceability during the embryonic and later stages of development.

Because there is an inner necessity for development to peter out in the individual, there must be a chain of individuals to keep alive the type in the continuity of the germ plasm. A reserve fund is created in this way that is not drawn on and that—involuntarily—helps preserve the species.

We should not simply pass over this "involuntarily." There is a great tendency to think teleologically in this context: it is not the individual that matters, but only the overarching unity of the species, whose preservation is paid for with the life of millions. The higher, the supra-individual, is also held to be higher in the sense of more valuable. The arsenal of the classic idolization of the universal is used to prop up a biology aimed at very specific teachings affirming the community and the state.

It is not the species, whose realization as a collective everything allegedly comes down to, that is primary, with the individual merely a means of life, of a chaotic power that breaks out into a myriad of individual destinies, both creatively blind and playfully seeing. Rather the species "becomes" by way of the individual; it is anticipated in the individual's development as the form "under" which the latter becomes, becomes something, *turns out*. Just as a final, unbridgeable hiatus between the shaped body and its type, which no relation of realization can erase, becomes comprehensible here, so does that between the individual and the species or the higher levels of generalization. The fact that the species, once it has become visible in a singular form in a thousand variations, appears to be incorporated into a hierarchy defined by an essential law may be indicative of the ideality of the actual forms, but never of their intrinsic purposeful meaning. In the becoming individual, the form (and thus implicitly also the hierarchy of form-ideas underlying the gestalt) in the mode of anticipation has the character of a final cause. Detached from the individual, on the other hand, the finality of the form levels (which, of course, can be detached from what is spatiotemporally real) is quite problematic; life only "turns out" under their guise and only on the occasion of life's becoming do they also become.

[215] Thus there is a disparity between the developing and fully evolved individual and the breadth of the type, the species, the phylum it ended up in. The unbridgeable margin between form and individual being as the evident, unexploited wealth of possibilities that the individual, due to the course of *its* particular development, had to forgo, but that nevertheless exists in its unexploitedness, in its abiding fulfillability—this hiatus itself between the form and that which is formed creates a tension that demands to be compensated.

Life is development—that is, the transition from unevolved potential to actualities [*Aktualitäten*], the narrowing of possibilities that were originally there and under certain circumstances may still become actual if interference in the organism demands regulation: life *is selection*. And yet the possibility remains potential, at the same time covered up by the actual, as long as the organism lives and insofar as the whole is represented in the particular and the relationship between the whole and the part is not affected by the process of aging. The decreasing prospect of fulfilling potential that comes with increasing age is contrary to the invariable potentiality of the individual (founded on the invariable relationship between the whole and the part).

As a result, there is a dual disparity for the becoming individual in its relationship both to the breadth of the form that gives it leeway and to the scope of possibilities it will necessarily have to forgo, as well as to the abundance of its own potentiality that allows it to realize the offered possibilities. It is this disparity that gives any factual development, whatever direction it might take and however many leaps it makes, the character of an individual development that could have also taken a different course—even if development along a particular path is essential to the individual. The path in fact taken must necessarily be incidental. Despite the general meaningfulness of taking a certain path, in actual development chance decides that it is this one and not another. The abiding breadth of the individualized form and the breadth of abiding potentiality that "properly" belongs to it do not rob the law according to which the individual realizes a particular possibility of its necessity. But they do rob the *result* of the realization of its necessity, [216] making the individual appear to be the victim of blind fate.

To be alive per se means to have been blindly chosen, selected. To be alive is to necessarily forgo one's possibilities and thus to participate in selection. The common view is that selection is brought about by factors external to life, such as the climate, nutrition, struggle with other members of the same species, selective breeding—as if life could exist as an as yet unmodeled mass, an as yet unregulated current, untethered and free. Then of course there is also the fact that only those individuals are alive that could just as well have turned out differently—a problem that has to be solved with the help of factors alien to life. There is no choice but to devise empirical theories of selection just as theories were devised for the phenomenon of adaptation.

For life, selection is an a priori mode of its real occurrence in bodily actuality. "For life" has the value of an inner structural law (it could be called the law of the "categorical subjunctive" constituting the inner mechanism of selection): "It could work, but it does not." (On this level of organization, the law only takes a physical shape. But it carries life into its highest sphere, the human, who consciously has to live according to it.)

For the observer coming to the problem from the natural sciences, things are much simpler. While he will not claim to be able to causally determine the development of this particular oak or that sea urchin down to the last detail, he will not tolerate any doubt in the fundamental possibility of such an explanation. He is convinced that every developmental step had to take place as it in fact did: given these light conditions, these chemical properties of the medium, etc., there was nothing else for the seed or germ cell with this particular genotype to do.

Does this contradict the theory put forward earlier? It does not. The effective disparity between the imposed path of development, describing a line that is of an entirely individual character, and the formed organism, which, despite its necessary uniqueness and one-sidedness carries *in itself* the whole wealth of possibilities opened up by the scope of its form-idea—this disparity is not contested by the natural sciences, but rather emphasized.—

[217] The fact that the organism in its germ cells possesses a relatively invariable wealth of potential that can be realized relatively independently of its development does not diminish the individual, but it does balance out the (nevertheless existing) contradiction between the diminished opportunities for fulfilling this potential with increasing age and the invariable potentiality of the individual (which is based on the invariable relationship between the whole and the part). Except for the compensatory sense in which the irreversibility of developmental processes is opposed by eternal renewal—in another individual—reproduction and heredity manifest their necessity for the individual by forming organs that serve them in the individual itself.

It may seem surprising that there has been no word yet in this context about sex differences, given that the interplay between male and female occurs in all of organic nature, in a thousand variations, as the prerequisite for reproduction. The renewal of living matter, it seems, cannot take place (at least in the long run) without a change in the plasm or nucleus. This compulsion to mix different substances can be comprehended by looking at the special physical conditions life requires in order to become realized. The current state of physiological chemistry does not allow for the deeper ontological and causal understanding this would require, however. There are processes at work here yet to be explored by exact biology, processes taking place in layers of being that can no longer be seen [*erschaubar*] but only measured.

An a priori understanding of the compulsion to differentiate sexually based on the compulsion to renew must consider the special physical conditions to which *the realization of life is tied*—must consider, then, something whose content can only be recognized a posteriori. This dependency of the phenomenal layer of life on other nonphenomenal layers of bodily being brings about the particular tension that is only upheld by the polar opposition between male and female. Especially in view of the research by Nicolai Hartmann, Fritz von Wettstein, and others, there can be no doubt that, *all quantitative conditionality notwithstanding*, the difference between male and female being or behavior from the simplest gamete [218] to the human is a qualitative difference, a difference of essence. Sexual differentiation is a modal, a condition for the realization of life in consideration of—and this distinguishes it from the other conditions—its particular physical conditionality.

The Open Form of Organization of the Plant

Nowhere, as far as experience reaches, does life appear in one form only. The form is always specific and can in every case be attributed to either the plant or the animal type of organization. There are indeed crossovers among unicellular organisms, just as reproduction and death can coincide in these beings. But the transition to multicellular organization seems to bring with it the compulsion to decide to be either a plant or an animal. We cannot dismiss out of hand the assumption that this compulsion to differentiate manifests an essential law of life itself.—

If this is the case and the organism fulfills its purpose in itself by achieving unity only by virtue of a life circle that moves through and back out of it again, we have to ask how this can be reconciled with the unavoidably closed nature of a body. Every body in space is a thing bound by and enclosed within walls, and even if the boundary surfaces in the different aggregate conditions cannot be determined with equal ease by vulgarly sensual intuition—they are there and can be verified in every case.

For the living thing there is a radical conflict here between the compulsion to close itself off as a physical body and the compulsion to open up as an organism. The solution to the problem is found in the living thing's *form*, whose manifestation in the specific gestalt of its type can be grasped by the senses without, however, appearing itself. Form here denotes the organizing idea (and an idea is essentially intuitively mediate) according to which the living body combines its self-sufficiency as a thing with its non-self-sufficiency as something that is alive.

Balance in the form of course presupposes conflict, which only occurs where organization is supposed to take place—strictly speaking, then, only in multilevel living beings (multicellular organisms). There can be organization in unicellular organisms as well; only that here the modeling of [219] the whole out of a cohesive protoplasmal basic substance means that the part occupies a fundamentally different position in relation to this whole than in multicellular organisms with specialized cells and tissue. Alfred Kühn's studies of amoebas, which under certain conditions take on the form of flagellates, show clearly the set of problems presented by the entire morphology of protists. Thus, for instance, we always speak of organelles, not of organs in the proper sense, in relation to unicellular organisms. When life chooses the path of multicellularity, it also chooses the conflict between organization and corporeality and so must balance them out in the form.

This balance can occur in two different ways: in an *open* or in a *closed* form. If it occurs in an open form, the living thing is a plant; if it occurs in a closed form, the living thing exhibits the characteristics of an animal. Thus, in ideal terms, the animal and the plant are strictly distinct from each other in their modes of organization, which is why there may be many properties in which they only differ from each other by degrees and in some areas may even coincide. A purely empirical distinction between plants and animals will always run into great difficulties because it will not be able to simply pass over the existence of transitional forms.

(The terms "open form" and "closed form" to distinguish between plant and animal organization come from Driesch, who does not, however, ascribe absolute significance to this opposition. He finds "open forms" in the animal kingdom among corals, hydroids, bryozoa, and ascidians, which he sees as analogous to plant morphogenesis; at the same time he suggests that many plants are not in fact individuals but rather colonies—so that his comparisons here are between two things of different degrees of order.)

A graph (see Fig. 5.2) will help us gain a clear idea of what is meant by open and closed forms of organization in the a priori sense.

A form is open if the organism in all of its expressions of life is immediately incorporated into its surroundings and constitutes a non-self-sufficient segment of the life circle corresponding to it.

Morphologically this is manifested in the tendency of the organism to develop externally and expansively in a way that is directly turned toward its surroundings. It is characteristic of this kind of development that it does not have the need to form [220] centers of any kind. The tissues responsible for mechanical solidity, nutrition, and stimulus conduction are not anatomically or functionally concentrated in particular organs but rather permeate

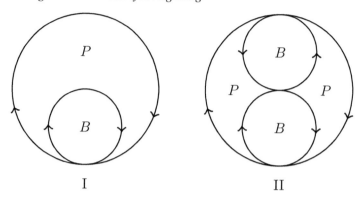

[220] Figure 5.2. *B* refers to the body; *P* to the positional field; the circles around the *B* and *P* zones to the circle of life. In diagram I, the incorporation of the body into the circle of life is represented by the arrows pointing in the same direction for both zones. Diagram I thus describes the case of the dependent open form. In diagram II, the incorporation into the circle of life is preserved, but the body acquires its closed form and independence from the twofold direction of the *B* zone, which moves in accordance with *and* counter to the direction of the *P* zone. As in Fig. 5.1 in Chapter 5, the body zone and its surroundings moving in the same direction express openness, while the two zones running in opposite directions express the closed nature of the body zone. There are no further points of comparison between the two diagrams. They are merely convenient visualizations of the characteristics of form.

the organism from its outermost to its innermost layers. The absence of any central organs tying together or representing the whole body means that the individuality of the individual plant does not itself appear as a constitutive but rather as an external moment of its form associated with the singularity of the physical entity; in many cases the parts remain highly self-sufficient in relation to each other (grafting, cuttings). This led a great botanist to go so far as to call plants "dividuals."

Ontogenetically, this dividuality of the plant form of organization can be seen in the preservation of phases in the structure of the individual. "If an animal germ proceeds in its development from a stage d to the stage g, passing [221] through e and f, we may say that the whole of d has become the whole of f, but we cannot say that there is a certain part of f which is d, we cannot say that f is $d+\alpha$. But in plants we can: the stage f is indeed equal to $a+b+c+d+e+\alpha$ in vegetable organisms; all earlier stages are actually visible as parts of the last one. The great embryologist, Carl Ernst von Baer, most clearly appreciated these analytical differences between animal and vegetable morphogenesis.... It seems almost wholly due to the occurrence of so many foldings and bendings and migrations of cells and complexes of cells in animal morphogenesis, that an earlier stage of their development seems *lost* in the later one; those processes are almost entirely wanting in plants, even if we study their very first ontogenetic stages. If we say that almost all production of surfaces goes on outside in plants, inside in animals, we shall have adequately described the difference. And this feature again leads to the further diversity between animals and plants which is best expressed by calling the former 'closed,' the latter 'open' forms: animals reach a point where they are finished, plants never are finished, at least in most cases."[6]

This essential unfinishedness characteristically manifests in the so-called embryo zones, which in more highly organized plants are found in the growing points of buds and in the cambium between bark and wood; it is from here that differentiation and growth can continuously begin. Such unshaped, shapeable material, comparable to the germ, always survives as a zone of the differentiated organism, providing it an immediate reserve for its own morphogenesis. The animal organism also possesses embryo zones in the form of germ cells—that is, systems of harmonious equipotentiality in the harmoniously equipotential system of its body—but they serve the creation of new individuals, not (in an immediate sense) the differentiation of its own lived body.

Unfinished is most certainly not the same as undeveloped. It is precisely mature development that determines the character of the essential [222] incompleteness of the whole that makes its appearance with this maturity. Roots, leaves, flowers, and fruits from the most primitive to the most differentiated forms emphasize—each in different ways that serve the purposes of nutrition, stimulus conduction, and sexuality—their integration into the surrounding medium. This integration has repeatedly been described as absolute devotement, as the disappearance and full absorption in the function-circle of the life of the species, and has given rise, particularly in the psychistic interpretation of life phenomena, to the notion of the ecstatic nature of the plant.

The dominant significance of reproduction for plants is nothing other than the expression of the throughness, of the transitional nature of the open form, which regulates the process of reproduction with its means in this way.

Pollen is transferred with the help of insects and the wind. The color and shape of the flowers, the saccharides of the nectaries, the waxes of the pollen, and the aromatic substances all serve to attract the pollen-transferring animals. Chemical substances serve to attract the spermatozoids for the purpose of fertilizing the egg.

The immediate integration of the organism with its form-creating surfaces is expressed in the overall metabolic cycle as well as in the cycle of reproduction. The fact that plants containing chlorophyll are able to synthesize, under the influence of sunlight, the highly complex organic substances from elementary inorganic bodies occurring in water, in the earth, and in the air, in addition to substances from organic compounds, does seem to be a specific characteristic of plant life (as this ability is only found in plants). But the general difference between plant and animal metabolisms only manifests in the gradual difference between predominant assimilation, weak thermogenesis, and oxygen excretion in plants and predominant dissimilation, lively thermogenesis, and carbon dioxide excretion in animals. Since the form-creating surfaces of the plant are without exception involved in the metabolic process as it is determined by the lived body's direct contact with the medium, supplying it with inorganic and organic substances and sunlight, there is no differentiation of tissue into organs for eating, digesting, and excreting. There is no need for a distribution of the metabolic stages [223]. In this too the plant "has neither core nor outer rind, being all things at once."[7] Parasitic, saprophytic forms, whose nutrition is tied to organic matter, are no different from "normal" colorful forms containing chlorophyll, in that both lack a specific digestive system (as an isolated process). Apart from that, the fact that the synthesis of valuable proteins as well as of living cellular material composed of basic inorganic compounds constitutes the most unique achievement of the open form in the area of nutrition justifies the instructive and scientifically well-founded assessment of the wild green forms as those expressing most purely the essence of the plant. Here the chemically definable material reveals what otherwise only becomes visible in its processing type.

It is equally legitimate to see the lack of locomotion as a characteristic of the open form. The vast majority of plants are sessile, in accordance with a maximum degree of integration into the surrounding medium. But it is not even this attribute that essentially distinguishes plants from animals.

Locomotion can be found in some plant organisms; even more common are sessile animals. While locomotion is almost entirely reserved to the closed form, this cannot be claimed equally emphatically about partial movement with a fixed location. All we can say is that movement phenomena in the plant kingdom are less common than in the animal kingdom. Among plants, movement phenomena are completely governed by the conditions given by the medium and the changes undergone by the functionally fit body. Flowers opening and closing, the day- and nighttime positions of leaves, the growth of stems and roots toward gravity or the light are in no way centrally mediated movements that can be attributed to instinctual or even volitional impulses. As Hedwig Conrad-Martius puts it, all of the plant's movements happen *to* it but never originate *from* it, just as the open form does not have a center from which it would be possible for movement impulses—whether instinctive, drive-driven, or volitional—to originate.

Due to the dominance of physiological ideas shaped in response to the functional schematic of the animal, the idea that plants also exhibit reflexive processes persisted up until very recently. References were made to [224] plant phototaxis, geotaxis, and chemotaxis, or, at minimum, the various forms of phototropism, thigmotropism, chemotropism, and geotropism were attributed to the reflex schema—that is, the schema of centrally mediated stimulus response. Once the existence of special cells sensitive to light, pressure, and gravity had been demonstrated, in particular by Gottlieb Haberlandt, the obvious next step was to speak of sense organs that perceive the stimulus and conduct it to the movement effectors. There is no doubt that such perception zones play a supportive role in the progress of stimulated movements oriented toward light, gravity, or elsewhere, although their comparison with sense organs responsible for mediating the stimulus *object* (regardless of how broad the scope) must be strictly rejected. Equally untenable is the equation of stimulus conduction with nervous processes. Leaving aside the fact that in many cases there is no evidence of a conductive relationship even if the closed organ cell aggregate would lead us to expect one, conductive relationships not only exist between the cells of the stimulus conduction system but also between parenchyma cells with flowing plasma. Locally triggered geotropic and phototropic stimulus effects can be distributed relatively widely on the plant body (see the work published by Haberlandt, Hans Nemec, and Hans Fitting).[8] Research by Anton Hendrik Blaauw and his students also produced surprising results and represents an important step forward in our understanding of tropistic reactions in plants. It shows, for instance, that the positive phototropic growth reaction of oat germs is very likely based on growth inhibition in the side facing the light. If we assume that growth-regulating substances are distributed equally throughout the tissue from the tip to the base of the germ and that these substances are photochemically decomposable, then growth will simply predominate on the shady side, causing the germ to bend toward the light. The explanation of thigmotropic growth reactions in vines leads in the same direction. Of course, it is impressive to see [225] the movements of a Cucurbitaceae vine or a morning glory, sped up ten thousand times on film, as it moves from a phase of "searching" for supports into a phase of turning back on itself where it finds a hold for regressive growth. Here too, however, there is not the least reason to assume that sensation-based or even only centrally mediated processes underlie these phenomena. Here too the corresponding reaction occurs not as a response to the situation existing objectively for the living being, but purely according to the laws of growth. We may perhaps say that the plant as open form would not be able to cope with the situation any better even if it possessed the gift of consciousness and arrived at the only existing possibility by feeling its way. Thus the inhibition of the growth-regulating substances accomplishes something for the plant that the closed form only succeeds in doing via the mediation of the situation itself—that is, its objectification.

Sensation and action (that is, centrally mediated movements that can be modified by associations) contradict the nature of the open form. Erich Becher's highly commendable efforts to demonstrate simple associations in the "carnivorous" *Drosera* produced negative results. Without a historical basis of reacting, without a memory, the movements of even these seemingly predatory plants are entirely automatic.—

Bergson's "the plant is a sleeping animal" is the creed of all romantics. Their desire to use introspective intuition to interpret the phenomena of life involuntarily leads them to look for something with which to empathize. Empathy, however, can only be extended to interiority, aspiration, intention, attitude—to a sphere of life closed around a center such as is essentially missing in plants. So they must have "fallen asleep," must be dreaming, and differ only from animals in that they lack waking consciousness.

Or we reverse the terms in a no less romantic and empathetic way and see in the plant in distinction to the animal something eminently positive, radiating without restraint, the embodiment of lavish self-surrender, ecstatic emotional impulse, a somnambulistic assurance that no longer needs consciousness—in form and [226] function the genius of nature. Such intuitions [*Intuitionen*] do not get us any closer to the essence of the plant.

It is, in fact, a betrayal of the essence of the plant (just as it is a betrayal of the essence of nature) to take it symbolically, as the embodiment of a principle communicating itself in it, as the expression of a force, a soul, a reality that is no longer the plant itself. Such literary methods not only rob nature of the greatness of its simplicity, but lead to a misconception of nature's true meaning, which intimates nothing but what it graspably is itself. The secrets of nature do not lie behind it or in it like a secret text written in code, but are out in the open for everyone to see. According to this law of essential phenomenality, the organizational idea of the open form can be shown in all expressions of plant life to be the founding unity of their essential characteristics, without having to appeal to psychological or psychoid forces. It is true that not a single expression of life can be deduced from this idea, but it allows us to understand every one of the expressions in their significance as determining the essence of the plant.[9]

The Closed Form of Organization of the Animal

The closed form mediately integrates the organism in all of its expressions of life into its surroundings and makes it into a self-sufficient segment of the life circle corresponding to it. If it belongs to the open form to make the organism, with all of its surfaces bordering on its surroundings, into a carrier of functions, the closed form is obliged to express itself in the living being sectioning itself off from its surroundings as much as possible. This sectioning-off has the purpose of mediately integrating the organism into the medium. The mediated contact not only allows the organism to maintain a greater closure than the plant, it also attains true self-sufficiency—that is, a position in which it is set upon itself, which at the same time provides a new *existential foundation*.

As a rule, we can only speak of mediate integration into a context where there are intermediate links between the context [227] and the integrated element. How does the body, whose outer surfaces immediately border on the medium, manage to insert intermediate links between itself and the medium—links that on the one hand do not belong to the body, but on the other must have a living connection to it in order to be equal to their task of naturally mediating integration?

Were the circle of life into which the body is integrated to coincide completely with the medium it borders on, we would have here an insurmountable difficulty. Thus either the significance of the medium as a component of the life circle must be reduced or the organic body must of its own accord attain another *level of being* for which the contact indissolubly given with its own corporeality no longer has sole significance. If a viable organization is to be possible according to the idea of the closed form, then integration into the circle of life must mean *something more* than contact with the medium, for the latter is immediately and essentially posited for the living thing *qua* thing.

In a nutshell, then, the organism's task is to insert a mediating layer between itself and the circle of life that takes over the contact with the medium (without of course abolishing it). What conditions must be fulfilled in order for this balancing out to constitute a revaluation of immediacy? The organism must have nothing at its disposal for the completion of this task other than its own body.

The organized body is given in its boundary surfaces. Closing off these surfaces would abolish the organs' function of mediating between the body and the medium. If the body is nevertheless to form a closed unit over against the medium, it must therefore have the boundary in itself—that is, disintegrate in itself into an antagonism. (We encountered this same law earlier in the necessity determining the essential characteristic of assimilation/dissimilation. There too the balance between the purely physical openness of the body toward the medium and the closedness inherent in genuine boundedness could only be found by way of the inner antagonism of the oppositional [228] process of assimilation/dissimilation.) It is this antagonism that allows the organism to consolidate into a unity. The oppositionality mediates it with itself into a closed wholeness, or organizes it.

Oppositionality as organizational principle is only possible if the *organs* oppose each other morphologically and physiologically and possess an antagonistic structure and function both individually and in groups. For this to be the case, there must be a highest consolidation within the organism that overlays and balances out the antagonism. Why? This consolidation is not reserved to the organism *as* a whole. The whole already includes the contraposition of a unity for itself and a unity in the variety of the parts that is mediated into the whole. The antagonism is here the ordering principle of the whole, making it the unitary form of all *variety*. If this variety, however, is separated into two opposing zones, there must be a *center* that technically sustains this opposition, otherwise its organizational purpose will be lost. The antagonism in that case would not be an organizational force, would not be the precondition for the unity of the consolidation, but rather the rupture that does not permit any unity. The unity of the consolidation allows the closed organism to combine the real peripheral unity and the real central unity, the antagonistic functional unity of the organs and the central organ, into the unity of the whole. It is in this way that this unity has to become real to itself and to join the real manifold unity (the functional coherence of all parts) for the closed form to actually exist physically.

This somewhat strange form of balance can be made more comprehensible by comparing it with the metabolic form of balance, where the organism as a whole itself forms the unity of the consolidation. The organism is engaged in the antagonistic process with all of its organs, has disintegrated into this process, and yet each component meshes with the one opposing it in such a way that the unity of the opposition emerges as its simple immediate synthesis. So it is that the metabolic organism disintegrates into two functional directions, each of which taken on its own lacks self-sufficiency; only together with the other does it constitute a meaningful function of the whole. The case with which we are concerned is different. Here the disintegration into the antagonism, as it were, concerns not only the [229] functional character, but also the organization of the organs. Thus two organizational zones standing in opposition to each other must be created so that the organism can separate into two relatively self-sufficient parts. This excludes the possibility of their immediate synthesis into the unity of the whole, despite the determination that the living body is unified by the antagonism.

In the case of metabolism, the organism becomes the unity of the whole by means of the antagonism of assimilation/dissimilation, of the opposition itself. Here, by contrast, the antagonism, because it is an antagonism of an organizational kind, must itself be mediated by something else. In the metabolic process, the body *immediately mediates* itself in opposition *to* the whole. The balancing out of the disintegration into zones in a center, on the other hand, is not itself the whole balance, but only the means to that end.

What is the nature of this disintegration into zones? We can find the answer by considering that it serves to balance out the organism with its medium. With its organs the body as body (not as a living whole) is in contact with the things in the medium. The living whole, on the contrary, encounters these things by means of its organs—that is, is in mediate and not immediate contact with them. In a real sense, however, the organs taken together *are* the organism. There would thus be no actual difference between body and living whole; the living whole in its organs would be immediately open to the medium if the antagonism of the organs and the formation of a center that this antagonism renders necessary did not really create something in the organism able to consolidate all organs. This representational organ—its function is to have all organs represented in it—is on a higher level than the antagonism and sustains it as a condition of organic unity. Thus it is only the *way* in which the organism relates to the medium that is antagonistic or, to be more precise, the way in which the organism, starting from the organism that it is, enters into relation with the medium.

Two types of relationship between organism and medium are possible: a passively accepting or an actively shaping relationship. In one case the organism accepts the medium and the medium shapes it; in the other the organism shapes the medium and the medium accepts this. Both [230] antagonistic relationships really take place and are balanced out by a center for the sake of the unity of the whole. The disintegration into zones appears as the opposition between *sensory* and *motor* organization and is mediated by centers of a primarily nervous nature.

The sensory-motor schema, the "function-circle," as Uexküll calls it, is the condition of the possibility of the reality of the closed form, the organizational idea of the animal. This idea allows us to understand all the essential characteristics of animal life in their unity: in morphological (and thus also ontogenetic) terms, the predominant formation of inner surfaces into organ and organ systems, drawing on the outer surface of the body as little as possible, which, in turn, is designated as the carrier of sense and movement organs, and in physiological terms, spontaneous movement, in particular the capacity for locomotion, circulation that is distributed across the organism's own organ spheres and divided into stages, respiration, nutrition (based solely on organic matter), and sensation.

It is not immediately clear, however, that the organism in a passively accepting relation "notices" stimuli coming from the medium or, in an actively shaping relation, "affects" the medium. In order to understand this, we first have to recall that these oppositional relationships in their morphological and functional dualism are meant to make possible the existence of a closed form. They do so because they induce the emergence of a central representation of the organism. This means that the living organism as a whole is no longer immediately the unity of the organs (mediated in itself, of course!), but is this unity only by way of the center. It is thus no longer in direct contact with the medium and the things surrounding it, but is so merely by means of its body. The body, then, has become the intermediary layer between the living being and the medium. Here we have the solution to the problem posed earlier of the mediate integration of the organism into the circle of life: the living being borders on the medium with its body, has acquired a reality "in" the body, "behind" the body, and thus no longer enters into direct contact with the medium. The organism has thus attained a higher level of being that is no longer on the same plane as that inhabited by its own body. It is the unity of the body mediated by the unified representation [231] of its members, a body that is therefore dependent on this central representation. Its body [Körper] has become its lived body [Leib], that concrete center by virtue of which the living subject is connected with its surrounding field.

There is an essential connection between the emergence of centers of organization in a body and the shift onto a higher level of being of this body as a living thing. Physically speaking, the emergence of a center leads to a doubling of the body: now it is also (that is, represented) in the central organ. This "center" [Mitte], which belongs to the essence of every living body, this core unity standing for itself over against manifold unity, which after all is a purely intensive quantity, is not, of course, filled by a spatial entity. It is a spacelike center, a structural moment of the positionality of the living body. But the nature of this body that encloses it in a spacelike way has changed, because it is now really mediated, represented in the body that it is. It is *set apart from itself* and dependent on itself as a body. Even in a purely physical way it is "its lived body." The spacelike center, the core, or the self thus no longer immediately "lies in" the body. To be more precise, the center inhabits a dual spacelike position [*Lage*] in relation to the body: in it (inasmuch as the whole body *including* the central organ is not its lived body and does not depend on it) and outside of it (inasmuch as the body depends on the central organ as its lived body).

In this way, the center, the core, the self, or the subject of having distances itself from the living body while being completely tied to it. Although it is a purely intensive moment of the positionality of the body, the center is set apart from the body, which becomes the center's lived body that it has. Now as far as the body physically is its own lived body, the center also enters into a special relationship to it as to a zone that is subjected to the center because it depends on the whole body (including the center). The whole body itself is not dependent on the center, but the zone that is represented in the central organ is. With this lived body the living thing exists as with a means, an intermediate layer given into its possession that at the same time connects and separates, opens and covers up, exposes and protects.

The formation of centers allows for a dependency of the body on itself, thereby creating a relation of opposition [232] between the carrier of life and bodily variety, which in turn becomes a relation of dependency of the centrally controlled lived body on the carrier of life. In this way "having" takes on concrete meaning. The self, although it is a purely intensive, spacelike center, now possesses the body as its lived body and thus necessarily has that which influences the body and that the body influences: the medium. The dual relationship in which the organism (as closed form) antagonistically stands to the medium—passive acceptance and active shaping—thus manifests as a dual mode of having or possessing. If the relationship is to the self, the self has the other in an accepting way; if the relationship is to the other, the self has the other in a seizing way. The self and the medium as the other relate to each other *across a chasm*. This bridge between the self and the other preserves the in-between and is the only form in which having or possessing is possible.

In its distance to its own lived body, the living body has its medium as its surrounding field. This set-apartness from its own lived body allows it to have contact with being that is set apart from the lived body. The body "notices" and "affects" this being. Across a chasm it has a sensorial and motor connection to the other. Contrasting this whole complex of the closed form to the complex of the "living thing as such" according to the principle of levels shows that everything that in the latter is still bound, only present in itself, only implied, and determining the structure of life in the closed form is released, has come to stand on its own, and been explicated. The plant too has stems, leaves, flowers, and fruits, but neither its self nor its having enter into an actual opposition to its body as a lived body. The plant self is only a characteristic of its living wholeness and cannot be positionally set apart from its body. But as soon as the formation of a center has brought about a real difference in the body itself, the whole also changes positionally, creating the basis for all phenomena linked to the existence of consciousness.

The path of the greatest possible self-sufficiency for the organism taken by the closed form leads to the contraposition of an open medium to the organism's enclosedness in the circle of life. This means that the animal generally moves from place to [233] place, offensively and defensively seeking, under constantly changing conditions, nourishment, prey, and opportunities for mating. While the positional field is finite—that is, its physicochemical dimension and type fit the animal organism or is closed—as the animal's opposing field it is boundless. There is no horizon line for the animal; it does not yet have the means to notice one.

The openness of the positional field corresponds in its essential law to the closed form of organization because they both define a state of affairs that can be found in all animal attributes: that of the primal *unfulfilledness* of the living being. To be primally needy is the same thing as to be mediately integrated into the circle of life. In the open form, self-sufficiency has passed into the entire circle of life, with the individual plant only a passageway. The animal depending on itself, on the contrary, with its closed form preserves its self-sufficiency vis-à-vis the circle of life, which it nevertheless belongs to with its entire organization. It is therefore by nature a needy thing seeking fulfillment—a fulfillment that is potentially guaranteed but in actuality is only attainable *across a chasm*. In its self-sufficiency the animal is the starting point and point of application of its drives, which are nothing other than the immediate manifestation of its primal unfulfilledness, its mediate integration into the circle of life. A maximum of closedness determines a maximum of dynamism in the animal's restless drivenness, peacelessness, and compulsion to fight. Whether instinct seeks to satisfy the drive for the animal or the animal itself (consciously) forces the drive's satisfaction makes no difference before the law stating that being an animal means being a fighter.

Finally, the law of the closed form dictates that organs are sectioned off from the outer world and at the same time strongly differentiated into relatively self-sufficient systems of circulation, nutrition, reproduction, stimulus conduction, etc. This sectioning off and differentiation is for one of course immediately connected to the form's closedness, which only has outward-facing sense organs and effector organs (organs of attack and defense). It also mediately depends on this closedness insofar as central direction or representation demands a greater separation of individual functions, their spatial distribution onto tissue systems that are as distinct from each other as possible, and their temporal distribution into particular stages. Representability presupposes the segmentation of [234] what is to be represented. An organism whose processes of breathing, stimulus conduction, and nutrition are as intertwined as they are, for instance, in a plant would not be able to implement a central representation of these processes and thus their regulation. Division of labor grows with the formation of central consolidation; they require each other, just as each draws an advantage from the strength of the other. When a growing and refined centralization goes hand in hand with decentralization, this only leads to an intensification of the organizational principle, the exceedingly artful results of which can be studied in certain arthropods and in the higher vertebrates.

The closed form as the principle of primal neediness or of having drives finds fundamental expression in the animal organism's inability to synthesize proteins, fats, and carbohydrates from inorganic elements, as do plants. Animals require organic nourishment; they have to live off of life. This inability does make the organism largely independent of its medium, but this cannot be the only reason. The animal's independence is already ensured by its other essential characteristics, which do not exclude inorganic nourishment a priori. Thus it seems likely that the existence of the closed form itself has another purpose, one we automatically associate with the essence of animal nature as the properly predatory: the purpose of intensifying life at the expense of life. All differentiation and sublimation sap life and can only be purchased with the loss of vital energy. The closed form constitutes intensification as it lifts the living body onto a higher level of existence. As a result, the living body needs life itself in order to live. The constitutive parasitism of the animal world, as it is expressed in the necessity of its organic natural basis, gives us a first sense of the enigmatic coherence that binds the levels of life in sensory laws. Parasitism is another possibility of which plants in their highly differentiated forms do not avail themselves. Animals, on the contrary, have to live off of life.—

Plants and animals cannot be essentially distinguished according to empirical attributes. The difference between them is ideational—and fully real. Open form and closed form are ideas [235] to which actual living bodies have to conform organically and under which living being falls when it takes the path of the organic. The dividing line between the plant and the animal kingdom cannot be found empirically; transitional forms exist alongside distinct ones.

It is the unity of life, as it is expressed in the general affinity of all vital processes, that, empirically speaking, makes the specific difference between plant and animal existence into a difference of degree. Certainly plants are consistently devoid of central organs to mediate and reverse the direction of excitations triggered peripherally, and animals are consistently unable to synthesize proteins, carbohydrates, and fats from inorganic compounds. But there are also animals without fully formed centers, just as there are plants without the ability to assimilate inorganic substances.

The bipolarity of the organic world does not prevent gradual transitions between the extremes, just as the extremity of red and blue on the visible spectrum does not cancel out the fact that orange, yellow, green, and purple continuously mediate these colors. Properties reserved only to animals or to plants do not exist; properties therefore cannot ground the essential difference between the two. The difference between plants and animals must be established before it can be correct to say that animals can never exist without organic nourishment; mushrooms, for instance, cannot do so either.

The organizational ideas of the open and closed forms are leeways that can never be fully filled by a particular complex of attributes. The hiatus that was shown earlier to be essential to life remains between the living thing and the form type under which it falls, even in this sphere of differentiated formation. The form of the inorganic thing is the gestalt given with the functional unity of all of its elements and as such is transposable, but for the thing itself, the form is not set apart from it. The thing is bound to the form itself because the form coincides with its boundedness. The form of the organic thing is the gestalt of wholeness arising from the relationship between boundedness and boundary, transposable and at the same time set apart from the thing itself. In this sense the organism is free from its form.

We will thus have to hold on to the infinite variability [236] of the individual form within a particular form type, to a whole scale of form levels. All attempts to deduce even just the individual phyla (not to mention species, genera, families) from the idea of plant or animal form are not doomed to failure because reality [*Wirklichkeit*] is infinitely superior to our paltry understanding and its concepts, as we say with impertinent modesty, but because it is futile to regard the idea as something whose restriction could lead us closer to the reality of the individual.

Ideas are not concepts used by experience to denote lesser or greater degrees of kinship on various levels of lesser or greater abstraction. Ideas rather generate a discontinuous variety of reciprocal super-elevation, without the possibility of moving from one level to the next following a principle of continuous progress. Not constructible from ideas but comprehensible in view of them, the concrete living thing conforms to the ontological relationship between being and form that is characteristic of life. Between the physical and the form, leeway remains.

[237] SIX

The Sphere of the Animal

The Positionality of the Closed Form: Centrality and Frontality

A living being whose organization exhibits the closed form has actuality as this body and as its lived body—that is, in its body. The central representation of all members and organs creates a zone of the whole body that is dependent on the whole; in this zone lies the spacelike center [*Mitte*], the core of the living thing. Positionally speaking, there is no possibility here yet of mediating between the whole body (including the central organ) and the lived body (the bodily zone dependent on the central organ). In positional terms these exist next to each other without canceling out their unity. The oscillation between the two situations of being, the alternation between being a body and being in the body in a spacelike way creates a dual aspect, but this oscillation, this alternation does not cancel itself out in itself but simply represents the same basic state of affairs. It is not that one and the same x is looked at this way and then that way, but rather that the dual aspect of body and lived body is the positional equivalent of the physical separation into a bodily zone that contains the center and a bodily zone bound by the center.

It is only in this ambiguous way (an ambiguity, in other words, that does not conceal unambiguity or that could be rendered unambiguous) that the living thing stands at a distance to its body, to that which the body itself is, to its own being. The living thing is itself in itself [*in ihm*]. The position is a dual one: being the body itself and being in the body—and yet it is singular, since the living thing's distance to its own body is only possible due to its complete oneness with this body alone.

The spacelike center, the core, means the subject of having or the self. In its set-apartness from its own lived physical body the self also forms the center around which the body [238] is closed, against which the body and the positional field surrounding it totally converge. While it is in no sense localizable, the self is not without a relationship to the spatial. Spacelike, it is the point contrary to which all the other points having the character of "there" stand, the point of a nonrelativizable "here." This "here" lying in the body, the placeless self (and thus the nonrelativizable place, the "natural," essential place) is, for the sake of this very distance to and at the same time oneness with its body, isolated as a something. As the absolute point of reference for the positional field and the body, embedded in them and set apart from them, the self is the simple mediation between being the body itself and being in the body in the pure, nonrelativizable here. Although it is the subject of having, it nevertheless in fact coincides with the object of having, the body, while remaining distinguishable from it. As the unity of subject and object, the self at the same time leaves the separation between subject and object by mediating between them in the pure here.

Thus the living thing whose organization exhibits a closed form is not only a self that "has," but a special kind of self, a reflexive self [*Selbst*] or an itself [*Sich*]. We may speak of a living thing of this kind as being present to the living thing that it is, as, by virtue of its set-apartness from this living thing, forming (not yet "having," which is why it is not yet an I!) an unshakeable point in this living thing in relation to which it reflexively lives as *one* thing. In the irreducible oscillation between being inside and being outside that distinguishes the positionality of the closed organism on the ground of simply being the body itself lies the boundary for the referentiality of the thing back to itself.

In and set apart from the organism that it is, the closed organism, the animal, is the unity of the alternation of the aspects as they are mediated by the "here." This unshakeable here does not elude the alternation; there is no back wall onto which it is projected (which, as we will see further on, is characteristic for the I), no strange point of coincidence of absolute distance (being in itself) and absolute proximity (being to itself). There is only that by means of which the alternation from inside being to outside being concretely forms the unity of being-itself. Controlling the organism that it is in its body, impulsively moving it from within, the organism stands "in" the "here"; its existence is positioned in the center of the fullness of its own body [239], is absorbed into this fullness as the center of the positional union of space and time. And because "here" only expresses the spacelike nature of this union, we must also include the timelike nature of the central position of animal existence: ahead of the being that it is, it stands in the "now." As a nonrelativizable here/now, the animal has and controls its lived body and with it the field given to it.

The barrier for the animal lies in the fact that everything that is given to it—its medium and its own physical lived body, *excepting* its being-itself, being the body itself—stands in relation *to* the here/now. Insofar as it is itself, it is absorbed in the here/now. The here/now does not become objective for the animal; it does not set itself apart from it. The here/now remains the mediating passing-through of concretely living execution; it is lived out, lived away. Even if in its distance from its own lived body the living body has its medium as a field set apart from and opposing its own lived body; even if it notices this medium and acts on it with the help of its physical lived body, which it equally—only not across a chasm but connected with it—has as something it notices and acts on, its having nevertheless is hidden from it. This having carries the living body but is not for it; the living body merely is it.

Positionally, an animal as single thing, as individual, forms a here/now against which its outer field and its own body stand concentrically and from which its own body and outer field receive influences. It notices and it acts; the difference between that which is alien and that which is its own is clearly given in zones. It is separated from the alien by the chasm that allows it to have and to notice what is given outside of its lived body. It exists in what is its own insofar as it immediately controls its body. The fact that it can control its body because it is set apart from it, is at a distance to it as its lived body, constitutes the positional character of the animal, carries its existence, but is not given itself, not noticeable. To whom would it be given? What point, what projection surface could this state of affairs itself be referred to, on the basis of what structural moment of living thinghood's distance from the thing of this animal body?

The here/now nature of the animal is not given to it, is not present to it; the animal is still fully absorbed in this here/now nature and carries therein the barrier, hidden to itself, set against its own individual existence. While it (as lived body) is present to it (the whole), the whole is not present to itself. The outer field and the physical lived body are present to the whole [240]. In the reference back of the lived body's own sphere to the whole sphere of being-the-body-itself, standing in the here/now, lies that special kind of self that cannot be addressed in the first, second, or third person—for it is not yet an I—but that is nevertheless fully reflexive in this reference. In other words, it is a self—that is, this reference renders it objective. Insofar as the animal is lived body, it is given and present to itself and, as the whole body standing in the here/now, it can exert influence on its body and bring about the "appropriate" success for its impulses. But the whole body has not yet become totally reflexive.

Every animal is potentially a center to which (to whatever varying extent) its own lived body and alien content are given. Physically, it lives present to itself in a *surrounding field* set apart from it or in a relation of *opposition* [Gegenüber]. As such it is conscious; it notices that which opposes it and reacts from out of its center—that is, spontaneously: it acts.

Spontaneity (like centrality, set-apartness from its own lived body, relation of opposition) is only one attribute of the positionality of the closed form. Theories of freedom or unfreedom do not belong in this context. Spontaneity is the simple expression of the essence of a being from out of a center that is itself no longer given, of a genuine beginning, commencing, initiating. If the animal is really [wirklich] essentially absorbed in the here/now and lives in this central position; if, moreover, the center of this position is not noticeable to the animal nor given as a vanishing point of its own interiority that lies "behind it," then from the animal's perspective its actions commence immediately, regardless of the share of drives, instincts, and involuntary and reflexive elements in each individual case. The animal lives in an essentially impulsive way, spontaneously moving its limbs, acting, and reacting to stimuli. At the same time, this structural moment of positionality also includes the possibility of choice, which can structurally precede the spontaneous act.

Here too we have to leave aside all ethical and metaphysical considerations. To choose means to be in a state of wavering. The possibility of doing something in this way or that, as is essential to the aspect of spontaneously beginning and carrying out an action, only means that the absorption [241] in the here/now presents itself "before" the beginning or, to be more precise, the fact that the individual is ahead of itself entails continual possibilities of action with the aid of its lived body. The presence of an undetermined abundance of possibilities means, in the transition to the spontaneous act, the obligation, the compulsion to choose.

The surrounding field, toward which the actions are directed, is open and without boundaries for the individual. For if it is true that the positional field is a surrounding field set apart from the body by a chasm, a hiatus, only insofar as the living center itself is set apart from the body, the form of objectivity of the surrounding field will also exhibit a corresponding element to the form of set-apartness. Now while the living center, the here/now, is the point that is set apart from the body, it does not itself appear as given; the living body stands *in* the here/now. Thus the animal's surrounding field cannot indicate the structure binding and carrying it—that is, its boundaries. It is in this sense that it is open. It remains finite, of course, because the animal has no means of breaching the zone of primary adaptedness, but this finitude does not appear as the structure of the positional field itself. It is only a condition for the existence of a positional field in the first place.

Facing an alien zone that as a whole remains opaque to it, which it can respond to but never come to terms with, the animal lives, positioned in itself, absorbed in itself, secure and vulnerable at the same time. This special position of *frontality*—that is, an existence oriented toward the surrounding field of alien givenness, opens up two divergent paths for animal organization.

The first option is for the organism to develop individual centers that are loosely associated with each other and that are decentralized to the extent that the execution of individual functions is independent of the whole, forgoing central consolidation. This is the path of circumventing consciousness in order to achieve the greatest possible protection against the field. The second option is for the organism to consolidate in a strictly centralist way under the command of a central nervous system and to seek to bring under its control the execution of its individual functions. This is the path of activating consciousness in order to achieve the greatest possible penetration into the field. Since the realization of the closed form concludes not with a center as such, but only with physical [242] cells and complexes of cells of a very specific structure and function, life must select one of these organizational paths. The idea of the closed form can be satisfied in either way, and in reality we find the greatest variety of transitional forms between the two extremes.—

Before turning now to the concretions of the closed form, it seems necessary to revisit the main points of what we just covered in order to forestall misunderstandings as well as the objections that can easily be brought against this line of argument. The dual aspect, it was stated earlier, is the positional equivalent of the physical separation into a bodily zone containing the center and a bodily zone bound by the center. In and set apart from the organism that it is, the closed organism, the animal, is the unity of the alternation of the aspects as they are mediated by the "here."

Is this not something entirely new, something that has been absent so far from our examination of the structure of the living thing-body? Does the introduction of the notion of aspect not signify a radical break with the direction we have been pursuing so far? Is it not a *metábasis eis állo génos* to say that what has until now only been described as physical variety is an alternation of aspects? Does not aspect, sight, require a seeing [*die Sehe*] (to use a term of Fichte's)? And where might this seeing come from where there is no eye, no knowing? Has the subjective aspect not been won surreptitiously by way of the concepts of the self and the subject, which are being treated as equivalents of the core, the space- and timelike center?

It is not surprising that all of these questions are forcing themselves upon us at this particular juncture. Bit by bit, our analysis has been establishing double-sidedness as the double-sightedness of the bodily objects of intuition that can be considered to be alive, and this is emerging clearly now. According to the original thesis, living being appears in the dual aspectivity of the appearing thing, and as a thoroughly objective aspectivity at that. Intuition can take a dual direction in relation to the thing because the thing as appearing object is double-sided (in a specific sense). It was by implementing this thesis that we were able to develop the objective characteristics of living things in context. From the level [243] of the essential characteristics forming the individual we arrived at the organizing characteristics and finally at the determination of these characteristics' specific differences, the open and closed forms. Here we find something remarkable: the closed form raises the existential *level* [*Niveau*] of the organized body; it becomes set apart from the body that it is in the body that it is, so that it comes to stand above itself (in the body that it is).

In this distance of the core of the body's positionality, in this setting-apart of its space- and timelike center, we gradually came to see the reason for its awareness. The core, the center, which positionally has the value of a self (as in the phrase: the flower itself as the carrier of its properties), of a subject of having, with distance (in the closed form of organization) does not acquire a new value and meaning, but is, as it were, only set free; it emphatically becomes what it is in itself: a point of view for seeing, the subject-point of an awareness.

Let the doubters try to bolster their accusation of a surreptitious attainment of the subjective standpoint, of a *metábasis eis állo génos*, by naming a difference between that which makes the I into an I, the subject of consciousness into a subject, seeing into seeing, and that which we have been describing as core, as space- and timelike center, as self and subject of having. We have made it easy for them, having begun this study with a detailed elucidation of the alleged untranslatability of the subjective standpoint and the absolute incompatibility between subjective and objective sources of intuition. We have intentionally pursued the arguments for an alternative position of the gaze in relation to the reality [*Wirklichkeit*] that presents itself to outer and inner perception and have attempted to uncover the motifs of this ontological-gnoseological dualism, which, if it were correct, would have to negate the phenomena of life and admit them only as conglomerates of physical and psychological being.

It is not due to a change in methodology if being here, in living bodies that are formed in a closed way, turns into consciousness and the core becomes an aspect center. We have remained faithful to our line of argument from the beginning and have not left one level in order to break into another dimension. And the attentive reader will not have to be reminded that [244] all of the concepts developed thus far essentially had the task of neatly elaborating the sphere of positionality that—neutralizing the very distinction between physical and psychic—may be considered, in a convergent position of the gaze, the sphere of existence encompassing both sides of living being.

Of course, the character of positionality changes with the form of organization. The closed form, distinguished by the core's distance from the body, has the character of frontality, of a contraposition, of an existence directed against the surrounding field of alien givenness, both open to this field and at the same time separated from it by a chasm and thus closed off. The closed form is quite properly the situation of consciousness, in which the living being operates out of an impulse center and remains, having, in a center of gazing or noticing. That is why it notices itself, albeit only as lived body—that is, to the degree that the body by way of centers has become dependent on itself, is represented in itself. The extent to which the living being is aware of itself (and of the outer field) is physically determined by the extent of the central representation of its own body, just as the possibility of the space- and timelike core distancing itself from the body is conditioned by the physical presence of centers.

What on the contrary can never be determined or conditioned by spatiotemporal bodies is the space- and timelike essence of the distanced core of positionality, as well as the subjectivity, "eye-havingness," and impulsiveness not only made possible but downright composed by these essential traits. The space- and timelike center of positionality is not to be found in the center of nerve net or brain. The nervous apparatus is only the means of interruption between the whole body and the body as sensory-motor antagonism that envelops the plenitude of organs.

While an interruption in the physical and positional core distance and the presence of nervous centers and subjectivity determine double-sidedness as the double-sightedness of animal existence and break it down into an outer and an inner aspect, its ascertainment is not based on all of this. The philosopher can grasp its coordination at a glance, for it belongs to the *one* (psychophysically neutral) sphere of positionality and constitutes the peculiar character of [245] frontality: a being-for-*itself* (as here/now a being-for-*myself*) that stands within its lived physical body before alien things.

The Coordination of Stimulus and Response in the Case of an Inoperative Subject (Decentralized Type of Organization)

As already mentioned, there are two objective possibilities for organizationally coping with the position of frontality: either decentralization, rendering consciousness inoperative, or centralization and the extensive development of consciousness. In both cases it is a matter of the organism reacting to what it notices.

The individual coordination of stimulus and response, which is *potentially* guaranteed for every living being—unicellular or multicellular, plant or animal—and actually for all living beings with the exception of animals, has to be secured on a case-by-case basis by animals. The sensory-motor antagonism of their nature is only effective when there is an interruption between noticing and effecting, an inhibition of the excitation brought about by the stimulus before it empties out into the organs responsible for movement. Incidentally, the primary inhibition of the excitation caused by a stimulus is already reflected in the nature of "noticing" and "effecting." Would there be any point in noticing something in order to have an effect on it if the conversion of the excitation from the sensory into the motor sphere could take place in an uninhibited way, on its own?

Noticing is equivalent to inhibited excitation; effecting to disinhibited excitation. The sphere of consciousness spans these two poles, and it is here that the transition takes place from noticing to effecting. This sphere is thus the spacelike inner boundary, the timelike interval between what comes from without and what goes out, the hiatus, the void, the internal chasm through which the response follows upon the stimulus. In this interruption, which demands the spontaneous intervention of the organism, the exerted stimulus has the value of something objectively given upon which a fitting response must follow: an answer, as it were, to a question. Clearly, the interruption means the possibility of failing to answer the stimulus. It is also clear that the likelihood of a correct answer will only grow with the narrowing of the leeway between stimulus and response.

[246] Total narrowing can occur, on the one hand, as a rendering inoperative of consciousness. In this case, the conversion of stimulus into response takes place in the lived body alone, with the centers reflexively

answering the questions posed to the organism by the surrounding field. Or, on the other, narrowing takes place as the fullest possible operation of consciousness by the corrective of experience on the basis of highly differentiated sense organs and a central nervous system corresponding to the wealth of possibilities of their connection. Circumventing consciousness in the first case, exploiting it in the second—this is how we could label the two methods life uses to cope with the situation of frontality.

The circumvention of consciousness can be schematically represented as the engagement of the body into the surrounding field with the help of its receptor and effector organs. This engagement, it must be noted, is not an absolutely rigid bond that would (with only the means of animal organization) make the animal into a plant, but rather a bond within certain boundaries that provide the living being with leeway. In this sphere, the surrounding field plays a very minor objective role, as nothing in it appears to the organism that would not require the latter to respond with action. The sense organs here have to perceive stimuli to the same degree that they have to tune them out. They are eyes and blinders in one. The fact that the animal only notices what it can make use of and for which it has an answer and, furthermore, that it is unconscious of its own movements significantly reduces the possibility of error without, however, entirely turning the organism into an instinct and reflex automaton. Completely rendering consciousness inoperative and stripping it of any meaning for the body would be to deny even the existence of the sphere of noticing.

But everything that is given here is related to action. The action plan of the animal is the net in which the world is caught. An entirely primitive priority of the practical structures the sphere of noticing here, both formally and in terms of content, according to motor categories: the sphere of noticing is simply put to the service of the search for food, self-defense, mating, egg deposition, etc. When a datum appears in the sphere of noticing, it presents itself as a signal, never as an object. Objects are only found in the sphere of action—that is, prey, nourishment, enemy, mating partner, lair. These are not objects of perception, but correlates of needs and drives, since on this level actions [247] are not connected to the network of noticing and the animal does not sense its own movements.

The relativity to action of the sensations—that is, the sensorially mediated contents of consciousness—is a substitute for the lack of objective unity in the surrounding field on this level of animal organization. It tightly binds together elements that for the animal are disconnected: the correlates of its sensations and the targets of its movements. If, like Uexküll, we understand the correlates of sensations, the stimuli-emitting properties of the object, as carriers of attributes and the properties of the object serving as targets as carriers of effects, we can make sense of what he writes here:

> The carrier of attributes and the carrier of effects always coincide in the same object—this is a pithy way of expressing the wonderful fact that all animals fit in with the objects of their environment. The object, which in its dual property of carrier of attributes and carrier of effects becomes a thing in the environment [*Umweltding*], also possesses its own structure which ties together these dual properties. Whether it is a matter of a dead object or of a living being, this "counter-structure" of the object is also always incorporated into the construction plan of the animal subject, even if there can be no effect of the counter-structure of the object on the structure of the subject. This fact alone vouches for the presence of a general universal plan in nature that equally encompasses subjects and objects.[1]

An outside observer noting that the environment of an animal is only filled with "things" that belong to this particular animal alone—that an earthworm is only surrounded by earthworm things, dragonflies only by dragonfly things—can understand this as a lack of *genuine* thinghood. These things are devoid of, precisely, objectivity, since in relation to the senses they are signals, in relation to the motor functions they are needs satisfiers, and they are fully absorbed into the function-circle (Uexküll) that binds the animal subject and its environment into a unity. Uexküll's diagram (Fig. 6.1) makes it clear how the sphere of noticing and the sphere of effecting are not unified for the subject and leave space between them for the counter-structure and the unity of the object.

[248]

For animals organized in a decentralized way, the unity of the plan replaces the unity of impulses. Just as the individual object can only make itself noticed in the surrounding field by means of a certain combination of stimuli characteristic of the particular animal's plan and with a purely signaling value, the significance of individual impulses for action is restricted

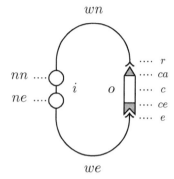

Figure 6.1. Uexküll's function-circle. *o* = object, *ca* = carrier of attributes [*Merkmalträger*], *ce* = carrier of effects [*Wirkungsträger*], *r* = receptor, e = *effector*, *c* = counterstructure, *wn* = world of noticing [*Merkwelt*], *we* = world of effecting [*Wirkungswelt*], *nn* = network of noticing, *ne* = network of effecting, *i* = inner world. I use the terms "sphere of noticing" and "sphere of effecting" rather than "world of noticing" and "world of effecting" and avoid using the expression "inner world" in the Uexküllian sense entirely. (Jakob von Uexküll, *Umwelt und Innenwelt der Tiere*, 2nd ed. [Berlin: Springer, 1921], 46. I have replaced the labels of the original diagram with abbreviations.)

and in some cases entirely suppressed. According to Uexküll, sea urchins, for example, can be thought of as veritable "reflex republics": "To be sure, there are centrally located reservoirs regulating the general excitation pressure, but the individual reflexes run their course completely independently. Not only each organ, but even each muscle fiber with its nerve center acts completely on its own. That in this situation something sensible nevertheless results is only the contribution of the plan.... When a dog runs, the animal moves its legs. When a sea urchin runs, the legs move the animal."[2]

A consistent characteristic of the decentralized form of organization is the receding of the sensory behind the motor apparatus, the shrouding of the world of objects except for sparse signals benefitting the smoothest possible function of the actions necessary for the body. A scant possibility of error corresponds to a scant capacity for forming associations or learning. An increase in the importance of consciousness for the coordination of response and stimuli, on the other hand, requires a development of this capacity as a corrective.

[249]

The Coordination of Stimulus and Response by a Subject (Centralized Type of Organization)

It is only from the human perspective that the nonalignment between the surrounding field's sphere of noticing and sphere of effecting in the case of the decentralized type appears as a lack. The animal does not notice these two spheres coming apart because it does not notice that, and how, it has an effect on its surrounding field. "In the whole array of invertebrate animals, from the lowest to the highest, the unity of the central nervous system lies exclusively in the construction plan. The functions merely form a continuous chain that nowhere in the inner world"—Uexküll means the body!—"closes into a circle. It is for this reason that these animals never attain the highest level of uniformity. Only jellyfish . . . receive their own movements back in the form of stimuli, at the expense, it is true, of the environment from which they receive no stimuli."[3] The surrounding field is only present sensorially in signals, not motorically. In terms of its motor functions, the animal merely exists; it is fully absorbed in its actions just as it is fully absorbed in the here/now core of its position. One could say that positionally it is enclosed in its surrounding field in a sensorial but not in a motor way. Drives, signals, and drive satisfaction are the contents of the positional field of lower animals.

If, for the individual coordination of stimulus and response, life chooses the way of consciousness and thus of the subject; if it chooses to go through the hiatus, the inner, middle void, in a space- and timelike way "right through the middle" of the crack of impulsivity, actions have to take place *on the basis of* sensations. It becomes necessary for the animal subject to use its sense organs to monitor the surrounding field as much as possible in order to discover the situation in which it finds itself and to be able to choose from among a greater or smaller number of action possibilities, allowing it to select a particular movement within a certain range of movements.

Compared with the simply signaling sense data of the lower animals, the differentiation of the receptors here affords a *surplus* of givenness, which is no longer related to certain individual actions and action chains, but only to a *type* of action. The development of the motor apparatus only keeps pace with this differentiation of the receptors to the extent that the centers of effecting are increasingly centralized—that is, subordinated. The motor functional plan of the higher animals does not contain new ideas [250] as compared to that of the lower animals. This is not surprising, given that there is a primacy of the motor functions in the latter that gradually gives way to a primacy of the senses in the ascent to the higher forms.

This primacy of the senses becomes final the moment the animal's actions come under the control of sensation. Only then does the surrounding field come to exhibit an objective sphere of effecting, does the animal subject become enclosed by the surrounding field in terms of its sensory-motor functions, does the surrounding field come to contain "things next to each other, after each other." Not only from the human perspective but measured against the idea of the closed form is the greatest possible realization, the highest level, the purest manifestation of animal being achieved with the total representation of the organism's own body. Here the living being attains the greatest degree of freedom, the highest fullness of power by confronting a sphere of action objects to which it is existentially exposed.

At the same time, because the growing role of the central impulse in view of many equally possible action objects increases the possibility of error, the living being forfeits some of the certainty of individual action. Making a decision in a particular case is of course more difficult for an animal whose highly differentiated eyes allow it to survey the situation in its surrounding field and distinguish the confusing abundance of colorful images and forms in its setting. The greater the presence radius, the more likely the mistake—as long as the living being is not able to discover the crucial element in the images themselves that would require a particular, and no other, action. The uncertainty arising from the increasing breadth of intuition is compensated by instinct and experience, the former based on anticipation as it is imprinted upon the individual by heredity, the latter on what the individual has itself experienced as it passes through life.

As double-edged as the increasing breadth of intuition and the possibilities of monitoring the surrounding field is the living being's control of the actions of its own body. The advantage for future actions of being aware of the success of an action in relation to certain objects in a certain situation is obvious. Implementation becomes more precise with increasing control over the object and the phases of action suited or not suited to it. At the same time, however, [251] the animal's control of its own movements has an inhibiting effect on their execution. The living being's attention is drawn from the object of the movement to the movement as object. Fragmentation is the unavoidable result: uninhibitedness is lost; the certain outcome of the action, which requires total commitment to the object, is called into question.

It is the antagonism between action and consciousness that nature has in mind when it keeps the movements of a living being's own body from the gaze of consciousness as long as possible. Even in the highest invertebrates—arthropods and octopods—movements of the limbs are governed by autonomous networks of effecting. "Even if the motor centers of insect limbs can already be found in the ventral nerve cord and the centers of mantle movement in octopods have moved as far as the esophageal ganglia, the receptor apparatus nowhere acquires even the slightest bit of information about the activities of the motor apparatuses."[4] The principle of maintaining certain zones of the body under autonomous systems, independently of the central control of the brain and thus inaccessible to spontaneous intervention, remains in place all the way up to the most complicated form of the human. To the extent, however, that the living being's own body is revealed to consciousness, instinct and habituation have to compensate in order to at least attenuate the effect of this antagonism on life, the cause of which, since it lies in life itself, they cannot eliminate.

The means of consciously coordinating stimulus and response and of gaining mastery over the lived body and the surrounding field are given in centralized organization. From a purely physical perspective, lived body and surrounding field have to find their representation in the center. What this representation means anatomically and physiologically is not our concern here. Nor is it yet our concern to understand the relationship the living subject has to the excitations of this physical central apparatus: whether it grasps the actual outer world *in* them as in the signs of a language that grasp the meaning and the thing, or rather regulates its contact with the outer world *with* them as a switchboard closes and interrupts circuits. We first must ask whether there is in fact a necessary coexistence between the centralized organization of the living being's body and the organization of its surrounding field into things. Only once this has been decided are we equipped to ask how the animal with the help of its physical central [252] apparatus, the brain, comes to represent other things outside of itself.

As has already been established, the closed organization of a living body positionally furnishes the possibility of conscious being. All conditions are in place here for the living being to notice itself—that is, its lived body—in relation to an outer sphere. As long as there is no physical center to represent the animal's own body with its limbs, glands, and digestive, sex, and respiratory organs, etc., the animal in its motor functions is still open. The idea of closed organization has not yet been completely realized, has not yet been extended to the area of motor functions, of action. Unable to notice its actions, in particular its movements in the surrounding field, the animal, because it cannot represent its own limb system, its lived body, forfeits the possibility of objectifying itself and thus the boundary between the zone it itself occupies and the surrounding field. As a result, the surrounding field also presents itself without boundaries or inner structure. As a pure sphere of noticing, it contains for the animal nothing but signs to which the animal reflexively reacts. The surrounding field is a pure signal field.

If, however, the organizational principle of the closed form is extended to the motor skills of the body and the circle of sensory-motor functions, whose unitary plan is the body itself, is closed again in the central organ, the animal comes to notice its movements in the surrounding field; it notices itself, its lived body, and the zone it itself occupies; the surrounding field with its own boundary recedes into the distance and becomes structured. Instead of losing itself in its actions and being fully absorbed in them without noticing anything about them, the animal now senses its gripping and letting go, its attacking and fleeing, its successful and unsuccessful movements. Now it is put in the position of steering its actions, of impulsively getting them going and arresting them, of controlling and modifying their course. Now it has gotten ahold of itself, just as it feels itself take hold of things in the surrounding field and feels this field interfere with it. The surrounding field presents itself as wieldy, no longer as a pure sphere of noticing but as a sphere of noticing *and* a sphere of effecting. It is a signal field and an action field in one.

What was once a field of pure instantaneousness, filled with signals that are glimpsed and then disappear and whose effectivity depends entirely on the drives and drive satisfaction, hunger and satiation of the organism, now becomes a field [253] of concrete presence. As an action field it offers "possibilities" and is a field of movements and grips still to be performed but also to be left undone. In the sensation of its own handiness, the living being has itself and its surrounding field in advance [*vorweg*]; it does not have instantaneous impressions but persisting and waiting givens. Just as this being, "choosing" and "wavering," occupies the inner center and void of the "not yet" in the eye point of noticing, in the impulse point of effecting, so the surrounding field is given to it with the same temporal character of awaiting. It is in this constancy that the structure of the thing quite properly lies, the form of reference for all of its sensory particularities.

If all intuitive thingness is characterized by the fact that it does not exhaustively manifest itself in what can be distinguished sensorially in the thing; if the thing therefore is more than the sum of its attributes, than the gestalt of its parts; if it is still more than it can be demonstrated to be (*nota bene*: for sensory intuition), then the requirement for the appearance of "things" in the surrounding field has been fulfilled here. Now the sensory data in the visual, olfactory, tactile, auditory, vibrational, etc., areas can be noticed as being related to a core, as being arranged around a core. As sphere of action, the content of the sphere of noticing objectively presents waiting opportunities for movement. In this noticeable wieldiness, this ability to deal with the surrounding field, to move, attack, and flee, to thrust and pull, shift and tug (as a property given in the sphere of noticing itself) lies the constancy and durability of intuitional things.

In addition to sense data, there must also be some kind of support, resistance, or background, itself not of a sensory, but of an intuitional nature, for their togetherness in a form, their shaped interpenetration, to be perceived as typically thingly. The unity of the intuitional thing is constituted by the, for instance, visual, audible, tactile contents of a structure assimilating and overlaying each other without concealing each other, remaining openly on display while equally revealing and disclosing the thing, all the while leaving intact a never-manifesting remainder, the "core" as the carrier of properties.

How could this be possible without a special schema? Are we to assume that there is some concrete intuitional element shared by a visual, a tactile, an [254] auditory quale (of intuition) that as structure, support, or background could form the basis for the unity of the individual thing? Indeed, there are commonalities between the sensory cycles of qualities, negative sensory characteristics of unity, a unity of intuition despite its specification,[5] even if it is still very much up for debate whether these characteristics can be considered, like Julius Pikler does, as effectively self-sufficient quantities.[6] But such shared characteristics, which are intrinsic to the sensory qualities, do not form the backbone of the thingly entity. This is why it is vacuous to claim that this nonsensory and yet intuitional "durability" of the thing is a work of abstraction based on many sensory, in particular tactile, lived experiences of pressure resistance.

That which appears as a structure of durability in the thing-formation is in truth its relationship to the motor skills of the living being that perceives it. The thinghood of held-together sensory content consists in this special schematization in relation to vital action. The ability to steer the movements of one's own body (as a result of the ability to sense them) and the thingly structure of the surrounding field correspond to each other. The centralized organization of a living body and the appearance of things in its field of noticing necessarily coexist.

In its sensation of the wieldiness of the things in its surrounding field, the living being is ahead of itself [*ist . . . ihm selbst vorweg*]. Because, however, the living being is in the mode of "having" its positional field and "itself" (as lived body), we must rather say that in its sensation of the wieldiness of things, the living being *has* itself and its positional field in advance [hat *sich . . . vorweg*].

Here we also have the reason for the appearance of "possibilities" in the field of noticing, as well as how we are to understand this (in contrast to the nature of sensation, inasmuch as it focuses on isolated, momentary contents [255]). The claim we just rejected, that the perception of things is derived from an abstraction based on many individual impressions, operates with the notion of possibility lying in the nature of the thing (the possibility of it turning, moving, changing, being destroyed, etc.). The idea here is that possibility is an abstraction. That, however, is not true. Possibility (we are not referring of course to the concept of possibility) is only abstract as an explicated state of affairs—for instance, the possibility of a cup breaking, of a human being becoming ill. The possibility immediately present in the sensory configuration of the formation, on the other hand, that which makes it into a "thing"—taking a seat on a chair, drinking from a cup or picking it up—this possibility implied by the phenomenon is not abstract. It rather comprises the starting points of living action.

Certainly, noticing and the sphere of noticing take on a different character here than they had on the level of the lower animals. There we find isolated, momentary contents, and the actions grasping them have the character of sensation. Here things, persisting and waiting complexes of a certain "approachability" appear, which leads to a change in the character of the actions with which the organism attends to them: they become a matter of intuiting. In the first case, noticing encounters sensible, substantive ultimate elements; in the second, sensible substances in a certain structure.

Of course, "sensing" and "intuiting" are ways of attending to the outer field that are only accessible to self-observation. Since, however, we saw earlier that the positionality of a particular type of living being has the same structure in its relationship to the outer field, it is not that we *may* use the same concepts here by analogy, but that we *must* use them. The consideration of positionality includes that which is accessible to self-observation insofar as it constitutes the relationship to the outer field.—

If it has been demonstrated that the centralized organization of a living body, the connection of its actions to the sensorium (thus rendering a closed loop of the sensory-motor functional play) corresponds to the appearance of things in the field of noticing, this raises the question of how these things become noticeable to the living being by way of the brain. This is an important question, for the living being is supposed to deal with these things based on its sensations and intuitions; indeed, it has to because its actions [256] no longer reflexively correspond to the stimuli but require steering in order to be "right."

The method according to which things and relationships to things in the outer field become noticeable to the living being is the same by which it becomes aware of anything that is alien or its own: by representation in the central organ. Just as areas in the brain come to represent the living being's own organs and functions, such areas emerge for the organization of the outer field corresponding to transmission by the sense organs. In the comparative anatomy and physiology of the central nervous system, brains can be classified in a hierarchy of increasing complexity, reflecting an ever-sharper grasp of things and their relationships to each other and the living being's own body, along with an ever-greater control of things as well as of the living being's own motor skills.

Naturally, representation in a spatial brain is itself spatial. Growing differentiation induces localization of functions; localization also implies typing, simplification, formulation. Such formulas or schemas provide a stable organization for the animal's fields of noticing and operation. The range of what can become noticeable to it and upon which it can have an effect is determined in advance for the type, if not for every detail. The "breadth" of the type is capable of great variation; in one case the schema could encompass a great many kinds of objects or movements and in another only certain objects and movements.

We should nevertheless not conclude that this method of spatial localization is entirely foreign to what it is meant to represent. Method and object, brain and the world of things correspond to each other particularly in spatial and temporal terms, albeit, of course, not in their measurements. In a certainly highly simplified form, the brain as a thing in space and time presents a small image of the surrounding field with the things in it. It first marks only simple differences in direction between left and right, up and down, front and back in the simplest contours, then gradually moves on to represent concrete matters of fact such as movements, images, and their location in the surrounding field. This is not to say, of course, that there is similarity in the sense of a strict depiction of external conditions in the brain; something that is on the left or the right *outside* of the brain does not necessarily have to be on the left or the right *in* the brain. And yet there is one fact that shows, more than all verifiable localization, the particular significance [257] of the spatial properties of the brain for the proper representation of the spatial properties of the outer field: the neutralization of the central organ in relation to changes in position by means of the development of organs of equilibrium and spatial orientation.[7]

Since the central organ is firmly connected to the body, the body's movements with their continuous shifting of the position of the central organ ought to interfere with the representation of the outer field, if a fixed position is in fact the *sine qua non* for the correctness of representation. According to Uexküll, who developed a precise understanding of the correctional process that compensates for this interference,[8] animals with an unstable sense of balance, whose position in life does not correspond to their position in death, require a constant corrective for the muscles in order to maintain their physiological position. This corrective is provided by the statolith organ, which automatically regulates the continuous relationship of the whole body to the center of the earth. It is true that eye mobility and, in particular, compensatory eye movement can somewhat offset the influence of the body's movements on the brain as well as extend the visual field; this alone may allow the living being to grasp intermediate spaces. But the form and position of an object remain uncertain under these circumstances, dependent above all on the function of the eyes and on the light.

This deficiency is only redressed by an organ of equilibrium of the structure of the semicircular canal apparatus, which constitutes a fixed system of coordinates in which the living being's own body and changes in the outer field are recorded like a fixed system of measurement in darkness and in light. Now the brain becomes "the shared field for all spatial measurements . . . which hold for both the eyes and the tactile organs as well as for the movements of the limbs."[9] The central organ has been neutralized in relation to changes in movement and shifts in the position of the living being's own body and of things outside of it. Physiologically speaking, this creates an absolute space, just as the regulation of the sequence of movements creates a physiologically absolute time. These are absolute only for the organism, but they provide it with fixed coordinates of its spatiotemporal position. With this restriction they correspond to the positional [258] character of the living being, which constitutes an absolute space- and timelike here/now. They correspond to this positional character in the same way that spatiality corresponds to spacelikeness, temporality to timelikeness.

Centralized organization, which essentially coexists with the organization of the outer field into things, also provides the living being with the means to become aware of this organization and to have an effect on it. This essential coexistence, however, does not guarantee the reality of the coexisting elements, of the organism and things. Self-contradictory as it would be to assume that an organism with the apparatus of a brain and sense organs would exist in a positional field without the things to which its whole body conforms, this absurdity is irrelevant to the question of whether these things exist, for the organism too is a thing in space and time. One cannot extrapolate from a thing to the reality [*Wirklichkeit*] or unreality of the world of things as such, of space and time. To conclude something along these lines from what has been set out here thus far is a sure sign of not having understood it.

A relation of coexistence between a certain physical organization and a field of perception and action of a certain nature, where the carrier of the organization lives, also holds for the visions of a sleeping world soul, for the fantastic creations of an artist-god. And yet if a living being sees things outside of itself and plays and struggles with them, it cannot but believe that they are real. There is always the possibility, of course, that it labors under an illusion.

The action of grabbing does serve as a criterion of reality insofar as that which is real [*wirklich*] must allow itself to be grabbed onto. But this method of assuring oneself that something is real [*Wirklichen*] remains *immanent to the sphere of coexistence of the organism and its surrounding field* and does not extend beyond it. It does not get at the coexisting *members* of the relation, which *themselves*, even if the intuitional conditions of the sensory-motor function-circle are fulfilled between them, if living being and things interact by seeking and finding each other, may "be" or not. Independent of each other—in balance with each other: it is in this form of relation that an organism and "its" outer field coexist.

Something that can appear to an animal and of whose actual existence the animal can become convinced by its own actions exists in the same sense as does the animal itself. Its embodied [259] being and the beings given to it outside of itself are on *one* level of reality. In this relation, no member outweighs another. Uexküll is wrong when he writes, "The environment as reflected in the counterworld of the animal is always a part of the animal itself, constructed by its organization, and processed into an indissoluble whole with the animal itself. . . . The environment is properly understood only as a projection of *its* counterworld."[10] It is true that nature does not force animals to adapt, but neither do animals form nature according to their needs. That would amount to zoological idealism, as it were, to the replacement of a world-creating consciousness with a world-creating organization, as Bergson did.

Following the principle of adaptedness set out earlier, coexistence can only be understood as the primary harmony and equiprimordial accord between spheres *separate* from each other. This separation, this hiatus, through which thing and living being mediate themselves (in sensation and action, intuition and action) to the immediacy of contact, forms the solid partition that Uexküll attributes to the stimuli of the environment. These stimuli serve the animal both as starting points and as a bulwark in order to be able to "enclose it like the walls of a self-built house and to keep the entire alien world away from it."[11]

Of course, one is tempted to find a way for this dual function of stimulus and response, dividing partition and immediate contact, to have room next to each other. That is to say, the function of contact is fulfilled in that which is positively accessible to the animal in sensory-motor terms; the function of separation in that which (thanks to its organization) remains concealed to it. This, however, is false. When sense organs, brain, and action organs make selections, they simultaneously posit contact and separation both in what they make accessible to the living being *and* in what they deprive it of. Concrete sensation and its ontic correlate are in themselves already a surrendering to reality *and* protection from it.[12] The presence of reality in sensory intuition and concrete action *is* to be set apart from reality and the exclusion of other possibilities in one and the same sense. Just as drawing means to omit and a clear image consists in the presence of certain lines and colors that by their existence exclude the presence of other lines and [260] colors, the current surrounding field corresponding to the organization is, in its contents and in its gaps, disclosed *and* concealed reality [*Wirklichkeit*].

This strange relationship of indirect directness, of mediated immediacy between the organism and the world, already expressed in the essence of the closed form and profoundly grounded in life's structure of being, does not rob the given sensations and intuitions of their character of actuality, does not simply make them into signs of an actuality entirely alien to and different from them, but does restrict their signification and that of their correlates—colors, shapes, sounds, qualities of touch, vibration, smell, taste, etc. that appear as absolute—to that of objective givens.

It is on this basis that we can more easily understand the relationships between the excitations of the brain and of the sense organs on the one hand and the consciousness of the living being on the other than from the point of view that identifies the bodily process of stimulus and excitation with the contents and structure of consciousness (even if this is done indirectly by way of a metaphysical theory of parallelism or reciprocity). Not to mention the absurd notion that the nervous processes are given to the subject (of consciousness) and that it is with this givenness that the subject arrives at lived experience—the experience of colors and sounds, of things in space and time rather than of nerve cells and their chemical changes, as if it were the task of the sensory apparatus to generate images of stimuli and that of the central organ to present these images to an inner observer.

Such nonsense, as well as the difficulties inherent in the theory of parallelism and reciprocity, are inevitable as long as dualisms alien to life corrupt our intuition of original positionality. Just as the brain and the sense organs cannot be held *directly* responsible for the existence of consciousness as such, its space- and timelike center and radius, they cannot be made responsible for the organization of its contents. An excitation of the retina, the optic nerve, the occipital lobe neither *is* the seen figure nor *means* this figure. It only corresponds to it, just as the nervous system responds to certain stimuli with certain excitations. This response posits, from case to case, the *interruption, inhibition, interval* (between stimuli and response), *which in positional terms is* the being [261] of a self in the central position—that is, its being "against something in the surrounding field" or its intuition of something.

The more differentiated the receptors and the brain, the more varied the evoked excitations and the more manifold the intervals and thus the structure of the positional field. Nervous excitations of the sensory (and motor) apparatus merely afford the living being the opportunities to occupy the central position as which and in which its conscious life plays out.

The Animal's Surrounding Field Organized into Complex Qualities and Things

For the progress of empirical animal psychology, whose methods for the most part strictly follow the logic of experimental observation, the problems of positionality invariably become relevant whenever it is the matter of interpreting the bodily behavior of the living being in terms of its state of consciousness. Without such attempts at interpretation, animal psychology is no different from the physiology of stimulus and movement or comparative biology in the sense of physiologically oriented life plan research as established by Uexküll. It is clear that an interpretive approach exchanges the safe path of causal explanation for an uncertain understanding of the conscious processes underlying bodily phenomena.

But interpretive animal psychology is not, or at least need not be, as faltering and arbitrary as the physiologists make it out to be. It is true that a particular measure of self-criticism is required in order to avoid the lazy anthropomorphisms of animal welfare calendars and tellers of fairy tales as they follow the law of the heart. Nothing positive is gained by self-criticism, however. What is needed is an objective disciplining of interpretation, beginning with a determination of the basic principles that must guide an understanding of states of consciousness.

Philosophers have thus far passed over this task because the Cartesianism of the two-world aspect prevented any reasonable solution. Not recognizing the zone of positionality that is posited with the existence of living bodies, they were also unable to see that a theory of interpretive principles [262] in animal psychology concerns categories of positionality.

As usual, however, research itself has not allowed itself to be influenced by the inner inhibitions of philosophy. Notwithstanding all epistemological concerns, it has attempted to grasp the animal not as a machine, but as a living center of action and to use concepts in line with life in its interpretations. Reflected here, of course, is the turn in recent psychology in general, which has abandoned the atomism of psychic elements and thus the allegedly verified doctrine of mechanical association as a hardly tenable generalization of certain marginal cases of memory and application of analytical and mechanical principles to psychology. This raises interpretation to an entirely different level.

In particular, the concept of instinct no longer functions as a simple antagonist to the concept of consciousness as it does in a mechanistic understanding of instinct as a chain reflex or equivalent thereof. Today there is the attempt to achieve an understanding of instinct that is more appropriate to life and to grasp it as a trend of behavior, certainly compatible with awareness, that is in place from birth and that possesses a particular "latitude." Within this instinctual latitude, actions take place without compulsion. Different animals exhibit different instinctual latitudes, with some animals being instinct specialists—that is, certain specific actions (nest-building, egg deposition, etc.) are fixed from the outset and are executed without prior experience.

In its particular type and latitude, instinct in respect to living behavior is what the whole body morphologically and functionally is for its organs:

the (itself non-isolatable) precondition, the framework, the selective principle for the parts of the world that become given to the animal, for its consciousness, for its surrounding field. Action determined by instinct thus does not necessarily have to take place without consciousness; instinct does not take the place of consciousness but rather forms and carries it. In the very same way, instinct as a general biological function determines what in an animal can become a reflex—that is, an unconscious, automatic reaction, thus withdrawing it from the area of living spontaneity.

For mechanistic thinking, the reflex was a component [263] of actions whose instinctive realization was made possible by chains of reflexes. This necessarily associated instinctive behavior with unconsciousness; consciousness was left with the area into which instinct could not reach and the mechanistic theorem had no ground on which to stand. The mechanist thus attempted to avoid the notion of consciousness as much as possible or to explain it as a "side effect" of mechanical processes in the reflex-mediating centers. The model of the reflex mosaic was to remain untouched.

Animal psychologists reject this primacy of the reflex and replace it with the primacy of general biological functions represented in the construction plan of the animal body and in its instincts. From this perspective, it is not a paradoxical outcome (as it must be for the mechanist) if, for instance, a slight rustling or scuttling sound has an effect on a lizard while a pistol shot in its direct proximity has none, or if mice find their bearings more easily in a "labyrinth" (a puzzle box with a maze-like layout) if the pathways are narrow and complicated than if they are wide and straightforward.

Methodologically, then, animal psychologists must ask biologically meaningful, instinctually appropriate questions and take into account the framework of consciousness if they want to learn about its structure. In situations that are alien to their instincts, animals will often appear to be "dumber" or "smarter" than they actually are, for now their actions appear against an artificial rather than a natural backdrop. If a dog is able to free itself from a cage with complicated machinery it makes a more intelligent impression than if it simply has to use its snout to push open a flap covering the exit. The "intelligence" of the experimental apparatus is wrongfully attributed to the laboratory animal. Of relevance is not that a door handle, ladder, hammer, pail, or rope was used, but only *how* it was used. Specifically human tools should only be employed in such experiments if they coincide with what is biologically possible for and consistent with the animal's nature.

The decision about particular sides of the structure of consciousness of course has to take into account what instincts are involved in the animal behavior in question. [264] Hans Volkelt's spider was verifiably hungry, but nevertheless behaved indifferently toward a fly if it did not appear in the spider's sensory field in the usual way—that is, if it did not fly into the web and, after a brief shock, attempt to free itself by violently shaking the threads.[13] If this sequence was not observed—if the fly, for instance, crawled onto the spider's web on the signal thread, right up to the predator's fangs, or even entered the web from behind—it was no longer a "fly," food, prey, but rather something unknown, "frightening," causing the spider to bolt.

These strange findings (as well as related phenomena reported by other researchers such as Hugo von Buttel-Reepen and Thomas Hunt Morgan) led Volkelt to champion the hypothesis of *complex qualities* in the structure of animal perceptual consciousness. The fact that bees whose hive has been moved only a few centimeters and thus remains safely within their visual range gather where the old hive entrance was and take a considerable amount of time to find the new one, and then only by accident; the fact that terns look for their nest precisely where it originally was even if its new location is only a slight distance away, of no relevance to these animals' excellent eyesight—these and other frequently observed dependencies of animals on a certain environmental *situation* are for Volkelt an undeniable indication that we have here a different *mode* of perception than we humans are familiar with.

Our perceptual world falls under the order of thinghood. For us, sensory givens are "attached" in their properties to relatively invariable and solid thing-bodies; they are like shells crowding round their cores. Despite their different aspects and varied constellations in relation to each other in different situations, the same things form the relatively lasting substances and carriers of our world. An intuition not sicklied over with the pale cast of thought can never call into doubt their objective existence and autonomy.[14] For us, if we notice them at all, a fly remains a fly, a nest a nest, even if they present different aspects on different occasions. In our perception, an appearance already presents itself as an "open gestalt," as a multiplicity suited to all manner of variations [265]. It is this conjunction of givenness and hiddenness that distinguishes the "thing" from the pure phenomenon, as is perhaps best illustrated by the rainbow. We are thus not surprised to find a thing in a different situation; because it is spun for us from "core" to "core," we do not lose the thread of the appearances.

Volkelt attributes the fact that animals so easily lose the thread and that they can be deceived by entirely irrelevant changes in their environment (see also his description of George Romanes's work on sand wasps), while on the other hand there are cases of astounding memory performance found only in animals (as is shown, for instance, by the spatial memory of bees, homing pigeons, and migratory birds, to say no more of the great trainability of many animals), to a kind of animal-specific perception. This kind of perception, he argues, is just as incapable of grasping thinghood as would be a form of consciousness privy only to atomized information about its surroundings based on individual sense data.

> We have seen, then, that animals are not adapted to the occurrence of individual things, but rather . . . and this is a baffling fact, that it is the *overall situation* and not the *individual* thing that determines the actions of animals. . . . So what is the structure of the modifications in consciousness that are elicited by the surroundings of animals? . . . Does highly developed consciousness not contain a known type of qualities . . . that allows us to explain in psychological terms the fact that the animal's actions are tied to whole situations as a reaction to a single quality encompassing the whole complex of the given? . . . These peculiar *single qualities of a multiple complex* are referred to in recent psychology as *"complex qualities."* . . . That there are in fact such properties only belonging to the concrete psychic whole revealed itself . . . in the striking observation that different complex wholes accord with each other in properties that do *not* inhere in their *parts*. In particular it was the transposability of *melodies* and the similarity of *figures* that led to this "discovery."[15]

Volkelt's engagement with Christian von Ehrenfels's concept of gestalt qualities in particular leads him to the following definition [266]: "The complex quality of a whole is not the sum of the parts really contained in the whole as such; second, the complex quality of a whole does not exist as a quality standing on its own next to the sum of the parts of this whole (as Ehrenfels thought)."[16]

This concept solves Volkelt's problem:

> On the animal level, the field of sense data is not organized into thingly formations; nor is this field split into atomistic elements. Rather in each case a broad field of sensory givens is . . . encompassed by *one* quality . . . containing everything. This quality "contains" parts . . . *only insofar* as an analysis would bring them to light, while the side of representation in human consciousness is already organized prior to any such analysis. . . . Therefore: . . . The actions of the primitive organism are attached to the occurrence of certain overall complex qualities encompassing much or even everything, including motor-visceral and emotional content. . . . That is what, in psychological terms, is inside the strange behaviors of "primitive life forms."[17]

According to Volkelt, sensory "melodies" and "configurations" dominate the animal environment in contrast to objective "things" surrounded by a horizon of infinite aspect possibilities that are given only to the human. It becomes very clear why such a rigidly circumscribed manner of orientation, why behavior that barely distinguishes between subjective-affective and objective content, has to be guided by instincts, which relieve the individual of the task of providing for itself, entrusting it instead to the process of living.

Volkelt's work seemed to reintroduce in the psyche the essential boundary between human and animal that nineteenth-century evolutionary theory had abolished. Köhler's intelligence testing of anthropoids, however, at least showed that Volkelt's hypothesis could not, as the latter intended, be attributed to the entire animal kingdom and that the essential boundary between the human and the animal psyche is located "higher" than Volkelt believed it to be.[18] Köhler's experiments allow us to salvage what is valuable in Volkelt's studies, which had of late been threatened with invalidation as a result of Fritz Baltzer's [267] repetition of the spider experiment.

It was Köhler's goal to find proof of primitive displays of intelligence in the animals anatomically, physiologically, and phylogenetically closest to humans—in anthropoid apes, in other words—in order to demonstrate a likely continuity between animal and man in the domain of psychology. Such an experiment requires a valid concept of intelligence that is not tailored to the human as well as reliable criteria for detecting it in the observed behavior.

Köhler satisfies these requirements. Intelligence, he says, has to show itself in the overcoming of a difficulty. This at least is what we call "intelligence," as long as we are certain that the difficulty was grasped as such and overcome by selecting a solution from a variety of possibilities. Only in that case does successful action demonstrate insight into the situation. If, furthermore, the course of action is characterized by an initial, tentative this way and that followed by an abrupt stop and ensuing implementation of the solution at one stroke, even the impression created by such a state of affairs points to insight, as it is expressed for instance in the words "Aha!" or "I've got it."

The simplest case of difficulty (not only by human standards) is provided by a detour a living being must take in order to arrive at a goal. By "detour" I mean quite intuitively any curve in the path that deviates from the shortest, most direct line between living being and goal. The direct path leads straight to the goal (as long as the stimuli in the surrounding field are otherwise in balance) in a natural frontal tendency that is the organism's primal orientation by way of the sense and movement organs. If the animal is hindered in its movement in this direction but nevertheless finds its way to the goal via a detour, it demonstrates "intelligence" ("it was able to help itself"), provided that, besides the obstacle, the remaining surrounding field is accessible and visually given to it. Varying this principle, Köhler's series of experiments progresses from simple to complicated tasks.

Preliminary experiments with chickens and dogs as well as with a fifteen-month-old child showed that even the simplest detour constituted a problem that could not be solved instantly or consistently. Even the child, who, it is true, was quickly able to [268] master the situation, still had to "find" the solution. The dog found the solution immediately if the desired object was thrown in a high arc over the impeding fence but failed if it landed directly in front of it on the other side of the fence—in this case the dog was not able to "tear itself away." Chickens need a long time to find the solution, which even then remains uncertain.

There is no need to go into further detail on Köhler's experiments in this context, especially as they quickly became very well known. They prove that anthropoids are capable of creating an *indirect* link between themselves and a goal that can be relatively richly differentiated into stand-alone subsections. In addition to the rather simple establishment of contact with the goal by means of ropes and sticks, the animals, especially the more capable ones, were able to implement rope-and-stick combinations, build structures with boxes, and finally combine distinct intermediate goals—in other words, perform actions that in their individual parts appeared to be futile in relation to the desired end effect (for instance, "sticking two reeds inside each other," "pushing the desired object in the opposite direction to the one one wants to approach it by," or "turning one's back to the goal").

Such segmented action, whose elements taken individually do not have a direct relation to the goal in order that the aggregate can attain such a relation, clearly shows all attributes of intelligent behavior. By performing action segments that in themselves run counter to its goal, the animal documents its grasp of its surroundings' field structure. It introduces things to fill in gaps in its path, combines them with each other, and correctly includes its own lived body (arm length, climbing ability, sense of direction) in the calculation—uses it, in other words, like a tool, something that does not occur to horses and dogs in similar situations, even the most simple.

This highly significant evidence, however, is not the most interesting thing about Köhler's experiments. Of greater interest is his observation of certain characteristic *weaknesses* in chimpanzee intelligence that do not seem to correspond to their abilities. The fact that the most capable of the examined animals, Sultan, discovered while playing that he could stick two reeds into each other at the precise moment when they happened [269] to be lying more or less behind each other in his line of sight, in no way discounts his genuine grasp of the utility of this extension, which afforded him the length needed to reach the desired object, inaccessible with only his arms and each individual reed. The latter connection he did not arrive at by accident but made himself. A genuine grasp of what he did with the two reeds is, however, called into question by the fact that he forgot about the solution under certain visual conditions (when his handling of the two reeds led them to be roughly parallel to each other).

The question of whether such a great dependency on a certain visual situation (as is the main issue here) can be attributed to a "weakness of form perception" [*Gestaltschwäche*] as Köhler believes, or whether this weakness of form perception is not in fact a symptom of a qualitative "lack" on the part of chimpanzee consciousness in relation to human consciousness, deserves serious consideration. Clearly one should not categorize failure in dealing with complicated forms (such as the rod-rope-ring-nail link or the rope wrapped around a pole) on the same level as when the animal betrays its complete lack of understanding of the "statics" of a box structure, a single box, or a ladder—a lack that even the most disastrous experience is not capable of correcting. "Simple gravitational physics," which after a few experiences already communicates itself to us in our visual perception of the thing and stops us from wanting to freely "stick" boxes to a wall, does in fact presuppose more than exact, good form perception—just as it cannot be attributed to issues of form awareness alone if a chimpanzee does not recognize a box pressed into a corner between walls as a box, a ladder propped straight up against a wall as a ladder.

These strange failures must rather be brought into relation with the fact—duly emphasized by Köhler himself—that the animals are evidently not familiar with the possibility of removing obstacles and are at a complete loss when faced with a task that can only be solved, even in the simplest way, by removing an obstacle, even of the simplest kind. Even the most capable animals were not able to properly remove obstructing rocks in a box or the box itself when it blocked their way to the desired object. They may be ever so good at [270] mastering the situation by positively using (incorporating) positively present things in the positively given, current field structure—and here they can cope with considerable complexity; they are certain to fail if the goal is only attainable in a negative way by removing something that is given.

The most intelligent living being in the animal kingdom, the animal most similar to the human, lacks a sense of the negative. This is the definite result of Köhler's studies, a scientist who was certainly not biased against animal intelligence or toward an unbridgeable essential difference between the human and the animal.

Here we have the key to understanding the failures in the observed behavior of the anthropoids that Köhler attributes to a "weakness of form perception" and at the same time the starting point for correcting the interpretation of his results. This correction brings them closer to Volkelt's findings while at the same time providing the latter with a theoretically unassailable foundation.

Genuine things as perceived by the human are characterized in their intuitive pictorial presentation [*Anschauungsbild*] by a surplus—a surplus of invisibility in relation to the real [*reell*] intuitive state of affairs; in other words, a surplus of negativity. We expect a genuine thing to have a back side, hidden sides in general, which form a fixed system of perceptual possibilities. A thing is only concretely present in a positive sense if what is intuitively present appears to be enclosed in or firmly tied to a fixed order of non-present elements. As phenomenological analysis has shown, this fixed order (which we have encountered more than once already), this structure of space- and timelike core and mantle, represents for the standard of sensory sensation, as it were, the absent itself. It is on this order that objectivity or the genuineness of things rests.

It is precisely this objectivity that remains inaccessible to animal consciousness, even in the highest animals. For animals with centralized organization, the thing in the surrounding field remains a correlate of the sensory-motor function-circle, starting point of stimuli, and point of application of its actions. As this durable formation of target surfaces, it shows relative constancy across changes in sensory aspect. Alternating impressions of different sensory types share one undertone: that of durability and tangibility. The play of impressions is based on this undertone; the abundance [271] of sensory data clusters around it and finds in it its unified point of reference.

Certainly, animals with a central nervous system do not simply perceive "melodies" or "configurations" organized into complex qualities, as Volkelt claimed, but rather closed, individual, relatively invariable, complex things. But these things do not have the character of objects, since they remain relative to the overall vital system of the living being, even if not only to the senses. There is no denying that substantial sensory congruence with the intuitive image of the thing, as the human has, can also obtain for animals. In the latter case, however, it has another "value." The animal faces the thing as a formation in the outer field, persisting, stable, workable, but not as an object, a matter in its own right. It essentially remains an action object and, in a perceptual sense, highly drive-bound. Anything that is not connected to a drive is weakly and superficially perceived and in some cases disappears entirely.

The animal perceives things whose core structure has motor significance and that corresponds to the animal's actions, is "meaningful" in relation to them. It has not yet awakened to the substantive character of the object, has not yet grasped the complete detachability of things from the sphere of perceptions and actions, has not yet noticed their inner self-sufficiency. It has not yet acquired a sense of the negative in any form whatsoever. Absence, lack, emptiness are intuitive possibilities closed to it.

In the relationship between living being and surrounding field characteristic of higher animals, neither the side of the subject nor the side of the field is set apart from the plane of consciousness, which itself does not become content. The subject is hidden to itself: it only has its lived body and is fully absorbed in the space- and timelike centrality of subjective life without experiencing it; from its own perspective, it is pure me, not I. So too the surrounding field in its boundaries is hidden to it, which (for the outside observer) is finite but not bounded (for the animal subject).

Consequently, any kind of intuition of homogenous emptiness in space and time *must* be denied the animal, which in turn allows us to understand the weakness of form perception Köhler observed in even highly developed animal consciousness as an essential predicate of the positionality of the [272] closed form. Failures in overcoming obstacles by removing given elements in the surrounding field and the phenomena of form perception weakness find their a priori explanation in a common root.

Volkelt's theory is justified only for animals without centralized organization who do not have a central representation of the network of effecting. "Melodies" and "configurations" are forms of order, corresponding to signals arising freely in the surrounding field, of a consciousness that encompasses neither things nor the animal's own lived body as field of action. But Volkelt's theory cannot claim validity for the whole domain of the conscious life of the animal. The higher animals, whose own lived bodies are represented in a central way, also experience things as correlates of their motor skills. Dependency on drives, weakness of form perception, the propensity to be easily deceived and disoriented, and a strong adherence to situations are thus necessarily related.

Intelligence

At the same time, lacking a sense of the negative creates a barrier for animal intelligence. Just as consciousness of the object as a thing [*Sache*] eludes the animal, so does consciousness of the state of affairs [*Sachverhalt*]. The animal only grasps field conditions [*Feldverhalte*] (which for the human are naturally given as states of affairs). Field conditions are structural relationships between existing elements in the surrounding field. The relationship between the goal and the living being's own body in the frontal orientation a priori to life (in the aspect of the frontal tendency natural to the animal) forms the measure for every other relationship between elements given in the surrounding field, the measure for certain difficulties, the principle for overcoming them. The animal cannot free itself from this relationship inherent in frontality; its entire perceiving and active life finds its orientation in it.

Köhler was certainly right to attribute genuine insight to the animals in his experiments. They grasp the difficulty in the given field structure and overcome it by choosing from among the possibilities it contains. It is only that this type of insight differs from human insight into states of affairs. In the case of the animal, it remains the ascertainment of form, a survey of a complex of given elements in the surrounding field.

[273] Human insight is often no more than this, either, such as when there is a knot to untie or some other kind of disorder to be removed. Even in the same situations in which the chimpanzees found themselves, a human would at first have to overcome the same "field problem." He would then, however, begin to reflect in such a way as to objectively address the field findings rather than treat them as a simple situation.

Humans, it is said, are able to do this by means of abstraction and conceptualization, which elude animals. This is true to a certain degree, but places the boundary of animal intelligence too high. Abstraction with the goal of conceptualization is based on an action Husserl called "ideation" and already implies a disengagement from the individual elements of intuition, an objectification of simple immediate givenness. In order to subsume a thingly formation under, for instance, the conceptual unity of "ladder," prelinguistic, schematically intuitional "ladderness" must first be grasped as pure gestalt, to which a thousand variations could correspond. This framework, which holds a wealth of concretely sensory, single gestalts but not just anything, is nothing but a sharply bounded void, a delineated negative, which—like every schema—is individually filled, is replete with certain gestalts. Because animals lack a sense of the negative even in sensory intuitiveness, ideation and thus conceptualization elude them.

Sensory abstraction as the grasping of similarities is, however, not affected by this. It requires only the ability to perceive form or intuit complexes, which in fact plays a vital role in animal consciousness. Thus Buytendijk's experiments with dogs and Johan Bierens de Haan's with monkeys have shown how easily animals can be trained to recognize transposable gestalts and how they can be brought to separately intuit round, pointed, square, etc., shapes as such. Because of its inability to inhabit an objective mindset, *animal consciousness tends to* disregard the individual thing—which of course never appears to it *as* something individual *with* the as-character, but rather remains connected to the structure of the overall field—and to focus its attention on the transposable features of a situation. [274] The individual entity of course also appears in the perceptual field as a component of the situation—that is, as in some way of motor significance and conditioned by drives.

In the past, animal psychology fostered the notion that animal consciousness was a scene of chaotically whirling sensations and perceptions, given a scanty enough order by the machinery of association based on contact or similarity. Accordingly, the animal could experience the individual but not the general. Once, however, it was recognized in psychology that a separation of this kind between purely sensory particulars and purely nonsensory generalities does not exist in concrete consciousness, that these are boundary cases construed by psychologists for theoretical purposes, the claims of animal psychology had to be revised. It became clear that every level of consciousness relates to the individual as well as to the general, that in primitive form they are not separate from each other and only come to oppose each other on the highest-known level of human consciousness.

On the lowest level of consciousness, the sphere of animals organized in a decentralized way, we find a form of intuition that is organized into complex qualities. This form provides no possibility at all of submitting its sensory "melodies" or "configurations" to the individual/general dichotomy. They are just as much both as neither. Color data, tactile data, olfactory data, and auditory data in rather rigid conjunction and in narrow limits of intensity and vividness determine the behavior of the organism not as individual data but as their complex quality, greater than the sum of its parts. The singularity of the elements has, if one will, entered into an indissoluble connection with the generality of the gestalt quality. The extreme particularity of the stimulus components is coupled with the high degree of transposability of the overall stimulus.

While biologists and psychologists once thought of the most primitive form of consciousness as a disjointed chaos of individual sensations that acquired a certain schematic order only by means of the organism's drive direction and its individual experiences (with the aid of the associative mechanism), modern studies and a priori analysis both show that on the most primitive [275] level, the difference between individuality and generality, concrete and abstract, cannot even be made. The complex-qualities mode of intuition saves the lower living being the trouble, as it were, of getting involved with the individual, concrete components of its situation, which it could only cope with if it were able to grasp what is typical and recurring about them. The orientation toward complex qualities is nature's expedient for living beings who are unable to grasp generalities and abstractions: it also denies them the consciousness of concrete individuality and gives them instead a consciousness that holds the undifferentiated middle ground between individuality/generality and concreteness/abstraction.

On the level of intuition organized into things—the sphere of animals with centralized organization—, which corresponds to the sensory-motor function-circle as a whole, the situation of the given field structure forms the perceptual framework, while the elements given in the field form the perceptual content. As a result, each individual thing's particular aspect separates itself from the formation that stands over against the animal subject as the durable correlate of its motor skills. The perceptual framework appears as the current backdrop of possibilities, of given possibilities of movement and gripping, bound together by the unity of the situation. The concrete and individual are thus embedded in an open unity, the finite, unbounded surrounding field, which in turn is not present in a completely objective way but, due to the animal's natural frontal tendency, remains indissolubly interwoven with the animal's conditions. The animal grasps neither the individual element against the backdrop of the field structure as individual nor the field structure as open unity in contrast to the individual field element.

For animals with centralized organization, then, there is on the sensory-motor level an analog of the opposition between concrete individuality and abstract generality, but not the opposition itself. Insofar as the individual concretum of the thing is gestalt and point of application, it is also there for the animal; insofar as it is subsisting actuality in itself, enduring matter, genuine object, it remains hidden to the animal. Accordingly, the general contextual relationships of things, their typical, invariable characters, only figure in animal consciousness insofar as they are motor equivalents. If it belongs to the essence of abstract generality to be a delineated [276] wealth of possibilities or an open unity (in distinction to the closed unity of the concrete thing), then animal consciousness is familiar with abstract generality—as the currently given wealth of possible movements or as "gestalt."

Genuine individuality and genuine generality, however, presuppose the ability to grasp the negative as such, the absence of something, lack, emptiness. There is thus an essential coexistence between the homogenous intuition of space and time, between hollow space and hollow time with gaps "demanding" to be filled with invariable elements, and the genuine objective perception of things and genuine ideative abstraction. This presupposition is only fulfilled in the human. Individual *and* general, conceptual and material generalities are known only to the human.

Köhler's discovery implied that, contrary to prior assumptions, intelligence was no longer to be a prerogative of the human, and this led to the concern that the essential boundaries between the animal and human psyches were to be abolished. The objection was raised from several quarters that Köhler's experiments did not demonstrate genuine insight; that their experimental design and the tasks posed to the subjects merely elicited accomplishments that could be attributed to the instinct specialization of tree-dwelling animals. Johannes Lindworsky, for instance, argued that the seemingly insightful handling of sticks, ladders, boxes, etc. could be ascribed to the fact that aids of this kind have for the animals the "functional value of branches" and that branches, twig combinations, and the like are, as it were, built into their system of instincts. An ingenious interpretation, certainly, but one that neither does justice to the true behavior of the animals as they solve the difficulties posed to them nor to their relatively extensive awareness. Genuine insight, whose occurrence, for instance, in dogs is comparatively easy to demonstrate, is certainly possible in higher animals. It is only that it is insight not into states of affairs [*Sachverhalte*], but into field conditions [*Feldverhalte*], insight into a certain structure or situation in the surrounding field.

Striking proof of this limit in animal intelligence was provided by Buytendijk and Géza Révész when they tried to train monkeys to grasp a particular sequence. They set up a row of small boxes, and at a certain interval (such as every second, third, or fifth box) they placed into them a piece of chocolate, a banana, or the like. Even very young children [277] were able to figure it out after only a few tries. The monkeys, on the contrary, could not be made to do so—they failed completely. Their consciousness is unable to grasp a matter of fact such as every second, third, or fifth, since it is only comprehensible as a state of affairs, not as a field condition, a structure of the surrounding field. To be sure, there may be cases in which comprehension of this kind appears to take place, such as when the sequence is highly structured and at the same time relates closely to the organism's frontal tendency.

Memory

The correctability of responses by the past plays a crucial role in the habitus of the highest as well as the lowest animals. This correctability is found only in animals, not in plants. Animals can learn; plants cannot. Experiments such as Erich Becher's, mentioned earlier, aimed at finding evidence of the power of association in the carnivorous plant *Drosera rotundifolia*, sundew, produced entirely negative results.

It is true that phenomena of functional adaptation are common to the entire realm of the organic. The same response is faster and better the more often it occurs. Likewise, all living beings experience fatigue. This immediate influence of earlier responses upon a current response does testify to a presence of the past in living substance, to memory as a general function of living matter, as Ewald Hering proposed—but the response is not qualitatively changed by the past, is not *corrected by* it.

Köhler was right not to include the feature of correctability in his concept of intelligence, intended to correspond to a particular habitus, even if this runs counter to the general tendency. Herbert Spencer Jennings, for instance, found that the protozoon *Stentor* reacts to the stimulus of powder falling on it first by simply turning to the side, then by reversing the movement of its cilia, and finally by swimming away in order to escape the stimulus. In a repetition of the experiment, it swam away immediately without applying the other responses first. This gives the impression that the animal is behaving in an intelligent manner; it knows how to adapt to the situation; it has learned from experience. But how do we know that the animal is actually aware of any of this? Its behavior [278] can be explained up to a certain point in terms of an association of the effect of the last appropriate response with the stimulus, without needing the help of consciousness. Reflection does not necessarily have to enter into it; to say nothing of the fact that the organization of a protozoon gives us no occasion to attribute its movements to conscious initiative. If, like Köhler, we reserve the concept of intelligence for accomplishments brought about by insight, it can only be applied to animals of the centralized type.

On the other hand, there is no denying that intelligence goes hand in hand with the development of the capacity to gain experience—that is, to correct responses as a result of the past. We must therefore think here in terms of the capacity of association as a characteristic moment of the animal's relationship to its surroundings as such, avoiding for this purpose the narrower concept of experience tied to consciousness and initiative and, following Driesch, apply the neutral concept of a historical basis of reacting.

Movements "which depend . . . for [their] specificity on the individual life history of [their] performer in such a manner that this specificity depends not only . . . on the specificity of the actual stimulus but also on the specificity of all stimuli in the past, and on their effect," are, according to Driesch's definition, actions.[19] Now there are machines, such as the gramophone, whose responses depend upon their individual past. They cannot, however, *revise* the *specific* correlations of their past; they retain it just the way they received it. The organism, on the contrary, "has the faculty of profiting from the specific combinations received in order to form other combined specificities." The historically created basis of acting thus does not resemble the gramophone record with its fixed system of engrams, "but consists of the elements of the experienced specificities."[20]

Driesch sees in this condensation of a past *shattered into elements*, as it were, the precondition for the formation of new combinations, upon which the active life of the organisms rests. He makes use of this [279] in connection with the phenomenon of the individual coordination of stimulus and response, discussed earlier, as the foundation for his third proof of the autonomy of life.[21]

Even if not all animal groups have been shown to perform actions—that is, their movement responses are correctable by means of the individual past of the organism—, there is nevertheless no doubt that actions are only characteristic of the animal. This fact must be based in the nature of the closed form and is clearly to be understood as a function of the timelikeness of living being, taking into account the special positional characteristics of closed organization.

Living being persists in becoming by being ahead of itself. It is present insofar as it is coming, as the basis of its foundation lies in the future, from the future, as it lives "in anticipation." Only in this "retrogression" is it posited being; only here does it show the positional characteristics of the space- and timelike union, confinement in the absolute here/now, self-sufficiency.

All living beings in general, whether plant or animal, *are* their past in a way that is inherently *mediated* by the mode of the future—that is, retrograde. Herein lies the difference to the inanimate formations. A mineral, a mountain, an entire landscape are also their past, but they are immediately formed by it; they are made of it. That which is alive, on the contrary, is more than simply what it was: it is that which is ahead of it. Its past thus immediately eludes it, and it actually fulfills a present. Living being runs ahead of and recedes toward itself in a genuine presence, and as such it persists.

It does not, however, persist *as* a becoming, and yet its becoming is the prerequisite for its persistence. How can this contradiction be realized?

It is resolved by the fact that living being *preserves* its past, its having-become, by *setting itself apart* from it. It persists as that which it has been without this defining and exhausting it. We can thus say that in this respect that which is ahead or is alive *is* still "its" past in the present insofar as this past lies *behind* it. Passing over into a new condition, life offers the possibility of its now confronting its then [280] and the present condition becoming distinguishable from former ones. Living being *in* the process of becoming is its past only because it *was* its past; it is what it has been only by having been it, or (since it remains connected to its past in the form of preserving) it has its past "behind" it.

We can thus understand the now as the separating/binding between coming and going, as the saddle, to modify a simile of William James's, on which living being rides out of the past forward into the future. That which exists in the *now* is therefore only mediately formed, through its *becoming*, by the *past*. The fulfillment of time broken down into three parts according to the modes of future, past, and present as being that, in a retrograde manner, is its past by virtue of its anticipatory, advance structure, is specific to life, whether plant or animal.

Living being in the closed form is present to itself. It stands in relation to the positional characteristics it exhibits in and of itself. In accordance with central representation, it has a surrounding field and its own lived body. The fact that it lives in anticipation, from the future—its positedness, in other words (it is tied to the center and stands on its own)—not only (immediately) forms the individual, but also (mediately) in such a way that the mode of its formation at the same time appears in that which is formed. This means that the closed form is on a higher level than the open form. The central core, point of reference for the living being centrally tied to it, becomes the current through-point of a *relationship* starting from and leading back to it, whose *execution* by the organism is what constitutes the life of the organism. A living being organized in a closed way is only alive insofar as it is actually absorbed in this central self-mediation, insofar as it comes to itself. As this reflexive *relation*, the living being is present and present to the living being that it is; in this reflexive relation it experiences "itself."

Because a living being organized in a closed way is present to and experiences itself in a *reflexive relation*, it belongs to its essence for it to experience its past or to have a memory—that is, a relationship to the past that can be conscious or extra-conscious. To have a relationship to the past, however, is to stand at a distance from it. Since its life, [281] in a positional sense, is based on being the current through-point of a relationship starting from and returning to itself; since it is only alive insofar as it is absorbed in this central self-mediation, in this coming-to-itself; since its presence to itself (the content of which is contingent upon the degree and type of bodily representation) is identical to this reflexive relation—the central passing-through of the mediation of the past must as such be at a distance, set apart from, and at the same time continually related to the past.

And thus emerges an in-between as current center, as the now of the living being's life, behind which lies the past, ahead of which lies the future. The position of frontality, characterized in a space- and timelike way, in actuality refers to the summit from which one can go backward into what has been and forward into what will be.

The presence of closed-form living being is executed in the relationship between what will be and what has been. The past is thus not present simply in the sense that it forms the current condition of the individual. *In this regard*, there is no difference between the plant and the animal. It is rather that for the animal, the past is in the mode of the mediate form of *per hiatum*—that is, is present to it. Extra-consciously, the past affects the animal's current condition without completely determining and fulfilling it; consciously it affects this condition by the subject going back to itself, finding itself to be the past that now influences its initiative.

We can thus understand, a priori, the correctability of animal movements as resulting from an individual past that still leaves open a surplus of possible and actual movements the individual can or does carry out. Otherwise there would be no leeway for "corrections." This surplus is only ensured if the current condition is not swallowed by the past, is not merely mediately formed by it, but occupies its own sphere, separate from what it has been and yet not completely isolated from it. This kind of setting apart between now and then, at the same time bringing the two spheres into contact with each other, is precisely synonymous with mediated presence.

Generally speaking, the extra-conscious or conscious influence of the past on the present condition can be understood to be a function of the positionality of the closed form. At least it rids the matter of the paradox commonsense philosophy [282] finds in the effectivity or existence (in conscious remembrance) of what has been. To resolve the paradox, this school of thought devised theories according to which memory is the material trace of prior impressions and the principle of association governs its re-evocation in the present.

And yet it does seem futile to attempt to understand the actualization of memory in individual cases without the notion of material traces or of a central mechanism (not to mention the fact that the findings of physiology and pathology speak a very clear language). Just as the configuration of the perceptual field depends on the filtering function of the nervous system, so, it seems, does the configuration of the historical basis of reacting. The specific combinations received are broken down into elements, which are then used for the formation of new combinations, and this corresponds to the analyzing filter function of the nervous apparatus.—Bergson's *Matter and Memory* put forward a closer interpretation of the facts surrounding aphasia, apraxia, etc., contradicting older approaches and influencing the work of the younger generation (Kurt Goldstein, Adhémar Gelb, Abraham Anton Grünbaum, Arnold Pick, and others); it would go too far for me to attempt such an analysis here.

While the existence or nonexistence of a historical basis of reacting or a faculty of association ought not to be taken as an empirical criterion of animal or plant nature, its occurrence is essentially tied only to the animal form of organization. This does not obliterate the strict difference between animate and inanimate formations in terms of their relationship to what they have been. The plant too is its own past in an only indirect way. As an open form, however, the plant has no possibility of entering into a relationship to the past it preserves in itself. A relationship of this kind is specific to animals. It can manifest as the extra-conscious influence of the past on animal responses (consider Pavlov's discovery of the "psychic" secretion of saliva and how this has been used to analyze dogs "psychologically"), or this turn of the living subject toward itself can take the form of a conscious influence on its initiative—but it always presupposes the distance of the subject, which distinguishes the position of the living being with a closed form of organization.

[283]

Memory as the Unity of Residue and Anticipation

On every level, life preserves its past only by virtue of the fact that it is ahead of itself. What it was has behind it something that exists, insofar as it is out beyond itself. Being based in the future is the possibility of having and keeping a past and, in those cases where the organizational form establishes a relationship between the living being and its positional characteristics, at the same time the possibility of the living being having this past at its disposal and thus making use of it. A gramophone record ties the apparatus to a particular kind of rendering, while the historical basis of reacting ties the organism to elements of the past only in the very wide and variable boundaries of combinatorics. This is because the organism's position in relation to its own having-become, which does not allow it to be immediately absorbed in it but only mediately makes it into what it has become, itself moves into the past along with the entire content of past life and, as form, remains interwoven with this content.

If life immediately passed over into what it has been as inanimate things do, if the historical basis of reacting was simply formed by the sedimentation of the occurrences brought by time, the current flowing into the living being and collecting there—then it would be strange indeed that accurate, "mechanical" reproduction requires the effort of monotonous repetition. If this were the case, there would in fact be such a thing as engraving or sedimentation true to the original, an impression method if you will, and the elements breaking down, as is characteristic of the historical basis of reacting, would be a matter of deterioration setting in after the fact, attributable to fleeting lived experiences and forgetting (itself rather mysterious).

Evidently, mechanical reproduction constitutes a boundary case of memory, which, as any kind of memorization or animal training shows, can only be realized by means of artificial concentration, by the suppression of other impressions and reactions. The mechanization of reproduction requires that the contents to be reproduced are forced out, in a sense, of their natural position [*Lage*] in the living whole. Normally, the relationship between living subject and memory goes beyond the former's futurity—that is, is mediated by an inclination and is taken up into memory itself as the form [284] of the retained content; in ordinary life everything that brings the living being into contact with what will be—drives, interests, volition—promotes and carries the memory-forming, preserving function. By contrast, animal training and memorization are aimed at restricting this mediation of inclination as much as possible and, by means of the monotony of repetition, at rendering effective the pure imprinting tendency alone.

Because all memory formation—that is, the emergence of a historical basis of reacting—is essentially mediated by inclination, there is in a primary sense no such thing as the complete retention and impression of experienced events in the chain of their actual coherence (which would gradually break down into elements only due to the fault of the subject—its hastiness, forgetfulness, and other deficiencies in its organization), but it is rather that *becoming the past and being destroyed into elements* is one and the same thing.

To put it figuratively: the living subject must penetrate into the finest meshes of the fabric of lived experience and break it down if it wants to take possession of the whole. The "gaps between the fibers of the fabric" correspond to the mediated relationship of the living being to itself. The fact that the past was not the way it was, but rather forms the historical basis of reacting in fragments and elements that can be shifted around in relation to each other, has its inner cause in the interruption or interval that constitutes the existence, mediated by self-realization, of a conscious living being. Precisely because it is the inner interval or interruption between that which is alive and itself, we could also say that self-realization, upon which the possibility of consciousness rests, works like a sieve whose pores pulverize the matter passing through it as it becomes the past.

It is a law for the animal to appropriate what in the course of its life befalls it. This appropriation, which rests on the animal's relationship to itself, on its distance to or interruption of its own life system, also necessarily entails the fragmentation of the appropriated content. As a result, the appropriation remains interwoven with the animal as form, as the form of being broken through, which both allows the elements of what has been to be shifted and rearranged as well as induces forgetting and the errors of memory. As indicated earlier, the actualization of this pattern is [285] tied to the function of the nervous system and corresponds to the decomposition process of memory formation (which is precisely a process of organization!), inasmuch as the function of the centers consists primarily in filtering the overall excitations of the nervous system and only in a second step in combining them into new unities.

At the same time, the appropriation of the past on the basis of its articulation/decomposition into elements allows us to understand why there are pre-memory, pre-experience "forms" in memory and experience, whose anticipatory function for the historical basis of reacting means nothing less than that they possess the contents corresponding to what has been. For if what encounters the living being can become sedimented into memory *only by way of its futurity*—that is, mediated by its being-ahead, if the decomposition of what was experienced into its elements rests on this "detour" (of an inner interval), then the anticipatory character must be immediately present in the contents of the historical basis of reacting—that is, as their form. This form determines what is and is not recorded by memory; it enables internal shifting, the life of memory, which is not dead mass restricting the organism's leeway, but rather enlarges it, by permitting the living being to rearrange the elements of the past and thereby learn from it.

It is thus characteristic of the animal organism that it is only able to form its historical basis of reacting within the framework of possible tendencies. An animal does not have experiences where it is not driven to go. Drive direction is memory's principle of selection, a "unity of apperception," to which the individual tendencies and drives are subject, as to categories—that is, selecting preforms.

(It is here if anywhere that the a priori forms of intelligent consciousness have their place. Strictly corresponding to the general character of the positional level in question, these forms are either—as with the animal—anticipations of certain areas of experience that can be integrated into the sensory-motor function-circle, or—as with the human—anticipations of certain objective areas of experience, although this does not exhaust their significance as forms of the objects of experience.)

[286] On every level, whether conscious or extra-conscious, the historical basis of reacting is a unity of residue and anticipation, or, to put it another way, on no level is the past, as it is retained by the living individual, a closed matter. Everything that has been is continually being transformed by the incessant anticipation of life. As long as biology adhered to mechanistic dogma in seeing association as the core phenomenon of memory, it could of course only attempt to explain the decomposition and rearrangement of what has been (the precondition of a historical basis of reacting) either, again, in terms of mechanical association or, once this was shown to be untenable, with the help of specific vital factors (dominants, psychoids).—

Entirely distinct from the anticipatory forms of the historical basis of reacting itself are the instincts, which belong to a more fundamental layer of organic being, the sphere of adaptedness. Although it is true that instinctive actions are determined prior to all experience and cannot be significantly corrected by experience, practice, and so on (as can be seen, for instance, in the way the youngest migratory birds fly ahead of the rest of the group even though they have never flown the route before; or in the way chickens immediately after hatching peck at grains with an accurate assessment of physical space), "prior" here has a temporal and not, as in the case of anticipation, a timelike/*non*temporal meaning. Instincts in the behavior of an organism manifest the primary accordance between it and the environment in time.

The fact that only animals have instincts while plants do not is due to the closed form of organization and the relative self-sufficiency of the animal individual in relation to its circle of life. An animal in its set-apartness is forced by its essence to act, to execute the appropriate responses to the environment's stimuli. This creates room for a primary equilibrium between individual and environment, while, in contrast, the plant's entire structure does not permit it to act: the plant's integration into its circle of life renders it dependent; it is fully absorbed in this circle as a part of it.

To look for psychic equivalents of the instinct and to see in it, for instance, something akin to "dreaming," as did some Romantic natural [287] philosophers, is not consistent with its nature. A sentimental interpretation would like only too well to recognize in instinct the manifestation of premonition and a sixth sense, of dream and telepathy. It fails to consider that instinctual life does not have anything immediate to do with consciousness and that it resides on an entirely different level. This means that instinct (and its differential, the reflex), while it can never be influenced by the initiative of the subject, is neither able to compromise nor to promote the development of consciousness.

SEVEN

The Sphere of the Human

The Positionality of the Excentric Form: "I" and Personhood

The limit of animal organization lies in the fact that the individual's being-itself is concealed to it because it does not relate to its positional center,[1] while its medium and its own physical lived body are given to it and relate to its positional center, to the absolute here/now. The animal's existence in the here/now does not relate to anything else, for there is no other to which it could relate. Inasmuch as the animal is itself, it is absorbed in the here/now. This here/now does not appear to the animal as an object, does not set itself apart, but remains a condition, the mediating passing-through of concrete, living execution. The animal lives out from its center and into its center but not *as* center. It experiences things in its surrounding field, things of its own and not of its own; it is able to gain mastery over its own lived body; it is a system that refers back to itself, a self, but it does not experience—itself.

Indeed, what subject is there to have such a lived experience on this level of positionality? To whom could be given this having and experiencing and acting, as it flows into the here/now and impulsively emanates from it? "To which point could this content relate, onto what object could it project itself, based on what distance of a structural moment of living thinghood from the thing of this animal body?" Insofar as the animal is lived body it is given to itself, relates to its positional center, and can exert influence on its body as whole body in the here/now, can respond with physical success to central impulses. But its whole body has not yet become completely reflexive.

Not yet. A progression is in fact conceivable, one that would lift the living, bodily thing onto a higher level of positionality, above the level of the animal—according to the same law [289] that determines the difference in level between the plant and the animal. Just as the plant's open form of organization contains positional characteristics without the thing thereby being "posited" in relation to its positionality and this possibility becoming realized in the closed form of organization of the animal, the essential form of the animal also reveals a possibility that can only be realized by something else. On the animal level, the living body is denied full reflexivity. Set in itself, its life out of the center constitutes the anchor of its existence but does not stand in relation to it, is not given to it. There thus remains the possibility of a realization. I wish to argue that this possibility is reserved to the human.

What conditions must be met in order that a living thing be given the center [*Zentrum*] of its positionality, in which its life is absorbed and by virtue of which it experiences and acts? A basic condition evidently is that the center [*Zentrum*] of positionality, whose distance from the living thing's own lived body grounds the possibility of all givenness, be at a distance from itself. To be given means to be given to someone. But to whom can be given that to whom everything is given if not to itself? This indeed is the only way for the spatiotemporal point of the absolute here/now to move away from or double itself (or however else one wants to conceive of this distancing). A pure here/now implies nonrelativizability, which however would be annulled if the center [*Zentrum*] were to be thus split. To put it in concrete terms: if there is an absolute here-and-now point, a positional center of a living being, then it would be impossible for this same center to, "in addition," be behind, in front of, before, or after it. It is tempting, however, to operate on this very assumption, since the positional center is supposed to be that to which something is given, that which experiences things, the subject of consciousness and initiative. Only an eye can see, and only an eye can see an eye. As it is not possible, however, to line up an indefinite number of eyes in one long row—since they all ultimately point to *one* seeing subject and it is here a matter, precisely, only of *one*—the ability of the eye to see itself, the givenness of the subject to itself, cannot be grounded in a (self-contradictory) multiplication of the subject's core.

[290] As long, however, as one thinks of the positional center [*Zentrum*], the subject, as a fixed entity that is completely given, that simply exists like any bodily feature, one cannot avoid this notion of multiplication and the impossibilities it brings with it. But this idea is as wrong as it is convenient: it forgets that it is the matter here of a positional character, whose existence is tied to an execution or a positing in the sense of the vitality of a being, a vitality that is determined by the constitutive principle of the boundary.

A positional center only exists as execution. It is the mediation through which a thing acquires the unity of a gestalt; it is the passing-through of mediation. As a moment of positionality, it is a subject that is not yet operative. In order for the subject to become so, an important shift must take place: the positional moment has to become the constitutive principle of a thing. With that, the thing is set into its own center, into the passing-through of the mediated unity of its being—and it thus arrives at the animal level. The law according to which the moment of a lower level, once it is grasped as a principle, renders the next higher level, while at the same time appearing in it as a moment (that is, is "preserved" in it), makes it possible to think a being whose organization is constituted according to the positional moments of the animal. This individual is set into the positedness in its own center, into the passing-through of the mediated unity of its being. It stands at the center [*Zentrum*] of its standing.

This satisfies the condition that the center [*Zentrum*] of positionality be at a distance from itself; set apart from itself, it makes possible the total reflexivity of the living system. The condition is met only in terms of positionality and without a contradictory doubling of the subject's core. Its life out of the center enters into a relationship to it; the reflexive character of the centrally represented body is given to itself. Although the living being on this level is also absorbed in the here/now, lives out of the center, it has become conscious of the centrality of its existence. It has itself; it knows of itself; it notices itself—and this makes it an *I*. This I is the vanishing point of its own interiority that lies "behind" it; it is removed from its own center in every possible execution of life and is the observer of the scene of this inner field; it is the subject-pole that can no longer be objectified or put into the object position. [291] This most extreme level of life sets the ground for ever-new acts of self-reflection, for the infinite regression of self-awareness, thus effecting the split into outer field, inner field, and consciousness.

Clearly, animal nature has to be preserved on this highest level, as it is only a matter here of the closed form of organization being taken to the extreme. There is only one way for the living thing in its positional moments to progress, and that is to realize the possibility of organizing the reflexive overall system of the animal body according to the principle of reflexivity and to also set in relation to the living being that which on the animal level only makes up life. A progression beyond this is impossible, for the living thing has now actually moved behind itself. While it still stays essentially tied to the here/now and has lived experiences without turning its gaze upon itself, taken instead by the objects in its environment and the reactions of its own being, it is capable of distancing itself from itself, of introducing a gap between itself and its lived experiences. This places it on both sides of the gap, bound in body, bound in soul, and at the same time nowhere, without place and unbound in space and time. This is the human.

In its existence set against the surrounding field of alien givenness, the animal takes the position of frontality. It lives separated from its surrounding field and yet in relation to it, aware of itself only as lived body, as the unity of its sensory fields, and—in the case of centralized organization—fields of action. It lives thus in its own body, whose natural place, the center of its existence, is hidden from it. The human, as the living thing placed in the center of its existence, knows this center, experiences it, and therefore is beyond it. He experiences being bound in the absolute here/now, the total convergence of his surrounding field and his own lived body against the center [*Zentrum*] of his position, and is thus no longer bound by it. He experiences the immediate commencing of his actions, the impulsivity of his stirrings and movements, the radical agency of his living existence, his standing between action and action, his ability to choose as well as to be carried away by affects and drives; he knows himself to be free and despite this freedom to be bound in an existence with which he struggles and that inhibits him. If the life of the animal is centric, the life of [292] the human, although unable to break out of this centrality, is at the same time out of it and thus excentric. *Excentricity* is the form of frontal positioning against the surrounding field that is characteristic of the human.

As the I that makes possible the full return of the living system to itself, the human is no longer in the here/now but "behind" it, behind himself, without place, in nothingness, absorbed in nothingness, in a space- and time-like nowhere-never. Timeless and placeless, the human can experience himself and at the same time his timelessness and placelessness as a standing outside of himself, because the human is a living thing that no longer stands only in itself but whose "standing in itself" is the foundation of its standing. He is placed within his boundaries and therefore outside of these boundaries that confine him as living thing. He not only lives and experiences, but also experiences himself experiencing. The fact that he experiences himself as something that *cannot* be experienced, that does not inhabit the object position, as pure I (in distinction to the psychophysical individual I, which is identical to the "me" as object of lived experience) is based solely in the particular way that the thing called "human" is set within its boundaries. To be more precise, it is the immediate expression thereof.

As the I, however, that grasps itself in this full return, feels itself, becomes aware of itself, observes itself desiring, thinking, doing, feeling (and also observes itself observing), the human stays bound in the here/now, in the center [*Zentrum*] of the total convergence of his surrounding field and his own lived body. He thus lives immediately, in the unbroken execution of that which he comprehends—by virtue of his unobjectified nature as an I—as his psychic life in the inner field.

To the human, the transition from being within his own lived body to existing outside of his lived body is the irreducible dual aspect of existence, a true split in his nature. He lives on both sides of this split as body and psyche *and* as the psychophysically neutral unity of these two spheres. This unity, however, does not cover up the dual aspect; it does not bring it forth from out of itself; it is not the third term that reconciles the opposition and leads into the opposing spheres; it is not itself a stand-alone sphere. It is itself the split, the hiatus, the empty passing-through of mediation, which for the living being itself is equivalent to the absolute dual nature and dual aspect of physical lived body and psyche in which it experiences this split.

[293] In terms of positionality, then, there is a threefold situation: the living thing is body, is in its body (as inner life or psyche), and is outside its body as the point of view from which it is both. An individual characterized positionally by this threefold structure is called *a person*. He is the subject of his lived experience, of his perceptions and actions, of his initiative. He knows and he wills. His existence is literally based on nothing.

Outer World, Inner World, Shared World

If being outside of itself turns the animal into a human, it is clear that the human must physically stay an animal, as excentricity does not enable a new form of organization. The physical characteristics of human beings thus only have empirical value. Being human is not tied to any particular gestalt and (to recall an imaginative conjecture by the paleontologist Edgar Dacqué) could just as well take on a variety of gestalts that do not correspond with our own. The human is tied only to the centralized form of organization, which forms the basis of his excentricity.

At a double remove from his own body—that is, set apart even from his being-himself in his center, his inner life—the human finds himself in a *world* corresponding to the threefold structure of his position: outer world, inner world, and shared world. In each of these three spheres there are things that confront the human as their own reality [*Wirlichkeit*] positioned in their own being. Because it appears in the light of a particular sphere, that is, against the backdrop of a whole, everything that is given to the human appears in fragments, as an extract or aspect. This fragmentation is essentially linked to the content's groundedness in itself, to the fact that it *is*.

The surrounding field filled with things becomes the *outer world* filled with objects, a continuum of emptiness or of spatiotemporal extension. Insofar as bodily objects manifest being within their boundaries, the empty forms of space and time in their direct bearing on these objects are manifestations of nothingness. (I do not mean with this sentence to again conjure up the old debate about the existence or nonexistence of empty space. It would be just as inappropriate, however, to try to either support or refute my claim using physics, epistemology, or metaphysics. What is being considered here is purely a matter of perception [294]. The pure where and when of the potential for being filled with entities is, in this precise regard, the latter's pure contrast or non-being—regardless of the ways in which physicists or metaphysicians, who necessarily go beyond this perspective, may uncover the provisional nature of these definitions.) The relative directionality of the spatiotemporal whole means that things in this homogeneous sphere can move in any direction, thereby constituting a situation that *strictly corresponds to* the position of the excentric organism. If this organism is outside of its natural place, outside of itself, non-spacelike, non-timelike; if it stands nowhere, stands on nothing, within the nothing of its boundaries, the bodily thing in its environment also stands "in" the "emptiness" of relative places and times. And the organism, by virtue of its excentricity, is to itself only such a bodily thing in its environment at a particular place and at a particular time and could just as well be anywhere else in this continuum of emptiness.

On this level, then, there is no longer a surrounding field in the strict sense of the word. The integration of the organism into the spatiotemporal whole with its relative directionality includes the integration of the surrounding field into this *one* emptiness. It is preserved here with all of its characteristics (its total convergence toward the absolute here/now, its set-apartness from the lived body, its boundlessness and finiteness), *in relation*, of course, to the organism *in its position*. This positional whole, however, itself stands in the outer world just like all other things. The excentricity of the living being's position [*Lage*], that is, of the irreducible dual aspect of its existence as *physical body* [*Körper*] and *lived body* [*Leib*], as thing among things at arbitrary points in the space-time continuum and as a system concentrically enclosing an absolute center in a space and a time of absolute directions, corresponds to the excentricity of that living being's structure.

It is for this reason that both worldviews are necessary: the first conceives the human as lived body in the middle of a sphere with an absolute top, bottom, front, back, right, left, earlier, and later commensurate with his empirical gestalt. This is the basic assumption of the organological worldview. The second conceives the human as bodily thing at an arbitrary point in the relative directionality of a continuum of possible processes and leads to the view embodied by mathematics and physics. [295] Although the lived body and the physical body in the material sense are one and the same and do not constitute two distinct systems, they also do not coincide with each other. The dual aspect is radical. In an equally radical way, surrounding field and outer world cannot be converted into each other, although they do not constitute materially distinct zones, either. Point by point, the surrounding field can be inscribed into the outer world, but it thereby loses the characteristics that define it as the surrounding field. An inscription of this kind renders a spatiotemporal area of particular dimensions that presents itself according to the laws of perspective. This is the physical equivalent of the positional field that houses the organism as bodily thing (the object of anatomy and physiology).

These aspects coexist alongside each other and are mediated with each other only at the point of excentricity, the unobjectifiable I. Just as this I constitutes the vanishing point of its own interiority, of its being-itself "behind" the physical body and the lived body and is the boundary that can only be approximated asymptotically, so the thing in the outer world exhibits the same structure as the appearance of inexhaustible being, as the structure of shell and core. "Behind" the sensory body in a spacelike way, binding the embodied fullness without becoming absorbed in it, temporally enduring the sequence of changes and eluding destruction, the substantive core signifies the "center" of the thing in its appearance without really [*reell*] being approachable. This is because the physical thing-body (that which really [*reell*] exists in space and time) is entirely appearance. Its center becomes a vanishing point, the X of predication, the carrier of properties. It is upon this structure finally that the one-sided appearance or the adumbrated presentation essential to all reality is based; it is the surplus of the given when it is grasped as existing actuality.—

The living being at a distance from itself is given to itself as *inner world*. This interiority is opposed to the exteriority of the surrounding field as set apart from the lived body. Strictly speaking, the term "exterior" or "outer" [*Außen*] cannot be applied to the world of bodily things as such. Only the surrounding field that has become integrated into a world, thus constituting an environment, is an outer world. The environment thus inversely correlates to the inner world, the world "in" the lived body, that which the living being is. This world too has its own ambiguity. The law of excentricity defines a dual aspect of its existence as *psyche* [*Seele*] and *lived experience*.

If a living being is absorbed in itself, whether naïvely or [296] self-reflectively, it experiences things; it becomes aware [*"wird inne"*] of its lived experiences and thereby executes psychic reality. At the same time, this execution is tied to this psychic reality, to the living being's being-a-self. The intensity and range of the execution determine the formation of the being's psychic life and vice versa. The depth and vehemence or coolness and mildness of feelings, volitions, and thoughts, the range of characters and degrees of talent—in short, all psychic dispositions and achievements both determine and are determined by lived experience. Lived experiences shape the human, unsettle him, and create new possibilities for lived experiences in the future, just as they are themselves made possible by the given properties of the psyche. The psyche on the one hand as the given reality [*Wirklichkeit*] of the dispositions, as an entity that develops and is subject to laws, and lived experience on the other as the reality of one's own self to be passed through in the here/now, in which no one can replace me and from which nothing but death (and not even that is certain) can remove me—these do not coincide with each other, even though in a material sense they are one and the same and do not constitute two distinct systems. The dual aspect is radical and corresponds to that of physical body and lived body, although it is not necessarily given in the same way.

The inner world, which is both in the position of the self *and* in the object position, as a reality to be passed through *and* to be perceived, is of a different ontic type than the outer world. For while in the latter it is also possible on the level of *appearance* to cover the entire spectrum from the pure conditionality of a merely carrying and accompanying environment to the pure objectivity of a world of things existing for itself, this is not possible here on the level of being itself. In the inner world, by contrast, there is also a spectrum of *being*. Here we can both "be in a mood" and "be something." It follows from the structure of the positionality of standing in the here/now (and the simultaneous excentricity of distance from this position) that being oneself exhibits a spectrum ranging from the pure state of abandon, of being consumed, at one end, to concealed, repressed lived experiences on the other. Sometimes psychic content functions like a clearly delimited thing able to exert force, such as in the case of a psychic trauma, a complex in the psychoanalytic sense, or a distinct, pleasurable memory. Other times, such as when we are completely consumed by pain, desire, or any other affect, our psychic being permeates and floods us; any gap between the subject of the act [297] of lived experience and the subject-core of the whole person disappears, and we are completely absorbed in the life of the psyche. Such conditions of our inner being are best described using metaphors of flowing movement. Between these two extremes of psychic reality lies a wide variety of intermediate forms. For instance: when I meet someone for the first time, I am given a certain impression, almost like a smell or a flavor. Sympathy and antipathy span a whole range of impressions. Or I "listen to myself" as if there were an inner voice speaking to me, trying to hear something where strictly speaking there is nothing to hear at all.

That fact that, in the self-position, I am the one who provides the material and the forms of my inner sphere and, furthermore, that this self-position is itself given to me, allows me to both discover and reshape my psychic reality. Even if the gradual increase of knowledge about the laws of the psyche has bolstered the belief in its resistance to the gaze of the psychological observer and has, empirically at least, pulled the rug out from under the idealistic notion that psychic content is irreducibly tied to consciousness—the claim that internal things and processes are influenced by the attention directed toward them cannot be entirely dismissed. In the act of reflecting, attending, observing, seeking, and remembering, the living subject also *brings about* psychic reality, which naturally has an effect on the objectified actuality of a wish, a love, a depression, a feeling. The gaze of the experiencing subject can cause its inner life to change as dramatically as light affects the sensitive layer of a photographic plate.

But it does not have to. The view that knowledge of a lived experience is never a simple observation but rather the formation of something—a view that can be found in neo-Kantianism and one that makes an empirical psychology impossible—is based on the assumption that all psychic content is exclusively of an experiential character. The lived experience of thinking, willing, feeling, etc., is psychological, the argument goes, ergo, psyche and lived experience coincide. Psychic *phenomena* are considered here (according to the general principles of idealism) exclusively as objects of self-reflection and are equated and identified entirely with the contents of consciousness. [298] Mitigated at best by Kant's more cautious reflections on the self-affection by the psychic thing-in-itself, on the inner form of intuition, and on the appearance status of the content gathered by self-observation, the principle of *esse = percipi* has been held to for the inner world even where it has long since been given up for the outer world.

Whether I am in the self-position or in the object position, as reality [*Wirklichkeit*] to be passed through or as reality to be observed, I *appear* to myself *while* I myself am the reality. There is a tendency to assume that in the self-position—that is, in the execution of lived experience—it is meaningless to speak of the appearance of the inner world, which is thought to show itself here immediately as itself. While admitting that it is only in the phenomenon that reflection on lived experience can *grasp* the self, there could still be no doubt that the lived experience itself is absolute or *identical* with the inner world. (This is a widespread assumption and fundamental to all forms of subjectivism and *Erlebnisphilosophie*.) The self-position could only have this kind of advantage, however, if the human were an exclusively centric, and not, as is the case, an excentric living being. For the animal it is true that in the self-position it is entirely itself. The animal is placed in the positional center and is absorbed in it. The human, however, is subject to the law of excentricity, according to which his being in the here/now—that is, his absorption in what he experiences—no longer coincides with the point of his existence. Even in the execution of a thought, a feeling, a volition, the human stands apart from himself.

How can we explain the existence of false feelings or sham thoughts or the fact that it is possible to work ourselves into something that we are not? How can we explain the existence of (good and bad) actors, the transformation of one human being into another? How is it possible that those watching him, but more significantly the individual himself, cannot always say with certainty whether even in moments of the greatest abandon and passion he is not merely playing a role? The testimony of inner evidence does not dispel doubts as to the truthfulness of one's own being. Such evidence does not overcome the incipient split that, because it is excentric, cuts through the human's being-himself, so that no one can know of himself whether it is still he who laughs and cries, thinks and makes decisions, or this self that has already split off, [299] the other that is in him, his counterimage or perhaps even antithesis.

When philosophy sets out to examine the principles of the psyche according to which the inner world is constituted, it should pay particular attention to this fundamental split, for it allows us to understand the instability or, to use a perhaps more appropriate image, the indifferent balance that is one of the chief characteristics of the inner world. This indifferent balance is the source of the inner world's greatest potential as well as that which can lead it to become ill and perish.

It seems that the empirical methodology of psychology has led to a somewhat simplistic treatment of the essence of psychic appearance. The "inner" sphere is conceived as being filled (more or less explicitly in analogy to the thing-filled outer sphere) with ideas, feelings, thoughts, and any other manner of psychic formations and processes, which are correlated with inner acts of perception and apperception. Sometimes I am the one seeing, at other times I am the one seen; sometimes I am a thing-in-itself, sometimes appearance (inasmuch as I know of myself and experience myself). This view is, at the very least, too narrow. Even when the human is not thinking about himself but naïvely executing his life and lived experiences, he is psychic appearance. And he is appearance (in precisely the same sense) because he is immediately himself "in-himself" ["*an sich*"].

This dialectical structure intrinsic to excentricity makes being-oneself an inner world, that which one feels, suffers, goes through, notices, *and* is; makes it the predisposition, temperament, and character underlying all acts and lived experiences and unconscious processes *and* that which counteracts these underlying elements, restrains, analyzes, observes, intensifies, and exaggerates them. The actual inner world is where one is at odds with oneself with no escape and no compensation. This is the radical dual aspect between the (consciously given or unconsciously effective) psyche and the execution in lived experience, between existence happening as necessity, compulsion, and law and existence executed as freedom, spontaneity, and impulse.

The human's being-himself also takes on the quality of a world in the sense that its constitution is not tied to particular acts. The human is present as inner world whether he is aware of it or not. [300] This inner world, however, is given to the human only in acts of reflection, and these acts of memory, perception, and attentiveness are themselves psychic. The fact that the human can perform such acts is ultimately due to the vitality of being. For living being is ahead of itself and thus forms a center [*Zentrum*] for acts of return to its own being.

By no means does the self grasp itself as the self *with acts of this kind*. It does not yet grasp itself, but rather the past, what it was. Animal subjectivity is also capable of simple reflection in this sense, in the form of a memory as ensured by the historical basis of reacting. In order for one's own being to encounter itself as a reality [*Wirklichkeit*] sui generis, it must be part of its nature to stand apart from itself. In such a relation to the zero point of one's own position—a relation that is not founded in acts, but that is given once and for all with the excentric form of being—the constitution of being-oneself consists as a proper world not tied to acts.

Excentricity, upon which the outer world (nature) and the inner world (psyche) rest, means that the individual person has to distinguish in himself between individual and "general" I. Normally, however, he only grasps this when he is with other persons, and even then, this general I never appears in its abstract form, but rather specifically in terms of the first, second, or third person. The human calls himself and others you, he, we—and not because he is forced by way of analogy or empathy to assume those beings that conform most closely to him to be persons, but because of the structure of his own mode of being. In-himself the human is an I—that is, owner of his lived body and his soul, an I that does not belong in the radius but nevertheless forms its center. He is thus free to experiment with laying claim to the timelessness and placelessness of his own position (by virtue of which he is human), for himself and for every other being, even those that are entirely alien to him.

Even within everyday experience, nothing refutes the famous theories of analogy and sympathy, according to which the human had to think up [301] the idea of a shared world and thus arrived at the certainty of the reality [*Wirklichkeit*] of other I's, more effectively than what can be widely observed in both individual and collective human development—namely, an original tendency to anthropomorphize and personify. Even dead things in a child's environment take on the character of living personality. The worldview of primitive peoples (insofar as we can assume them to be primitive—that is, early forms rather than degenerate forms) displays similar characteristics. It is only the disenchantment brought about by rational culture that makes the human conscious of dead things. Seen from this perspective, the pantheistic animism and animation of the world characterizing the worldviews of later cultures can be read as attempts to paralyze this consciousness and as a flight into childhood.

The assumption of the existence of other I's is not a matter of transferring one's own mode of being, the way in which a human being lives for himself, onto other things only corporeally present to him—in other words, an extension of his personal sphere of being—but rather a restriction and limitation of this sphere of being, which was originally not localized and resisted localization, to "human beings." The *process* of limitation, as it unfolds in the interpretation of alien centers of life appearing before us in an embodied way, must be clearly distinguished from the *premise* that other persons are possible, that there is such a thing as a world of persons. Fichte was the first to point out the importance of this distinction. Every positing as real of an I, of a person in his individual body, presupposes the sphere of you, he, and we. The fact that the individual human "thinks up" the idea or, rather, is imbued with it from the very beginning, that he is not alone and not only has things but other beings with feelings as companions, is not born of a specific act of projecting his own form of life onto others, but is one of the preconditions of the sphere of human existence. Certainly, finding one's way around this world requires constant effort and attentive experience. For despite the structural sameness between me and the "other," as a person he is simply an individual reality (as am I), whose inner world is, initially, as good as completely hidden from me and can only be uncovered by very different kinds of interpretation.

[302] The excentric form of his positionality ensures the reality of the *shared world* for the human. He does not become conscious of this world *on the basis of* certain perceptions (although in the course of his experience certain perceptions on occasion will certainly imbue it with color and life). This is associated with the fact that the shared world differs from the outer and inner worlds in that its components—persons—do not constitute a specific substrate materially different from that which is contained in the outer and inner worlds. Its specificity is rather life in its highest—that is, excentric—form. The specific substrate of the shared world is thus based only on its own structure. The shared world is the form of the human's own position, conceived by him as the sphere of other humans. We can thus say that the excentric form of positionality generates the shared world and guarantees its reality.

This assertion naturally runs the risk of being misinterpreted in a variety of ways. It seems impossible for a world not to have its own specific substrate, its own characteristic "matter." That would seem to imply that a pure form could occur as a world. If the unity of the person materially allows for no other differentiation than into bodily and psychic being and personhood is based on its *mode* of being (of life), it seems, according to the standard dualist and empiricist understanding, to be illegitimate to speak of a stand-alone shared world.—Blindness to the indissoluble layer of being that is specific to living things and cannot be ascribed to body or soul makes itself felt particularly when it comes to the highest development of these layers of being in the "person."

And furthermore, are we not conceding to the theory of the projection or transference of the individual mode of being onto other bodies when we say that the shared world is the form of the human's own position, conceived by him as the sphere of other humans?—Indeed, this view would not be particularly convincing if it did not include the *existence* of this sphere as the *condition* for grasping one's own position as such, as well as this precise form of positionality as a sphere. The existence of the shared world is the condition of the possibility that a living being [303] can grasp itself in its position—that is, as a member of this shared world. Just as we may not apply the standard schemata that have become habitual for us through the use of our sensory organs and their study by physiology and physics (in particular the schema of the face-to-face position to the outer world and that of receiving messages from it) to the living relationship of the person to the outer world, so we may not use these schemata to explain the relationship of the person to the shared world. Moreover, to make the reality of a world depend on whether our relationship to it corresponds to the categories that form our relationship to nature or to the psyche would be to anticipate a dualism for which there is no longer any good reason.

The shared world does not *surround* the person as nature does (albeit not in the strict sense, given that one's own lived body belongs to nature). Nor does the shared world *fill* the person, as can be said in a similarly inadequate way of the inner world. The shared world rather *carries* the person while at the same time being carried by and formed by the person. Between me and me, me and him lies the sphere of this world of the *spirit*. If it is the distinguishing trait of the natural existence of the person that he inhabits the absolute center of a sensory, figural sphere, which itself relativizes this position and divests it of its absolute worth; if it is the distinguishing trait of the psychic existence of the person that he grasps his inner world and, at the same time, executes this world as he experiences it, then the spiritual character of the person is based on the we-form of his own I, on the totally unified being-encompassed *and* encompassing of his own lived existence according to the mode of excentricity.

We—that is, not a select group or community that can refer to itself as "we," but rather the sphere designated by this word—is strictly speaking the only thing that can be called "spirit." Understood in the purest sense, spirit is different from the psyche and from consciousness. The psyche is real as the internal existence of the person. Consciousness is the way in which the world presents itself as determined by the excentricity of personal existence. Spirit on the other hand is the sphere created and existing along with this particular form of positionality and thus does not constitute a reality, but is realized in the shared world even if only *one* person exists.—These distinctions of course are often overlooked, [304] and spirit, psyche, subject, and consciousness are treated as equivalents or synonyms. This gives rise to fateful notions of the spiritual nature of the world and to animism; to subjectivist, but also objective-idealist presuppositions that classical paradigms are called on to renew even after they have died a natural death.

The shared world is real even if only one person exists because it represents the sphere guaranteed by the excentric form of positionality upon which every selection in the first-, second-, or third-person singular and plural is based. It is for this reason that the sphere as such can be distinguished from extracts from it as well as from its specific foundation in life. It is the pure "we" or spirit. It is in this sense that the human is spirit and has spirit. He does not have spirit in the same way as he has a body and a soul. These he has because he is them and lives them. Spirit, on the other hand, is the sphere by virtue of which we live as persons. It is where we stand, precisely because our form of positionality sustains it.

Only insofar as we are persons do we stand in a world of being that is independent of us and at the same time amenable to our actions. It thus holds that spirit is the precondition for nature and psyche. This sentence is to be understood in a specific sense: not as subjectivity or consciousness or intellect, but as we-sphere is spirit the precondition for the constitution of a reality [*Wirklichkeit*], which, in turn, only figures as and comprises reality as long as it also remains constituted for itself, independently of the principles of its constitution in one aspect of consciousness. Precisely by turning away from consciousness in this way, this reality fulfills the law of the excentric sphere as described earlier.

If one wanted to speak metaphorically of the spherical structure of the shared world, one could say that it devalues the spatiotemporal diversity of human standpoints. As a member of the shared world, every human stands where the other stands. In the shared world there is only *one* human, or, to be more precise, the shared world only exists as *one* human. The shared world is absolute punctiformity, in which everything human remains originally linked together, even if the vital base is broken down into individual beings. The shared world is the sphere of the "each other" and of absolute disclosure, where all that is human [305] encounters each other. As such it is truly indifferent to singular and plural; it is infinitely small and infinitely large; it is subject-object; it guarantees actual (not only possible) human self-knowledge in its mode of being-each-other.

Following Hegel (usually in a very superficial way), we speak of subjective, objective, and absolute spirit. The use of such set terms always has its disadvantages. It is important to understand that the terms "subjective" and "objective" cannot be applied to the spirit as sphere. This should not be construed to mean, however, that it is here a matter of absolute spirit. Irrespective of that which is carried by spirit or expresses itself as spirit, the sphere of spirit can only be defined as subjectively and objectively neutral—that is, as indifferent to the distinction between subject and object. Nevertheless, this does not mean that the predicate of absoluteness can be applied to this layer. It is the sublation or the bridging of the gap between subject and object—which nevertheless continues to be valid for the human—in the sphere of spirit that gives rise to the temptation to speak of spirit as the absolute (which is something different again from the concept of absolute spirit). If we recall, however, that spirit is only the sphere given by the excentric form of positionality of the human and that excentricity signifies the form of frontal positioning against the surrounding field characteristic of the human, then we can begin to understand the original paradox of the human situation: that as subject he stands against himself and the world and at the same time is at a remove from this opposition. In the world and against the world, in himself and against himself—neither of the opposing determinations dominates the other; the gap, the emptiness between here and there, the crossing-over remains, even if the human knows this and by virtue of precisely this knowledge inhabits the sphere of spirit.

It is spirit that makes it possible to objectify oneself and the outer world one faces. That is to say, objectification or knowledge is not spirit, but presupposes it. Precisely because the excentric living being, due to its frontality (which has developed naturally and is given with its closed organization) and its [306] being-positioned over against its surrounding field, is at a remove and is placed in a shared-world relationship to itself (and to everything that is), it is able to become aware of the impossibility of breaking out of its existential situation, which it shares with the animals and from which animals are also unable to free themselves. The subject-object relationship mirrors the "lower" form of existence—in the light, however, of the sphere by virtue of which the human living being constitutes and possesses the higher form of existence.

I would like to respond here to two possible objections. The term *Mitwelt* [shared world] is generally understood in a narrower and more specific sense than the existential sphere at issue here to mean the social environment of the human. *Vorwelt* [the world of the past, our predecessors] and *Nachwelt* [the world of the future, posterity] are also distinguished from *Mitwelt* in the narrow sense and *Mitlebende* [those living with us, our contemporaries] are treated as a current totality very much distinct from past and future generations. And yet it cannot be denied that any such empirically tangible specification refers to a special kind of sphere that can neither be equated with objective nature or the psyche nor be identified with a synthesis of the two. The final element of this sphere is the person as living unit who, from an analytical, objectifying perspective, can in fact be decomposed into nature, psyche, and spirit (or into units of sense and meaning as correlates of intentional acts), but can never be composed out of them. The other exists as a member of a social setting, as fellow human being [*Mitmensch*], only because of the special structure of the personal sphere. It thus seems appropriate, despite the generally more restricted use of it, to employ the word *Mitwelt* to refer to that from which the more limited definition properly derives its meaning.

Second, the identification of a shared world reserved only to the human implies a rejection of theories that categorize the social relations of animals, their cooperation and conflict, together with human forms of social interaction. A zoologist, however, can refer to the shared or social world of an animal species, can speak of bee, termite, and ant colonies, of group formation, sociability, stable herds, etc., without thereby (necessarily) claiming that the fact of a corresponding social world is there *in the perspective* of the animals themselves. Genuine biologists avoid making such judgments out of methodological considerations alone [307] and limit their research to phenomena in the field of socially determined reactions. Since these phenomena occur in the spatiotemporal world, biologists can of course speak of studying the social worlds of animals.

The theory that the animal's social horizon and scope are given to it as a world in the same or in a similar way as they are to the human is quite another matter, however. Such an assumption is false. For just as the surroundings of the animal's own existence cannot appear to it as a world—that is, in genuine objectivity—otherwise it would not be an animal—its *relationship-with* does not present itself to it as a world, either. It does not become conscious of it *as* a relationship-with; the relation remains hidden from it. The animal stands in this relation, but does not grasp it. Its form of organization is concentric, not excentric, and thus does not allow for the possibility of developing and grasping its position in its relationship-with.

If philosophy then reserves the *shared world* for the human, this does not mean that it wishes to ignore the facts of social life in the animal kingdom or to deny their special significance. It is only the proper evaluation and interpretation of these facts that are up for discussion. Of course, it is tempting to speak not only of the animal's surrounding field but also of its *shared field* [*Mitfeld*] and to entertain the possibility that the animal in its social behavior relates to such a sphere of a shared field. But that would be jumping to conclusions. The animal's closed form of organization does not allow for a constitution of its own shared field as distinct from the surrounding field. Those of its own kind or "fellow animals [*Mittiere*]" do not form a particularly distinguished or circumscribed setting for the animal. These other animals are completely amalgamated into the animal's surrounding field as a whole and are thus treated as equivalent in meaning to it.

There is no doubt that all animals "have a nose for" others of their kind with which they "objectively" stand in a relationship-with. It is a question for biologists how much of this sense is due to pure instinct, how far perception reaches, and whether there is room here for trial and error and thus for acquiring experience. The fact, however, that a relationship-with, from the simplest types of occasional sociability, such as during mating or fighting over prey to the highest forms of state-like associations, is intrinsic to animal life, [308] is immediately obvious from the essence of the closed form of organization and is indeed confirmed by experience.

Because it is alive, all living being stands in a relationship-with—that is, in a relation of accompaniment [*Mitgehen*], coexistence [*Nebeneinander*], and cooperation [*Miteinander*], which is entirely different from a bare relation of opposition (a limit case of which only the human with his sense of objectivity can conceive). This insight follows conclusively from the analysis undertaken in the preceding chapters of this work. Above all, the relationship-with governs the living being's relationship to its environment, regardless of whether it is made up of dead or living things. Only the human has a true relationship-against [*Gegenverhältnis*] (in an objective, not in a hostile sense). His world is also necessarily carried by environmental characteristics, just as in the organization of his own existence that which is higher and specifically human is carried by that which is animal. This world too necessarily appears, and is internally comprehensible, as a milieu, as an unstructured "atmosphere," as an abundance of circumstances surrounding and carrying the human. The myriad things with which we have to do every day, from a piece of soap to a mailbox, are only potential objects; as elements of our dealings with them, however, they are components of our surrounding field, members of our relationship-with them.

This vital zone of approachability and familiarity, where true relations-with prevail in a way characteristic of the situation of the animal (without, of course, being embedded in a world), thus obviously has nothing to do with the shared world as such. If the human can speak, in a direct rather than merely allegorical way, of brother donkey and brother tree, it is because he has grasped the continuous commonality of all that is alive and can emphasize the relationship-with that is the very hallmark of the positionality of the vital, in which he too sees himself as connected in his own way with all that is alive. The sphere in which you and I are truly connected in the unity of life and gaze into each other's unveiled faces is, however, reserved to the human. This is the shared world, where relations-with not only exist but where the relationship-with has become the constitutive form of a real [*wirklichen*] world where the emphatic I and you merge into the we.

[309]

The Fundamental Laws of Anthropology

THE LAW OF NATURAL ARTIFICIALITY

How does the human live up to his situation? How does he realize the excentric position? What fundamental characteristics must his existence as a living being assume?

Even this question points to the human's contraposition over against his own vitality and life situation. It is a question that *follows* from the fact of excentricity. Rather than simply an arbitrary problem the philosopher applies to the human (as to all things on earth and in heaven), it is (merely an articulation of) the obstruction with which the human necessarily has to wrestle if he wants to live. And the question of philosophy, as, essentially, every question the human has occasion to ask himself a thousand times in the course of his life—what should I do, how should I live, how can I cope with this existence—signifies the characteristic expression (all historical determination notwithstanding) of human brokenness or excentricity, which not even the most naïve, unbroken, content, tradition-bound, and close-to-nature era in human history has been able to elude. There have certainly been periods (and will be again in the future) when this fissure was not spoken of and the consciousness of the constitutive homelessness of the human being was covered over by strong ties to land and family, to hearth and ancestors. But these periods were not at peace, either, unless they sought peace. The idea of paradise, of the state of innocence and the Golden Age, which every human generation has known (today this idea is called "community") points to what the human lacks and to his knowledge of this lack, by virtue of which he stands above the animal.

As an excentrically organized being, the human must *make himself into what he already is*. It is only thus that he fulfills the way in which his vital form of existence forces him to stand in the center [*Zentrum*] of his positionality and at the same time to know of his being-positioned, rather than simply becoming fully absorbed in it, like the animal that lives out from its center and relates everything to its center. This mode of being of standing in one's being-positioned is only possible as *execution* from out of the center [*Zentrum*] of this being-positioned. Such a way of being can only be carried out as [310] realization. The human lives only insofar as he leads a life. To be human is to "set apart" being alive from being and to execute this setting apart, in virtue of which the layer of vitality appears as a quasi-self-sufficient sphere—a sphere that in the case of the plant and the animal remains a non-self-sufficient moment of being, a property of being (even where this sphere is the organizing, constitutive form of an ontic type—that is, for the animal). Consequently, the human does not simply live what he is to the end; he does not "live himself out" [*sich ausleben*] (the word understood in its radical immediacy),[2] nor does he only make himself into what he already is. His existence is such that it forces this distinction within itself, but at the same time extends beyond it. The philosophical explanation for this "oblique position [*Querlage*]" of the human is the excentric form of positionality, but that is not much help. Whoever is in this position finds himself in an absolute antinomy: he has to first make himself into what he already is, to lead the life he lives.

In very different forms and emphasizing a wide range of values, human beings have become conscious of this fundamental law of their own existence. This knowledge is always tainted with the pain caused by the inability to achieve the naturalness of other living beings. Human freedom and foresight came at the price of animal certainty of instinct. Animals exist directly, without knowing about themselves and about things; they do not see their own nakedness—and yet the heavenly father feeds them. The human with his knowledge has lost that directness. He sees his nakedness, is ashamed of it, and must therefore live in a roundabout manner via artificial things.

This view, often cast in mythical form, expresses a deep insight. Since the human is forced by his type of existence to lead the life that he lives, to fashion what he is—because he *is* only insofar as he performs—he needs a complement of a non-natural, nonorganic kind. Therefore, because of his form of existence, he is by nature *artificial*. As an excentric being without equilibrium, standing out of place and time in nothingness, constitutively homeless, he must "become something" and create his own equilibrium. And he can only create it with the help of things outside of nature that originate from his creative action *if* the results of this creative activity take on a weight of their own. In other words, he can only create [311] it if the results of his action become detached from their origin in this action by virtue of their own inner weight, whereupon the human must recognize that he was not their originator but that they were only realized *on the occasion* of his action.

If the results of human action do not acquire a weight of their own, if they cannot be detached from the process of their creation, then their ultimate meaning, the production of equilibrium—existence as it were in a second nature, repose in a second naïveté—has not been achieved. The human wants to escape the unbearable excentricity of his being; he wants to compensate for the dividedness of his own form of life, and he can achieve this only with things that are substantial enough to counterbalance the weight of his own existence.

The excentric form of life and the need for completion constitute one and the same state of affairs. Need, however, should not be understood here in a subjective or psychological sense. This need is presupposed in all needs, in every urge, every drive, every tendency, every volition of the human. In this neediness or nakedness lies the motive of all specifically human activity—that is, activity using artificial means that is directed toward the unreal. In it lies the ultimate ground for the *tool* and for that which it serves—that is, *culture*.

It is rare that these deeper relationships are clearly spelled out. Under the influence of purely empiricohistorical thinking, cultural historians and sociologists as well as biologists and psychologists approach the problem of the emergence of culture with the preconception that it can be solved empirically. The emergence of an individual culture or a particular cultural environment cannot, of course, serve as the model for the "genesis" of the cultural sphere as such. Once this limit has been recognized, however, it quickly mutates into an absolute prohibition against addressing the question of culture from a philosophical perspective. This is probably the main reason we still entertain two opposing answers to it—answers whose underlying principles have long since been shown to be dubious: a spiritualist and a naturalist explanation for the origins of culture.

The spiritualist theory attributes the unavoidable artificiality of human activity, its goals and means, to the spirit. This can be meant in a more objective sense, as in the old teaching that culture comes from God. [312] Or it is meant in a more subjective sense, in which case culture is attributed to special dispositions of the human, to his intelligence, his consciousness, his psyche. Neither of these theories, strictly speaking, explains anything. Both simply identify the special property that differentiates the human from the animal and go no further. This would only suffice as an explanation if it made clear how spirit came to emerge in the first place from the natural foundation specific to the human. That is not the case here, of course—not only is this not made clear, but there is no intention even to do so. The spiritualist theory merely displaces the problem.

The naturalist theory goes further in that it aims to deduce mind as the specific foundation of cultural activity from the natural layer of human existence. There is both a positive and a negative version of this same basic idea.

The positive version is the better known of the two and informed Darwin's theory of evolution. It posits that there were primitive humans (possibly in a variety of species and families), whose cerebra developed to the point that they were motivated to struggle for existence using weapons other than the natural ones of their bodies. They were supported in this by the correlative evolvement of upright gait, which kept pace with the development of the cerebrum (intelligence) and the ensuing development of the fore extremities into hands. Thanks to its opposable thumbs, the hand, "the outer brain of the human animal," was able to create the tool as its natural extension. In the past it was thought that humans developed opposable thumbs at a relatively late stage. Hermann Klaatsch and others, however, have convincingly argued that this characteristic is relatively old and that humans owe their superior standing (particularly in comparison to apes, the animals closest to us) to the preservation of a certain primitiveness. It is because we have not had to become specialists in our organization like the different species of apes, these scientists argue, that we have attained our characteristic predominance over animals. Be that as it may, both of these views suggest that humans can be said to have "sucked culture out of their thumb": intelligence and dexterity are thought to have brought about the use of tools as well as culture.

[313] Put in psychological terms, this version of the naturalist theory posits a human fear of annihilation by the forces of nature as the underlying cause of culture. Prehistoric humans are thought to have lived in fear and trembling, to have seen themselves as defenseless, and to have been naturally anxious to acquire means of self-defense. They sought after protection, which they hoped to find in their tools. The constant improvement of these tools is then what gradually led to culture, whose ultimate meaning lies in the preservation and furtherance of life. Even culture's highest sublimations,

which seem to threaten, suck up, and inhibit life, are (in a pragmatic sense) attributed to this purpose.

The *negative version* of the naturalist theory considers the development of the cerebrum and the correlative physical properties to constitute a threat to life, an illness. This theory regards the human as a sick animal that has been thrown off its natural course, off its vital balance. The human in this conception is the victim of the parasitic development of an organ. This cerebral parasitism, possibly caused by a disorder of the internal secretion process, gave him the poisoned chalice of intelligence, insight, knowledge, and consciousness of the world—this consciousness, the spirit, may be nothing more than a grand illusion, the self-deception of a biologically degenerate creature sucked dry by brain polyps. The only purpose of the crutches or artificial limbs of tools and culture is to secure his survival. And even that has a negative side. For even if this unreal world supports a form of life that has become too weak, it is equally the expression of this weakness, is itself sick.

Today the psychological rendering of the negative version can be found in theories influenced by Freud and psychoanalysts following him. Culture, in this view, is based on the repression of drives. To be human means to repress. If for the human, unlike for the animal, the sphere of consciousness also encompasses his own existence, this also generates censorship of the self. The self defends itself against this and tries to avoid the censorship of consciousness. If, then, the human wants to know about himself only that which suits him, he has to repress into the subconscious whatever is unsuitable. This includes the drives restrained and curtailed by every convention and custom, which robs them of their natural [314] effect so that they come to discharge in pathological ways. One form of pathological drive discharge is neurosis; the other is sublimation—that is, a deflection of the drive's energy into the spiritual realm. According to this theory, religion and the formation of philosophical, artistic, and political ideas arise from the sublimation of drives brought about by repression. They are thus nothing other than the indirect gratification of drive impulses inhibited from being directly satisfied. Freud writes that it "is simply that civilization is based on the repressions effected by former generations, and that each fresh generation is required to maintain this civilization by effecting the same repressions."[3]

A genuine theory of the origins of culture and of its biological function must go further, however, and inquire into the necessity of the collision between culture, customs, morality, and instinctual impulses. This necessity is brought about by a development of drives, particularly the sex drive, which, compared to animals living in the wild, is hypertrophic. Alfred Seidel, for instance, hypothesizes that this hypertrophy is the result of domestication.[4] Domesticated animals have been observed to exhibit increased nutrition and sex drives and, in the case of males, breaches of the rutting season. The human as domesticated animal, Seidel concludes, paid for the loss of a life in the wild with hypertrophic drives and the *compulsion* to repress them.

But hypertrophy can also be found in other human drives, especially the drive for power. The will to power, the struggle to be more, the tendency to be on top all show the same primary disturbance of the vital balance between the organism and its drives. For the human, striving and gratification are incommensurate with each other so long as gratification is sought in the same sphere to which the striving belongs. (Here one must note that culture can be conceived of as originating from hypertrophic drives either directly—such as in Schopenhauer's notion of the will, which delivers itself from itself, and in Nietzsche's concept of the will to power—or indirectly—such as in Freud and Alfred Adler with notions of a flight from drives or overcompensation. [315] Nietzsche's conception otherwise belongs to the positive version of the naturalist theory of origins, which itself reduces the meaning of culture to an increase in power.)

According to the *positive* version of the naturalist-vitalist explanation, culture is the direct expression and manifestation of life desiring—and called upon—to somehow enhance itself. According to the negative version, culture is the indirect expression of life condemned to—and seeking to escape having to—enhance itself. Of course, neither of these theories focuses on demonstrating how this process took place in time; their emphasis is rather only on the natural conditions of existence specific to the human, which gives this creature of nature his natural-unnatural position.

It is not necessary here to give a detailed critique of the theories of naturalism and pragmatism, especially given that others have done so on numerous occasions. These theories fail either because they do not explain the non-instrumental, nonutilitarian meaning of cultural objectives—that is to say, the spiritual sphere's own weight—or because they err on the other extreme and lose sight of the element of utility and the functional, objective meaning inherent in all cultural activity, and so get lost in pure psychologism. On one side we have a biologistic-utilitarian, on the other a psychologistic view of the world of spirit. The former makes the human into a healthy animal, the latter into a sick one. Both see him primarily as an animal, whether predatory or domesticated, and try to deduce the epiphenomenon of the spiritual expressions of his being from biological processes. This is their cardinal error, which strikingly demonstrates their inability to see, from *one* perspective, the human as human and at the same time as a creature of nature as long as they merely combine ideas derived from the natural sciences with ideas from the humanities. Both theories absolutize a symptom of human existence and attempt to explain with it everything else characterizing the human. Some reduce the human to the "all-too-human," to the sex drive, the nutrition drive, or the drive for power. Others construe a drive for redemption. Yet others see everything in terms of intelligence and calculation. And so everyone turns in circles, unable to escape the empiricism [316] of biological or psychological symptoms. They repeat the basic mistake, which Marx's historical materialism makes on a large scale, in their understanding of the unit of history, the individual human being. And if today all the insight into the economic, psychological, and biological determination of cultural work has led us to lose our faith in it, then it is the fault of these theories for not teaching us to see the foundations upon which this system of reciprocal determination rests.

Not a hypertrophy of drives or a tendency to enhance the self in the form of "the will to power, to be more, to be on top" following Nietzsche, Simmel, and Adler nor overcompensation or sublimation due to repression is the true cause of cultivation. Each of these is itself a consequence of the given form of life that alone makes up that which is human about the human. The fact that the human cannot satisfy his drives with the means naturally at his disposal, that he cannot find peace in what he is, that he wants to be more than he is and that he is, that he wants to count, that he is irresistibly drawn to irrealization in artistic forms of action and in manners and customs, is ultimately due not to drives, to the will, to repression, but to the excentric structure of his life, to the form of his existence itself. The constitutive lack of balance of his particular kind of positionality—and not a disturbance of an originally normal and harmonious system of life that one day can again become harmonious—is the "occasion" for culture.

The expression of his nature that corresponds to the human's essential dividedness, nakedness, and existential neediness is artificiality. Given with excentricity, artificiality is the detour to a second native country where the human finds a home and absolute rootedness. Positioned out of place and time in nothingness, the excentric form of life creates its own ground. Only to the extent that it creates this ground does it have it, is it carried by it. Artificiality in action, thought, and dreams is the internal medium by which the human as living natural being is in accord with himself. With the forced interruption by fabricated connecting links, the life circle of the human, into which he, as an organism of needs and drives standing on its own, is irrevocably forged, raises itself into a sphere superimposed on nature, and comes full-circle there in freedom. The human, then, lives only insofar as he leads a life. As such, [317] the life of his own existence is always breaking apart into nature and spirit, bondage and freedom, *is* and *ought*. This opposition persists. Natural law stands against moral law; duty fights inclination. Conflict is the center of existence as it necessarily presents itself to the human in terms of his life. He has to act in order to be. But the *vis a tergo* that acts upon him from his drives and needs does not suffice to keep him moving in the entire fullness of his existence. He rather requires a *vis a fronte*; it is only a force in the "ought" mode that is in keeping with his excentric structure. This force is the specific appeal to *freedom* as a stance in the center [*Zentrum*] of positionality and the motive for the human as a spiritual being, as a member of a shared world.

Because of the excentricity of his positional form, the human is a being who makes demands on himself. He "is" not simply and does not simply go on living, but counts for something and as something. He is moral by nature, an organism that tames itself by making demands on itself, that domesticates itself. He cannot exist without mores, without allegiance to unreal norms that have in themselves sufficient weight to demand recognition (which they do not require for themselves). In this way the essential fact of his positionality becomes what is called a *conscience*, the source from which ethical life and concrete morality flow. At the same time, his positionality becomes his censor, an obstruction that again and again triggers the conflict with and detachment of his "lower" nature with its drives and inclinations.

Considering now the spiritualist and naturalist explanations of the "origins" of culture as found in tools, mores, and work, the view seems to be justified that claiming that the problem cannot be solved is to contest the possibility of the problem itself. Thus spiritualism insists on the original facticity of spirit, spirituality, consciousness, and intelligence and declares any attempt to reduce this original fact to the principles of its constitution, that is, of its origin in a nontemporal sense, as nonsense. This point of view eliminates itself.

Naturalism, on the other hand, has at least had the courage to repeatedly take a fresh look at the problem. Its solutions, however, all take that form of explanation [318] that since Fritz Reuter has been referred to as the definition of poverty as *pauvreté*. Or is it something else when culture is reduced to sublimation and overcompensation, sublimation and compensation to complexes and the repression of drives, repression to hypertrophic drives, and these in turn to the domestication of the predator that is the "human"? He is only tame in his civilized form, after all—but who domesticated him if not himself? Or sublimation and overcompensation are understood as the energies that create and form culture and society, but are then traced back to the influence of culture and society. The "censor" at the boundary of consciousness and the unconscious presupposes mores and can thus at best serve to maintain these, but not to generate them. (A claim, incidentally, that prudent psychoanalysts who keep within the boundaries of experience do not make. Philosophy must object only to the metaphysical misuse of psychoanalytic ideas.)

The other version of the naturalist theory of origins is no better. Either it operates with the fear of annihilation—that is, the notion of the struggle for existence, of competition, and the selection of the fittest—or with the will or the drive to be more, the tendency of organic nature to want to ascend and to enhance itself.—In response to the first argument we might ask why it is precisely the human who has this exceptional fear that drives him to artificially protect himself, given that he is physically in no way less equipped than many other species and better so than thousands of them? Why are his needs such that they cannot be satisfied naturally (at least not in full)? Is it because he knows that he must die? Why is it precisely the human who becomes privy to this knowledge (as we are meant to assume that it is denied to animals)? Certainly the fear of death, the concern for one's own life, is something specifically human and generates culture in a much deeper way than we admit today. But it is hardly the ultimate foundation upon which the human tendency to think and create is built. The fear of death is in fact only a symptom of the basic structure of excentric positionality, which precedes all specifically human achievements. Everything that is alive, provided it is organized in animal form, becomes fearful when it notices its living environment being threatened or restricted. [319] But only the human is concerned *for* his own existence or even for the existence of other beings. Animal life does not know what it is to be afraid of coming dangers because it does not live ahead of *itself* [*lebt* sich *nicht vorweg*]. To be sure, it is ahead of itself—that is, it genuinely lives in the present—but it does not live in the future like the human, who knows of himself because he is ahead of his own being-ahead-of-himself, anticipates his anticipation. While true fear and true concern are not necessarily based on the knowledge of future things, they are only possible for those living beings for whom the temporal mode of the future has at least opened up. Fear alone cannot explain cultivation and the "invention" of the tool. If one wants to appeal to concern and fear, particularly the fear of death, as explanations, one must understand that such fear in fact presupposes the human form of life.

What about the second argument of the will to power, of the drive to be more? Even as it figures in Nietzsche and in pragmatism, this argument does not suffice to explain the irrealization of human action. All social animals exhibit domination tendencies, so there must be an additional element in the human. Some say it is intelligence; others point to the hypertrophic development of the drives (perhaps—under the influence of intelligence and the brain—in the form of a compensatory hypertrophy made necessary by the latter's predominance). But [human] intelligence must qualitatively differ from animal intelligence even just to explain the production of spiritual weapons and, above all, those spiritual entities that are neither weapons nor tools. Wolfgang Köhler's experiments have demonstrated the likelihood of highly developed animals successfully producing primitive tools if the obstacles in their way are powerful enough to inhibit the normal process of the drive-satisfying reaction. These experiments also show the difference between animal and human tool production: the animal does not know *what* it is doing. It seems to be able to remember its artificially assisted action and can generally reproduce it on command, but it does not notice the circumstance in which its action results. It does not occur to the animal that this invisible matter or possibility could become detached from the visible result in which it is contained. We can thus conclude that all tendencies toward power over one's own kind—[320] instead of "power" we should rightly use the more spiritual term "rule"—that all tendencies to rule, then, in no way explain the origins of culture, unless a specifically human intelligence is already presupposed.

The fact that the human is the apostate of nature and a troublemaker, that he needs to count as well as to achieve, and that life's tendency to enhance itself in him takes on the excessive form of a drive to power—none of this should be regarded as the foundation of the origin of cultivation, but must itself be understood as a symptom of excentric positionality. This positionality forces the appearance of a will to power and precedes it. For the human must act in order to live. Of course, the exigency to perform, which is grounded in his excentricity, does not exert itself in one blow. A single action is not enough for him; instead, he is caught up in the restlessness of unremitting activity. This creates the impression (and presumably also, secondarily, the psychological tendency as a reflex of this exigency) that what has already been done is constantly being outdone. The completed achievements, of course, do not simply disappear again, and so they continuously accumulate, with each new feat towering above all the previous ones.

It is in order to *find* balance and not to leave it that the human becomes a being who constantly strives after novelty, wants to outdo, seeks the eternal process. Excessiveness—falsely absolutized as a tendency of life to enhance itself—is the necessary form taken by the human attempt to compensate for his own dividedness, lack of balance, and nakedness. The human labors only to procure that which nature owes him for giving him the highest form of organization.

In conclusion, I would like to point again to the fact that of all the reasons for the failure of the pragmatist, biological theory of origins, the primary one is that it traces the nonpurposive spheres of cultivation, mores, and culture (in the narrow sense) back to the purposive means of action—that is, tools—and then is not able to explain the inner weightiness of the resulting works—that is, their claim to validity, their substantive character, except by appealing to the so-called heterogony of ends (Wilhelm Wundt) or to Freud's notion of the goal displacement of sublimation, particularly in the case of the sex drive. The substantive character [321] of tools, even the simplest—ladder, hammer, knife, and so forth—is usually overlooked, along with the fact, essential to their existence, that they are detachable from the process of their invention. The human does not invent anything that he does not discover. The animal can find [*finden*], but not invent [*erfinden*], because it "thinks nothing of it" [*nichts dabei finden*]—that is, does not discover anything. The results of its action do not disclose themselves to it. But how much more is required for a living being to intend mores and purposeless works. Something quite different must come into play here, something that has its origin in the special existential form of needing supplementation. Only because the human is naturally partial and thus stands above himself is artificiality the means to find a balance with himself and the world. This does not mean that culture amounts to an overcompensation for inferiority complexes, but points to a prepsychological, ontic necessity.

THE LAW OF MEDIATED IMMEDIACY: IMMANENCE AND EXPRESSIVITY

All of his initiative would not be able to help the human artificially restore the balance ontically denied him if the results of his initiative were not detachable from himself. Even the simplest tool is only a tool insofar as it is a state of affairs, comprises a state of being. The most primitive weapon, the most unsophisticated instrument is only serviceable under that condition. If we assume that the things we deal with and use acquire their full meaning, their entire existence only from the hand of the creator and are only real [*wirklich*] as they relate to our dealings with them, then we only see half of the truth. Just as essential to these technical aids (as well as to all works and rules originating from human creativity) is their inner weight, their objectivity, which appears in them as that which could only be found and discovered, not made.

Everything that becomes part of the sphere of culture thus exhibits both a connection to its human authorship and (to the same extent) independence from it. The human can only invent to the extent that he discovers. He can only make what "already" exists, just as he himself only becomes human by making himself so, and only lives by [322] leading a life. His productivity is only the occasion for his invention to become an event and to take form. The correlative relationship discussed earlier between the a priori and a posteriori elements that generally dominates—even makes up—the situation of the living being and its adaptation to its surroundings is repeated here on the level of conscious making, which only becomes creative when it succeeds in adapting in a specific way to the objective world. The secret of creativity, of hitting upon an idea, lies in the fortunate touch [*in dem glücklichen Griff*], in an encounter between the human and things. The quest for something specific does not precede the invention, for whoever is looking for something in truth *has* already found it. He is subject to the law of that which is, according to which a discovery is merely the fruition of a striving that is assured fulfillment. That which precedes both seeking *and* finding is rather the correlativity between the human and the world, which points back to the identity of the excentric form of human positionality and the structure of the reality of things (which itself exhibits an "excentric" form).

No one would claim, however, that this suffices to characterize the essence of invention and the "fortunate touch." To invent also means to convert possibility into actuality. It is not the hammer that existed before it was invented, but the state of affairs to which the hammer lends expression. The gramophone was, as it were, ready to be invented when it was established that sound waves can be transformed mechanically—a state of affairs that was not brought about by human creativity. But the invention still had to be carried out—that is, a form for this state of affairs had to be found. The creative touch is an achievement of *expression*. This lends the act of realization, which is dependent on the materials provided by nature, its artificial character.

Corresponding to its inner essence and external appearance, every achievement of expression breaks down into content and form, into a *what* and a *how*. The present discussion does not call for a treatment of the basic types of expression; it is rather the universal law that shows itself to be operative in every one of its types that I wish to work out here. My book *Die Einheit der Sinne* provides an in-depth exploration of the basic forms of giving expression while taking into account the aesthesiological problems involved. I recall that work here in order to preclude erroneous interpretations of the following. [323] For it was not my intention in *Die Einheit der Sinne* to abolish the specificity of expression, its fundamental categories, for the sake of a universal law of expression or something similar, or to deduce the former from such a law. Expression and thus culture as *manifestation* in a tangibly intelligible form are only possible according to one of these categories. (This in turn is different from the demonstration of these categories of manifestation or realization by way of the specific empirical facts of a historically evolved culture, such as Euclidean geometry, verbal language, or the "pure music" of post-Reformation Europe (as is done in *Die Einheit der Sinne* following the aesthesiological method). On occasion this has been absurdly misunderstood to mean that this particular music, geometry, or language was being presented as an a priori category. That is not the case, of course. Neither the form of expression nor the content can be a priori, only the way in which a content finds its "form"—a process that can solely be made visible by means of exceptional examples.)

In the present context we are concerned with the necessity of expression as such, which is prior to individual modes of expression, with the essential connection between the excentric form of positionality and expressivity as the mode of life of the human. Everyone has had the experience of feeling compelled to express himself, of having to speak his mind, and we attribute this to our being born for a life with others. This need to communicate varies from person to person and must in turn be distinguished from a different kind of need to express oneself, whose psychological significance is often underestimated and that cannot be attributed in the same way to sociality: a need to show things in facial expressions; to show or depict in general things we have experienced, disturbing feelings, fantasies, or thoughts. The way in which and the degree to which artistic achievements develop depend on the intensity and orientation of this need, which seems to initially be motivated by the tendency to preserve and structure what is fleeting in life by giving it form.

The need to communicate and the need to give form thus themselves point to existential forces that only act through them. Whether these forces are directly or indirectly connected to the sociality of the human form of life, whether other aspects of this form are not also involved cannot be explored here. One thing is certain from the preceding analysis: the excentric form of positionality gives rise to the shared-worldliness or sociality [324] of the human, makes him a *zoon politikon*, and at the same time gives rise to his artificiality and urge to create. The question is whether excentricity also implies, in an equally originary way, not this or that kind of *need* to express, but an essential feature of human life that would have to be referred to as the expressivity of all human manifestations of life. An essential feature of this kind of course also takes on the character of an exigency for the human, who is not only absorbed in his life but also opposes it, leads his life as he lives it.

To be excentrically positioned means that the living subject has an indirect-direct relationship to everything. A direct relationship is one where there are no connecting links between the members of the relation, an indirect relationship one where there are such links. We will call an indirect-direct relationship that form of connection in which the mediating link is necessary in order to produce or ensure the immediacy of the relation. The notion of indirect directness or mediated immediacy is thus not meaningless, not simply a self-refuting contradiction, but a contradiction that resolves itself without reverting to zero—a contradiction that is meaningful, even if analytical logic cannot follow.

I have attempted to make clear that the living as such possesses the structure of mediated immediacy. This structure arises from the nature of the boundary posited as real. Given that the positing of the boundary as real is the constitutive principle of all organic formation, the excentric form of organization also participates in this structure. This "abstract" participation of every organization in the structure of mediated immediacy, which is essential to all life, must be distinguished from the specific meaning that this structure has for each individual level of organization. Even the fact that the levels differ from each other according to the principle of openness and closedness has a bearing on the relations of the organic body to other bodies.

Furthermore, the organic body has a particular positional character. In terms of positionality, then, there is differentiation. In the case of plants, there is no positionally grounded relationship between living subject and medium. While a (direct) [325] relationship expresses itself in the organism, it is not there *as* a relationship. It is so, however, for the animal. It is true that the relationship between the animal and its surrounding field expresses itself in the organism according to the law of the closed form as indirect relation, regardless of whether it is decentrally or centrally organized. For the animal itself, however, the form of organization makes a difference in its positionality. In the case of decentralization, there is a direct pairing of stimulus and response. The relationship between the subject and its surrounding field is immediate. In the case of centralization, the coordination of stimulus and response is effected by the subject. The relationship between the subject and its surrounding field is mediated.

But what can it mean in terms of positionality, for the living being, that a relationship mediated *by the living being itself* exists between it and its surrounding field? Since the living being is still hidden from "itself," this relationship can only appear to it as direct, as immediate. The living being stands at the point of mediation and constitutes it. In order to notice the mediation, it would have to stand next to it without thereby losing its mediating centrality.

As already indicated, this excentric position is realized in the human. He stands at the center [*Zentrum*] of his standing. He constitutes the point of mediation between himself and his surrounding field *and* he is posited at this point; he stands there. This means, for one, that while his relationship to other things is indirect, he lives it as a direct, immediate relationship just like the animal—*inasmuch as* he, like the animal, is subject to the law of the closed form of life and this form's positionality. Second, it means that he knows of the indirectness of his relationship; it is given to him as mediated.

It would seem to follow from this with stringent necessity that the human as an excentric creature occupies two fundamentally different relationships to the outer world, to the alien world as such—a direct "and" an indirect relation. This conclusion is false, however, and overlooks a crucial premise: the identity of the one who *stands* at the center [*Zentrum*] of the mediation. As Fichte already recognized, this identity exists only insofar as it is achieved by him who is assumed to be identical with himself. It is only this positing of oneself that constitutes the living subject as I or excentric positionality. [326] Consequently, the human is no longer hidden from "himself." He knows of himself that he is identical with himself as the one who knows. The human occupies *one* relationship of mediated immediacy, of indirect directness, to external, alien things, and not two neatly separated relationships running along next to each other.

If we were to assume—incorrectly—that the human related to his surrounding field in both an immediate *and* a mediated way, it would be impossible to decide which of these relations dominated and shaped his life. The two relations would be incessantly competing with each other: now it would be one way (immediate), then it would be another (mediated). The human would never occupy an unequivocal position vis-à-vis his surrounding field, but only a position that oscillated back and forth between opposites.

Yet an oscillation of this kind, a competition between the contradictory relationships of the living subject to his surrounding field, is itself not even possible, as it would mean the identity of the subject with *himself* going to the devil. If this identity is to be preserved *even just in principle*, the struggle of one relation against the other that is contradicting it would come to constitute the foundation of the human position; both relations would, in ideal terms, have to take place in one and the same sense—which would amount to their mutually canceling each other out and making any position impossible. It would be a matter of a contradictory demand whose even partial fulfillment would render zero.

Now, why does directness and immediacy dominate over indirectness and mediatedness? Why is the human said to occupy a relationship characterized by mediated immediacy or indirect directness? Why can it *not* be said, and why is the apparently valid formulation that the human exists in his surroundings in direct indirectness, in immediate mediatedness, incorrect?

Because that which is positionally true for the animal holds analogously for the animal as a human being—that is, insofar as he is not subject to the same law of the closed form as is the animal, but rather *fulfills* the form of excentricity reserved to him alone. Seen positionally, the mediated relationship between the animal and its surrounding field cannot for the animal have the character of mediatedness, because the animal *is* [327] the mediation between it and the field and is thus centric, completely absorbed in this mediation and still hidden to "itself." The situation of the human is different. He is the mediation between him and the field, but only becomes completely absorbed in that field insofar as he still also "stands in" this mediation. He is thus above it, which means that he constitutes the mediation between *himself* and the field. It is not that he, to put it roughly, like the animal constitutes the mediation between him and the field "down below," while "up above" he is separate from this, stands above it, does not participate in the mediation, and watches himself, as it were, as if he were another, as sometimes happens when dreaming. If that were the case, *he would not be the other to himself*, would not be himself, would not be the mediation between *himself* and the field as immediate absorption in the relation.

In order for the human to carry out this mediation and to maintain it at his level of existence, it must actually pass through him as he stands in the field. His *standing above it* must guarantee the living immediacy between him and the field. The fact that he is set apart from himself, by virtue of which he can say "I" to himself and exists as I, hence forms the relationship between him and the field in such a way that the relationship reveals this setting apart.

And that is in fact the case. The human lives in a surrounding field that has the character of a world. Things are given to him as objects, real [*wirkliche*] things that *in* their givenness appear as detachable *from* their givenness. Their essence includes the surplus of their own weight, of existing-for-themselves, of being-in-themselves, without which we would not speak of them as real [*wirklichen*] things. Nevertheless, this surplus moment, this surplus weight becomes manifest in—their appearance, which of course belongs to reality [*Wirklichkeit*], but does not reveal all of reality, and only objectively presents the side of the real that is turned toward the subject in a real [*reell*], that is, direct way. As a result, the subject can only grasp reality through the mediation of this appearance—in the mode of immediacy, because the surplus weight of being-in-itself, of being more than appearance, immediately appears "in" the immediate presence of the appearance.

If we replace the idea of standing-above, which has been used to describe the excentric positionality of the human, with the idea, also employed earlier, of standing-behind-oneself, the situation of the human in the world suddenly becomes [328] clear, and old notions about this situation take on a new life. *The situation of the human is that of being immanent in consciousness.* Everything he experiences he experiences as a content of consciousness and *therefore* not as something in consciousness but as something existing outside of consciousness. Because the human is organized excentrically and therefore has come to stand behind himself, he is set apart from everything that he is and that is around him. Doubly set apart from his own lived body, placed in the center of his position and not simply living out from this center like the animal, the human knows of himself as body and soul, of other persons, living beings, and things in an immediate way only as appearances or contents of consciousness. It is mediated by them that he knows of the appearing realities.

In the direction of knowledge itself as experience, observation, perception, comprehension, and understanding, the human's relationship to knowledge must be immediate, direct. He has no choice but to grasp matters in their naked immediacy. Because they are for him, they are in themselves. For the subject itself standing behind (over) itself *constitutes* the mediation between itself and the object, so that it may know of the object. Or, to be more precise, knowledge of the object is the mediation between the subject and the object. In this way, mediation as it is executed extinguishes the human as the mediating subject standing behind itself. The subject forgets itself (*the human* does not forget *him*self!)—and the naïve directness convinced of having seized the thing itself comes about.

Just as the mediated relation given between the animal and its surrounding field cannot have the character of mediatedness *for the animal itself*, because it, the animal, is the one executing the mediation between it and the field (and, thanks to its centricity, is absorbed in this execution), so the relation the human mediates to his surrounding field also takes on the character of immediacy for him. For him also the executing center or the I is engaged by the execution; more than that, it becomes pure execution, pure passing-through. His excentricity, then, due to which he stands behind (above) himself, cannot prevent the consciousness of immediacy and direct contact. The gazing subject (the center of the position) and the subject standing at the center are, after all, one and the same.

For this reason, excentricity, even if it forgets itself in the execution [329] of knowledge (that is, of mediation), is not extinguished. It is by virtue of excentricity that knowledge immediately comprehends something that is mediated: the reality in the appearance, the phenomenon of actuality. Appearance is not to be thought as a leaf or a mask concealing the real behind it and detachable from it, but is like a face that conceals *by* revealing. In this concealing revelation lies the specificity of that which is there in the appearance itself—albeit "not quite" there, but still behind, hidden, for itself and in itself. This is the only way that an actuality can enter into a relation with a subject *as* actual: as thrown over against the subject, as object—that is, as appearance [*Er-scheinung*], manifestation of . . . , in short, as mediated immediacy. Otherwise the character of actuality, objectivity, is lost, as is the case for the animal. The animal's centricity does not allow it to grasp appearances [*Erscheinungen*] as appearances [*Er-scheinungen*]. It perceives images without the character of objectivity.

At the same time, excentricity means that the human doubts the immediacy of his knowledge, the directness of his contact with reality as it exists for him with absolute self-evidence. Just as only excentricity makes it possible for him to enter into contact with the real, it also enables him to reflect. Thus he becomes conscious of that which executes acts of perception and knowledge—that is, of his consciousness. In this way he discovers—not the psychic, which is itself a reality of his life—but the indirectness and mediatedness of his immediate relations to objects. He discovers his immanence. He recognizes that he effectively only has contents of consciousness and that no matter what he does, his knowledge of things interjects itself as something between him and these things.

If the knowledge with which he establishes contact, the eye with which he sees, is an intermediate thing, the knower, the subject-pole, can no longer enter into a direct contact with reality. This necessarily leads him to lose faith in consciousness as it is for itself and for him. He despairs of the act's self-evidence, of intentionality, of the value of the opinion consciousness evinces in its acts. The standard response to this is that of course the subject believes it is grasping reality [*Wirklichkeit*] and has ahold of reality itself. But that is only true for the subject. In fact, it is in a realm of sensations, representations, and contents of consciousness. The eye necessarily forgets itself [330] when it sees. Knowledge is, in essence, precisely a ray that leaves its point of departure, that extends outward and goes beyond itself: ekstasis. This is the reason for the necessary semblance [*Schein*] of immediacy, of whose groundlessness reflection convinces us.

And so the evidence of intending consciousness stands opposed to the evidence of reflecting consciousness. The doctrine of immanence confronts the doctrine of transcendence. The idealism of consciousness battles realism. There is motivation to decide which view is correct. Theories are devised in order to make natural experience, the everyday view of life with its "believe," its *doxa*, comprehensible to the idealists and teachers of immanence or, on the contrary, to explain to the realists the possibility of questioning their convictions, of the objection of immanence. Or, as in the case of Kant, there is a justification of scientifically trained experience's claim to objectivity. The reality of the inner world (which can be studied by psychology) serves as the basis for attempting to prove the reality of the outer world. Or there is an attempt to reconcile the two positions, and arguments from so-called critical realism (or idealism) are used to demonstrate that both realists and idealists have a one-sided view of the matter and that true reality or reality as such can only be reached indirectly, either by means of inferences or of emotional acts (drives or acts of the will or feelings).

What is certain is that the human necessarily finds himself in this dilemma because of his excentricity. The epistemological question is not an arbitrary one. The notion that the epistemological problem is "alien to nature," that the human brought this difficulty upon himself simply by having a false understanding of himself, is mistaken. But it is just as certain—as I have shown—that both the idealistic interpretation of immanence, according to which the indirectness of the relationship between subject and reality [*Wirklichkeit*] is carried by an intermediary, an intermediate layer, as well as the realistic denial or interpretation of immanence (which also assumes a mediating in-between that has taken on a life of its own) are wrong. Chapter 2 of this book shows where the idealistic theory leads. The strength of the new proof of reality lies in the fact that it understands the *subject's situation as immanent to be the indispensable condition for* [331] *its contact with reality* [*Wirklichkeit*]. Precisely because the subject is confined within itself and imprisoned within its own consciousness, thus doubly set apart from the sensory surfaces of its body, does it keep the distance required by reality as the reality that is to reveal itself, the distance *appropriate* to being, the space in which reality [*Wirklichkeit*] can come to appear. Precisely because it lives in indirect relation to that which exists in itself, its knowledge of that which exists in itself is, for the subject, immediate and direct. The evidence produced by acts of consciousness is not deceptive but rather legitimate and necessary. The evidence produced by reflection on the acts of consciousness is just as reliable and necessary. The breakdown into the two views of immediacy and mediatedness is given by the excentric positionality of the human. But he is not able to break down the two views themselves. It is not that both are right, presenting the philosopher with an insoluble antinomy. Nor is only one of the two right, allowing it to be played off against the other as the only one with any weight.

It thus becomes clear that both the monadological conclusion, according to which all consciousness is self-consciousness, and the naïve-realistic conclusion, according to which all consciousness is direct contact with reality [*Wirklichkeit*], are wrong. The first view reifies the mediating in-between of knowledge into a box of consciousness from which there is no escape. The second holds only to the intentional character of knowledge acts.

Phenomenology (not as method but as doctrine) uses this view to argue against the epistemological posing of the problem.

The failure of attempts to go beyond the *res in mente* to reach a *res extra mentem* by means of intellectual or emotional processes follows with equal necessity. The human does not infer from the contents of his consciousness a reality announcing itself within these contents. But neither does he need the testimony of inhibited volitional impulses (as in Dilthey and Maine de Biran) or of feeling, instinct, or intuition in order to be sure of reality. His own form of life, which is prior to all contemplative (intuiting, feeling, seeing, intellectual) and active (striving, pressing, desiring) behavior [332] furnishes him with the guarantee of the objectivity of his consciousness, of existence, and of the accessibility of reality. Because he is *in* his consciousness and has, immediately, only images of actual being in nature, psyche, and spirit, he comprehends the actual world in and with these images in a way that for him is immediate. Just as being existing for itself, by virtue of being, is mediated immediacy—that is, is actuality with the potential for appearing—so it *needs*, in order to be able to reveal itself to a subject with its character of actuality, a subject standing in itself and, as it were, imprisoned in itself. The subject's immanence in its own consciousness, which introduces twice as much distance between thing and subject as exists for the animal, is the only guarantee for the establishment of contact between thing and subject. Only indirectness creates directness, only separation brings about contact.[5]

This seemingly paradoxical insight causes difficulties for the standard view, which draws on the well-worn assumptions of idealism, realism, and their various hybrids. This becomes particularly evident when we come to the implications of the theory of the mediated immediacy of consciousness, to the question of the subjectivity or objectivity of sensory qualities. The aesthesiological inquiry of my *Die Einheit der Sinne* laid the groundwork in this regard. Some of what I put forward there and that is comprehensible in context but, when taken out of context and regarded on its own, seems to have no connection to traditional questions—particularly the discussion about the sensory modality as relational modality between spirit and physical, lived body and the possibility of perception[6]—only acquires its full meaning with the concept of the indirect directness of consciousness.

But have we not been waylaid from our original goal of discovering the essential connection between the excentric form of positionality and expressivity as a mode of life [333] of the human? What does the human situation of immanence, his imprisonment in his own consciousness, have to do with expressivity? The answer is surprising enough: immanence and expressivity are based on one and the same state of affairs—that is, the double distance between the center [*Zentrum*] of the person and his lived body.

As long, however, as one conceives of consciousness as a box whose walls hermetically close off our entire being and life, our knowledge and will, our cognition and action from the outside world, it is impossible to see this connection. In that case, all stirrings of life coming from the subject, particularly the motor reactions arising from feeling, drive, or will, would occur in the same internal region as the representations we have of them through our consciousness. They would thus not genuinely be movements with an expressive character, but would simply look like they were; expression would only be their "meaning." The essential connection between immanence and expressivity is equally incomprehensible to a view that thinks of consciousness merely as a lens placed in front of the eye of the body and the eye of the spirit, or in place of the head, as a box whose varied contents make up the matter of our knowledge while our emotional life takes place outside of it. This view tears apart the connection between immanence and expressivity. It understands the human's situation in life such that everything centripetal upon which his knowledge, intuition, recognition, and the abundance of his contemplation rest is made up of contents of consciousness, while everything centrifugal—urges, drives, tendencies, the will, everything that discharges itself in thrust and grasp—is in contact with reality [*Wirklichkeit*] itself. Only by taking the roundabout way via the emotions, interests, and actions would the contents of contemplation acquire reality—according to, as it were, a priority of the practical that occupies the middle ground between the primacy established by Kant and that established by pragmatism. This view, however, loses sight of the indirectness and brokenness of everything that is centrifugal and thus of the specific structure of expressivity characterizing human manifestations of life.

The adequacy of expression as a stirring of life that actually does bring to the outside what is within *and* its essential inadequacy and brokenness as the translation and shaping of a living depth that itself never surfaces—the law of mediating [334] immediacy can render this seeming paradox comprehensible and show it to be binding for human existence, as it can the seeming paradox of the consciousness of reality based on immanence. Every stirring of life in which the spiritual center [*Zentrum*] of action or the person participates must be expressive. That is, it is for itself, from the viewpoint of the subject, an immediate, direct relation to its object and finds its adequate fulfillment *only insofar* as the intention of the drive, urge, longing, will, aim, thought, or hope *does not* stand in a direct relationship to that which in fact follows from it and constitutes the ultimate, satisfying result. Thus the only reason that the de facto inadequacy of intention and actual fulfillment, which is due to the total heterogeneity of spirit, psyche, and bodily nature, does not prove to be the undoing of intention, does not condemn it to eternal disappointment, does not render its belief in fulfillment a merely subjective illusion, is that the relationship between subject and object, as a relation of indirect directness, absorbs, legitimizes, and *demands* this fissure.

There is a general tendency to overlook the problem inherent in the essential difference between spirit, psyche, and body and to simply put up with the fact that the human has to realize his ideas, desires, drives, and plans in a wayward medium. "The thoughts lie peaceful in their ample room; but in the world, things clash against each other."[7] The objectification and realization of those tendencies moving the person and coming from him would be impossible, however, if the relationship between his center [*Zentrum*] and the medium were direct and undiverted—regardless of whether the medium were a wayward reality [*Wirklichkeit*] or mere "stuff" subject to the mastery of the person.

In the first case, the personal subject would have to make a compromise with matter in order to assert and realize its aspirations—but would not be able to. The rays of its intentions would immediately encounter the real medium and suffer a diversion (one the subject could perhaps even predict). This diversion would mean the refraction of its intention, and the original goal would never be reached. Realization would by its nature mean the human attaining results that were not wanted. And even if he were told a thousand times and had to tell it to himself; even if he understood it to be an internal necessity that [335] intended fulfillment and objective fulfillment diverge; still no force and no insight of his own could bring him under such circumstances to speak of the fulfillment of his intention. All initiative would be futile. And even if the human did have the feeling of satisfaction and evidence of fulfillment, he would only be the poor victim of an illusion.—Those espousing tragicism and pessimism, of course, have an interest in this being precisely so. They shrug their shoulders at the resignation into which the human is forced and take the easy way out by reminding us that in the real world, things clash against each other. The physical existence of the human is used to explain the brokenness of human plans, the character of compromise in real culture. The fact that this dualism, which acquired its specifically philosophical form with Plato, has had enormous consequences in the history of ideas hardly needs to be recalled here.[8]

In the second case, the personal subject does not have to make any compromises with matter. For matter in this conception is completely lacking in any wayward reality that could dash the subject's aspirations. The medium of its creativity does not force a diversion of the ray of intention. The person can do what he wants. But if this were the case, there would also not be any such thing as fulfillment. There would only be a kind of translation or transposition from out of interiority into exteriority. Objectification would forfeit both the joy of fulfillment *and* the pain of the negative. The resignation of creativity would become a sadness without object. All human aspirations could effectively reach their goals; they would simply progress from fulfillment to fulfillment.—Those espousing optimism of all sorts have an interest in this being precisely so. Here we find the tendency to forge alliances with idealism, which gives the subject power over nature and turns nature into the subject's self-made counterprojection and counterimage, dissolves nature into sensations, and reduces it to a substrate of human creative whimsy. On this basis, clearly, the restlessness of human initiative in its appearance as world history can only be understood in terms of development and progress, just as [336] the panarchy of the idea of progress is only true on this basis.

A genuine fulfillment of intention, an immediate relationship between the subject and the object of its aspiration, adequate realization, is only possible as a mediated relationship between personal subject and attained object. Fulfillment ought to come from *there*, not from here, and its nature is that it can also fail to occur. Only where reality falls into place of its own accord is intention fulfilled, are aspirations attained. The compromises with reality [*Wirklichkeit*] that the subject enters into in order to make its wishes come true and to render harmless the refraction of the rays of intention in the wayward medium of the psyche and the body are thus not the precondition, not the means, but *themselves already a compromise* of genuine fulfillment. For a reality [*Wirklichkeit*] with which the subject has made a pact before approaching it with its aspirations is no longer the original reality in its in-itself. It has rather already been subjected, made compliant to the subject as a result of the subject's observations, experiences, and calculations. It is already the counterimage of the subject's possibilities for success, the medium that has disclosed its diverting, refracting nature. What, as technicians and artists, scholars and teachers, politicians and doctors, merchants and lawyers, we call compromises with reality: all this is a necessary consequence of that original compromise in the world that we are and that surrounds us, the compromise between our personal center [*Zentrum*] and reality [*Wirklichkeit*] in itself, on which the possibility of true fulfillment rests.

Only the original, *un*prearranged encounter of the human with the world, only the attainment of an aspiration by way of a fortunate touch, only the unity of anticipation and adaptation deserve to be called "genuine fulfillment." Fulfillment for the subject is immediate and adequate *and* in itself the bridge between the essentially different zones of spirit and reality precisely because reality demands the distance to be kept that the center [*Zentrum*] of the personal subject has by virtue of its excentric position, its double set-apartness from its own lived body. Now in the case of consciousness, this mediated immediacy becomes manifest in the structure of the object that it grasps. As the appearance of a reality, as a diversity that is bound to a core and can be directly experienced, the object is itself mediated immediacy. It corresponds for itself to the structure [337] of the consciousness of it. Should this law of correspondence not also hold for the aspiring behavior of the person? Is the genuine fulfillment of an intention not also objectively mediated immediacy, in that it is a fulfillment of the striving that led to it and not merely an element of reality like everything else that is real? Just as correspondence is the only possibility for consciousness to enter into what for itself is an immediate relationship to the object, to directly encounter fully self-evident actuality in the object, and to nevertheless not be deceived, to experience more than mere appearances, because none other than the real object in itself makes manifest mediatedness—so the same holds for aspiring behavior.

If the result of an aspiration of whatever kind exhibits the character of mediated immediacy, it presents itself in some way as a "what," as content in a form. The possibility of setting the "what" apart from the "how" of the accomplishment of the intention reveals this character. It is only due to this possibility that the subject is potentially able to reach the goal to which it aspires in the creative act of a primary encounter with reality,[9] *despite* the deviation and refraction of its intention in the medium of reality. Although the goal of the aspiration never coincides with the endpoint of its realization and the human in a certain sense never gets to where he wants to go—whether he is making a gesture, building a house, or writing a book—this diversion does not render his aspiration illusory and does not preclude its fulfillment. The *distance* between the goal of the intention and the endpoint of its realization is the "how" or the form, the mode of realization.

Every stirring of life of a person that becomes intelligible in deed, utterance, or facial expression is thus expressive, somehow *expresses* the "what" of an aspiration, whether it wants to or not. Every deed, utterance, or facial expression is necessarily realization, objectification of the spirit. It is not that we have content here and form there, in the manner of the professional who selects certain methods to achieve his goals. Anticipation and calculation of the form are only possible where the human already knows about the reality in question and his intentions are assured fulfillment. For this very reason, the form understood as the distance between the goal of the intention and the endpoint of the realization cannot be [338] anticipated, cannot be taken away from the content and placed onto it; it rather *arises from* the realization. The form happens to the content, which itself is only the goal of the aspiration held onto during the process of its realization. And because there is this kind of continuity between intention and fulfillment, despite the refraction of the ray of intention in the medium of psychic and physical reality—a refraction that could not be anticipated beforehand and that is essentially never given for itself—the subject has the right to speak of attaining its aspirations.

It is precisely for this reason that the subject has the right and the duty to try *again* to succeed. This is because a successful result at the wrong place is also the wrong thing (not because of the flaws of earthly existence and of unwieldy matter, but for internal reasons). According to its content it may be an adequate realization, but where is the content? It cannot be isolated from the form; it is fused with it, and no one can say where the content begins and the form ends as long as the former, caught up in the aspiration, attains and holds onto fulfillment. It is only when we observe the successfully realized work, the completed gesture or utterance, that we notice the difference. As soon as it is realized it breaks apart into the "what" and the "how." The discrepancy between what was attained and what was aspired to has become an event. The animating striving has escaped from the result, which has grown cold and stays behind as a shell. Thus alienated, it becomes an object of observation when before it was the invisible space of our striving. And since striving does not stop, since it demands realization, it cannot be satisfied by that which has become *as* form. The human must get back to work.

The expressivity of the human thus makes him a being who even in the case of continuously sustained intention continues to push for ever *new* realizations and in this way leaves behind a *history*. Expressivity is the only internal reason for the historical character of his existence. It is not that the human has to be creative and only is insofar as he creates. Activity alone, eternal restlessness, is not enough to introduce difference into progress. Since the law of natural artificiality already imposes creativity onto the human as essential to his nature, it is conceivable that if intention and fulfillment were *either* directly *or* indirectly related to each other, this could lead to a [339] purely external progress that had nothing to do with historicity. In the first case, progress would be like a perfectly straight chain of adequate fulfillments, in which each successive link outdid the previous ones. In the second, progress would mean endlessly treading water, one and the same illusion endlessly appearing in different guises, and the restlessness of humankind from generation to generation would be worth as much as the running of a squirrel on the cylinder turning beneath its feet.

Both of these possibilities are precluded, however, by the mediated immediacy of human existence grounded in his excentricity. The process in which he has his essential life is a continuum of discontinuously deposited, crystalized events. Something happens [*geschieht*] in this process, and so it is history [*Geschichte*]. It holds the center, as it were, between the two possibilities of a process whose meaning lies in progressing to the next stage and a circular process equivalent to an absolute standstill. The notion that the meaning of history is to be found in a goal it is hastening toward is just as false, then, as the opposite idea that history is just one great *nunc stans*. The true motor of the specifically historical dynamism of human life is to be found in expressivity. With his deeds and works, which are meant to give the human the balance nature refused him—*and actually do give him this balance*—he is at the same time also thrown off balance again, only to try once more with luck and yet in vain. The law of mediated immediacy constantly knocks him out of the position of rest to which he wants to return. It is this basic movement that brings forth history. Its meaning is the recovery of what was lost with new means, the creation of balance through radical change, preservation of the old by turning toward the future.—

Language is one of the most frequently cited essential characteristics of the human. Rightly so, as the research shows—although "language" is too narrow a concept for what lies at the heart of this characteristic: that is, expressivity. Nevertheless, aside from the fact that language dominates in the real life of the human, it also occupies a special position in the layer of expressivity: language makes explicit the correspondence between the structure [340] of immanence and the structure of reality—both zones constitute mediated immediacy and relate to each other in terms of mediated immediacy—upon which expressivity in all its guises rests. Language makes the expressive *relation* in which the human lives in the world into the object of expressions. Language not only becomes possible because of the situation of immanence, of the double distance between the center [*Zentrum*] of the person and his lived body, but, by dint of the excentricity of this center, it also expresses this situation in relation to reality. Due to its excentricity, the excentric center [*Zentrum*] of the person, the executing center of the so-called spiritual act, is able to *express* the reality that "corresponds" to the excentric position of the human.

In this way the essential relations between excentricity, immanence, expressivity, and contact with reality converge in a surprising way in language and its elements—that is, in *meanings*. Language, an expression to the second power, is therefore the true existential evidence of the position of the human: standing at the center of his own life and therefore beyond it, outside time and place. In the peculiar nature of propositional meanings, the basic structure of mediated immediacy has been purged of all materiality and appears as sublimated in its own element.[10] At the same time, the law of expressivity, which governs every stirring of life of the person that demands fulfillment, is proven in language: language does not exist, only languages. The unity of intention holds together only in the fragmentation into different idioms. And we can safely hazard the assertion that the quest for a proto-language is condemned to failure for more than empirical reasons. Such a quest demonstrates ignorance of the law of the concretion and objectification of spirit, according to which intention (exceeding all restrictive form) is only achieved when, in the process of objectification, a form (one that is not in itself necessary) *"happens to"* the intention.

Realization and fulfillment of an intention means the refraction of its ray in an alien medium. The fact that this [341] does not destroy it, even though the diversion was not foreseen and cannot be calculated, shows that intention is reality's equal. A language—could say nothing. The ability of intentions to refract as a condition of their fulfillment, their elasticity, which is also the basis for their differentiation into different languages and individual types—these assure their force of (and possible fidelity to) reality.

Because the human in human beings always strives for the same thing, he always changes and becomes other. And because this gives rise in him to the subjective desire for the ever new and ever different, for upheaval, adventure, and pastures new, he thinks that unprecedented means are always required for its satisfaction. It rarely happens to us humans that we seek a donkey and find a kingdom. We find what we are looking for. But what we find undergoes change, and now and then a kingdom turns into a donkey. It is a law that human beings ultimately know not what they do and only find out through history.

THE LAW OF THE UTOPIAN STANDPOINT: NULLITY AND TRANSCENDENCE

Dos moi pou sto.[11] This phrase is written above all of human existence. The excentric form of this existence drives the human to cultivation and creates needs that can only be satisfied by a system of artificial objects, which it stamps with the mark of transience. Human beings attain what they want all the time. And as they attain it, the invisible human within them has already gone beyond them. The reality of world history testifies to his constitutive rootlessness.

But the human also experiences this rootlessness in himself. It gives him the consciousness of his own nullity and, correlative to this, of the nullity of the world. In the face of this nothingness, it awakens in him the realization of his own uniqueness and singularity and, correlative to this, of the individuality of the world. And so he awakens to the consciousness of the absolute contingency of existence and thus to the idea of a ground of the world [*Weltgrund*], of necessary being resting in itself, of the absolute or God. [342] This consciousness, however, is not of unshakable certainty. Just as excentricity does not permit an unequivocal fixing of one's own position (that is, it demands such a fixing, but constantly rescinds it again in a continual annulment of its own thesis), it is also not given to the human to know "where" he and the reality corresponding to his excentricity stand.[12] If he wants to choose once and for all, there is nothing left for him but to make a leap of faith. The concepts of and the feeling for individuality and nullity, contingency and a divine ground of one's own life and of the world have in the course of history and in the broad diversity of cultures certainly varied in their aspect and in their importance for life. But they all contain an a priori core given with the human form of life as such, the core of all religiosity.

One can argue whether religion by its very nature presupposes a need for salvation or whether it merely performs a de facto function of salvation for the believer. The notions held of the divine vary along with those held of the sacred and of the human. One thing, however, is characteristic of all religiosity: it creates a *definitivum*. That which nature and spirit cannot give the human, the ultimate "it is so," religion wants to give him. Only religion provides the ultimate bond and integration, the place to live and the place to die, security, reconciliation with fate, interpretation of reality, a home. This means that between religion and culture, despite all historical peace settlements and the (rarely sincere) assurances such as are popular today, there is absolute enmity. Whoever wants to come home, to be secure, must sacrifice himself to faith. Whoever follows the spirit, however, does not return.

For someone thus positioned, excentricity is an insoluble contradiction. It is true that excentricity incorporates the human into an outer world and a shared world and allows him to internally comprehend himself as reality. But this contact with being comes at a high price. Excentrically positioned, he stands where he stands and at the same time not where he stands. He both occupies and does not occupy the here in which he lives and to which his entire environment relates in total convergence, the absolute, nonrelativizable here/now of his own position. He is positioned in his life; he stands "behind" and "above" it and thus forms the center of his environment, a center that has been removed from the circular field. But the notion of an excentric center remains an absurdity, even [343] if it is realized. Because existence for the human harbors a realized absurdity, a transparent paradox, a comprehended incomprehensibility, he needs something to hold on to that will liberate him from this situation. This dependency on a support for his own existence outside of the sphere of reality means that reality—outer world, inner world, shared world—which is essentially correlated to his existence, necessarily becomes in need of support itself. In relation to this reality-transcending point of support or anchor, reality then joins together into *one* world, into the universe [*Weltall*]. It is in this way that reality suffers its objectification as totality and thus its setting apart from something that *is* without being from this world. Having become a something, reality becomes a *this* and separates itself from a sphere that is not this, that is something else. It stands there individually as *one* world, for a horizon of possibilities of also being able to be otherwise has opened up.

In this world that is real in this way and not in another, the individual too is individuality. The human no longer signifies to himself simply an indivisible, seamless being, but rather an irreplaceable life in the here and now that cannot be substituted by anyone else. The irreversibility of the direction of his existence takes on a positive meaning. This is normally explained by the preciousness of a lifetime that is cut short by death. But death, in the face of which the human lives, does not give him the vantage point from which to see the uniqueness of precisely his own life. Just as the world as an individuality only sets itself apart from the horizon of the possibility of being able to be otherwise, the human's own existence only sets itself apart as individual from the possibility that he could have also become another. This possibility is given to the human on the basis of his life form. The human is to himself the backdrop of humanness as such, against which he stands out as "this one and no other." The single individual stands in the shared world as pure I or we. The shared world not only surrounds the individual like the environment, not only fills him like the inner world, but stands through him [*steht durch ihn hindurch*]. The human *is* the shared world. The human is humanity—that is, as an individual he can be substituted and replaced in an absolute sense. Every other could take his place, just as he and the other are joined together in the placelessness of the excentric position in an originary community with the character of a we.

[344] The *substitutedness* and *replacedness* of every individual by every other in the form of the we, which is prior to the formation and expression of the feeling and behavior of solidarity, prior to the concrete community, constitutes the background from which the individual sets himself apart as individuality. He is basically the same as the other; he stands where the other stands and the other occupies his place. This is why the other in outer-worldly and inner-worldly reality *can* occupy the position that every human possesses in his absolute here, or—"he could have also become the other." The individual human's actual replaceability and substitutability give him the warrant and certainty of the contingency of his being or individuality.

This contingency is the reason for his pride and for his shame. Even the fact of the irreplaceability of his own living substance, in which he is different from everyone else, does not offset the fact that he is replaced in the we, that he is replaceable by every person he meets. This is why the human in all his preciousness is ashamed. The nullity of his existence, its total permeability, and his knowledge of the fact that we are all basically the same because we, each for ourselves, are individuals and thus different from each other are the reasons for his shame (and only in a derivative sense the object of metaphysical shame and the beginning of humility). It is of course an indirect reason, mediated by the inner reality of psychic being. This gives rise to the ambiguity in which the human is torn between the urge to reveal himself and to be admired and the urge to hold back. This ambiguity is one of the basic motives for social organization. Straightforward relationships between human beings do not follow from human nature, are not part of the essence of the human. The human has to create them. Without the arbitrary establishment of an order, without doing violence to life, he cannot lead a life. This is the ultimate grounding of the claims made in *The Limits of Community*, which have met with a certain resentment-driven incomprehension on the part of some sociologists and social policy makers.[13]

[345] Of course sociality (understood here in a broader sense than in the work of Ferdinand Tönnies) does not derive its justification and necessity from this factor alone. The artificiality and indirectness of human existence also play an important part. Even if a purely communal form of living (again

understood in a broader sense than does Tönnies) seemed tolerable to the human himself, he would not be able to realize it. But social realization *ought* not to go in this direction, since respect for the other for the sake of the originary community of the shared world demands distance and concealment. It is precisely this originary community that sets society's limits. Thus human beings have an inalienable right of revolution if the forms of sociality destroy their own meaning, and revolution occurs when the utopian idea of the final destructibility of all sociality takes root. And yet this idea itself is only the means for renewing society.—This is not to be understood as a theory of restoration or an apology for the anxious bourgeoisie, but as the recognition of an essential law to which the human is subject regardless of social fashion; as an articulation of the essential law of social realization, an articulation that completely withholds judgment on particular social and political ideas.—

Consciousness of the individuality of one's own being and of the world and consciousness of the contingency of this overall reality [*Gesamtrealität*] are necessarily given together and require each other. His own lack of anchor, which both bars the human from finding an anchor in the world and becomes apparent to him as the conditionality of the world, suggests to him the nullity of reality and the idea of a ground of the world. The excentric form of positionality and God as absolute, necessary, world-grounding being are essentially correlated. It is not the image that the human has of God that is important, no more than it is the image the human has of himself. The anthropomorphism of the determination of the essence of the absolute corresponds necessarily to the theomorphism of the determination of the essence of the human—as Scheler observed—as long as the [346] human holds onto the idea of the absolute even only as the ground of the world. To give up this idea would mean to give up the idea of the *one* world. Atheism is easier said than done. Even Leibniz was not completely consistent with his idea of pluralism and was not able to forgo the concept of a central monad.

And yet it is possible for the human to have this idea. The excentricity of his form of life, the fact that he stands nowhere, his utopian standpoint, forces him to doubt divine existence, to doubt the ground of this world, and thus the unity of the world. If there were an ontological proof of the existence of God, the human, following the law of his nature, would leave nothing untried in an attempt to shatter it. As the history of metaphysical speculation shows, the same process that leads to the transcending of reality would have to be repeated in regard to the absolute: just as the excentric form of positionality is the precondition for the human to be able to grasp a reality in nature, psyche, and shared world, it also constitutes the condition for his insight into this reality's nullity and lack of anchor. It is true that the human standpoint is located opposite the absolute and that the ground of the world forms the only counterweight to excentricity. But in an existential paradox, the truth of this excentricity, for this very reason and with the same internal justification, demands to be removed from this relation of complete balance and thus to deny the absolute and call for the dissolution of the world.

A universe can only be believed in. And as long as he believes, the human will always "go back home." Only in belief is there such a thing as "good," cyclical infinity, the return of things from their absolute otherness. Spirit, however, sends the human and things away from itself and beyond itself. Its sign is the straight line of endless infinity. Its element is the future. It destroys the cycle of the world and, like Marcion of Sinope's Christ, opens us up to blessed strangeness.[14]

[349] APPENDIX

I hope the following two comments will forestall any misunderstandings or false expectations:

1. In my view, the times of vitalism are over forever. Vitalism—and I am thinking primarily of Driesch here—claims that it is impossible to understand life processes with an analysis based in physics and chemistry and therefore introduces a factor (entelechy) to compensate for what this analysis fails to do. The notion of entelechy itself, however, eludes all analysis. Such an approach is methodologically untenable, quite apart from the question of where research in physics and biochemistry happens to currently come up against limits.[1] Research must have free rein, and it must be given the methodological opportunity of using its own logical means to remove obstacles in its path. At what point a so-called insurmountable difficulty is not merely a temporary barrier but rather a genuine boundary is, first, its own matter to determine and, second, an epistemological matter. Indeed, even during Driesch's lifetime, developmental physiology was propelled

beyond his experimental findings by the work of Hans Spemann, Fritz Baltzer, and others. The extraordinary progress made in biochemistry, specifically in the elucidation of chemism—of the molecular mechanism, in other words—of cells, genes, and viruses, however, did come later. Neovitalism operates with a far too simple understanding of mechanism and machine when it combats the so-called machine-theory of life.[2] This was already brought up as an objection against Driesch around the turn of the century, and the further progress of research bore this out (Ludwig von Bertalanffy, Erwin Schrödinger, Pascual Jordan). Today harmoniously equipotential systems no longer present biochemistry with fundamental difficulties. "Mechanical" thinking has learned so much and become so refined that it is no longer helpless in the face of questions concerning regeneration, reduplication, and differentiation (consider, to name only one example, cybernetic models).

[350] This gain in operational possibilities for reducing vital processes to chemical and physical processes does not, however, necessitate a denial of their uniqueness, which consists in the fact that they appear as gestalts and in this respect—that is to say, phenomenally—possess a special quality that presents itself. This quality is an effect that, as organization, itself in turn affects the processes. Such systemic functions have regulatory significance for constituted systems. Under certain conditions, the phenomenal effect of the gestalt can itself have a biological function: the way an organism looks can become significant for its life cycle (see Adolf Portmann, cited earlier). For the gestalt as such to have an enticing or repulsive, a masking or a fascinating effect, its counterpart has to have corresponding receptors. Looking a certain way presupposes an entity able to look.

I agree with Gerhard Frey that the vitalism represented by Eduard von Hartmann and Driesch was an "amalgamation of operational and phenomenological concepts." I would not, however, call this amalgamation "arbitrary,"[3] because the organism as a whole constitutes the starting point of the analysis, which, in order to verify its results, has to in turn be assessed in relation to the whole of the organism (for instance, in relation to its nutrition or spatial orientation—that is, to purposeful accomplishments "for" itself). Unquestionably, the amalgamation of these two modes of observation became possible only with the development of exact methods in the natural sciences, in particular in causal biology. Without an operational approach in relation to living objects, without at least the idea of a "Newton of a blade of grass," there is no possibility of an amalgamation and hybridization with the phenomenological mode of observation. This, however, has consequences for the history of philosophy. Aristotle, the creator of the concept of entelechy, cannot be called a "vitalist" because his view of nature lacked the counterweight of an operational concept of nature. He was concerned with *physis* as a reality manifest in its phenomena. Not until Descartes did the situation change. Even so, one should not apply the concept of vitalism to thinkers such as Leibniz, Schelling, or Hegel either, even if speculative idealism responded to the critique of teleological judgment, which already evoked and rejected the possibility of a Newton of a blade of grass. Was Schopenhauer a vitalist because he spoke of the will in nature? Things are different for Eduard von Hartmann, who confronted an established causal research tradition—we can call him [351] a vitalist despite his speculative background because he wanted to intervene into theory construction in the natural sciences.

The phenomena of life are autonomous as phenomena, which is not to devalue them in relation to an actual or deeper reality accessible to operational analysis. Light as an electromagnetic state of a certain wavelength is no more real than the green "corresponding" to it, including the prerequisites for a functioning retina. Appearance [*Erscheinung*] is not semblance [*Schein*]. The reader should thus not be surprised to see appearance and being—here only in the sense of existing—used interchangeably. Our focus here lies on the phenomenon as such, without reflecting on the horizon of consciousness within which it becomes constituted and without taking into consideration transphenomenal "being"—in other words, the ontology of an in-itself, regardless of whether we consider such a thing possible or not. This study is concerned with presenting the conditions under which life becomes possible as appearance. It is up to the natural sciences to ascertain its conditions of reality.

2. The notion of levels of the organic naturally evokes levels of development. After all, our intuition of plant, animal, and human has been decisively determined by the discoveries of paleontology, whether we are neo-Darwinists or not. Our understanding of anthropogenesis, in particular, has become significantly more detailed since Eugène Dubois's discovery in Trinil, Java. Gustav Heinrich Ralph von Koenigswald's findings in

the same area and at Dragon [Bone] Hill near Peking, the discoveries in South Africa, and most recently in Tuscany make a monolinear derivation of *Homo sapiens* and of the species' relatively late emergence unlikely. The field of population genetics currently promises to provide new answers to the question of evolutionary mechanisms, with particular attention being paid to the factor of mutation. However we interpret the facts of the evolutionary history of life, their empirical nature is beyond doubt, and we should have learned by now not to consider the appearance of contemporary races of *Homo sapiens* in the geologically most recent layer of the Holocene to be a predicate of value, let alone evidence that with them evolution has reached its goal. We do not even know if it has come to an end. An increase in the degree of cephalization and cerebralization can be observed in vertebrates, making visible an orthogenetic differential whose most recent progression is found in today's human. That is all there is to say in zoological terms. The turn that was introduced with this progression, with the formation of a cultural impact [352] of this living being and its environment, no longer belongs to the zoological field.

Abilities such as language, systematic action, the invention of tools, or the creation of institutions of nonstable character have evidently led to an accumulation of power that has delivered the organic world and—as the last centuries have made apparent—inorganic nature into the hands of the human. But we do not know how far this appropriation, which is attended by a reverse appropriation of the subject of power, goes. Has nature, has evolution come to grips with itself? Has it, in a Schellingian and Hegelian sense, come into its own in the human? Does this inhibit evolution or does it rather propel it in a new direction (toward systematic breeding, which thus far is a fantasy of dilettantes but is becoming a serious possibility the more we learn about the chemism of genes and is foreshadowed in the utopias of Orwell and Huxley)?

The notion of an increase in power is particularly suggestive, and because the highest degree of cerebralization to date is connected with a disproportionate increase in power (disproportionate because it is based on insight), there is a tendency to measure all of evolutionary history—which in protists and invertebrates exhibits evolutionary branches directed in entirely different ways—according to the standard of vertebrates and to subject it to a value scale of high and low. If we apply the criterion of self-sufficiency in relation to the milieu, the *milieu interne* leads us to rank warm-blooded animals over cold-blooded ones. Placing mammals above nonmammalian forms requires yet other criteria, unless the increasing proximity to primates and to the human is simply allowed to tip the scales. In his *Creative Evolution*, Bergson argues that evolutionary history has two peaks, contrasting colonial insects as the culmination of instinctive behavior regulation with the cerebralized forms that can only get anywhere by trial and error, experience and learning. It is conceivable that biology—in particular by way of its fledgling branch, ethology—could differentiate further and discover other possibilities of progression in the organic forms.

But can instinct and insight be contrasted on the same level? As distinct types of behavior regulation, perhaps. But insight lays claim to truth, to which I myself lay claim when I assert that the arguments presented here are valid. But what we referred to as the appropriation [353] of nature, initiated by the emergence of *Homo sapiens* and grounded in its monopoly on objective insight, is not exhausted in the new type of behavior it establishes. This appropriation does not justify the pious assumption that the history of life was designed to accomplish this breakthrough—a theory of preformation espoused by Schelling and Hegel and today by Pierre Teilhard de Chardin. But even understood in purely epigenetic terms, as an abrupt change in the genes, it brought about a turn that cannot be reversed and established a new sphere (which Teilhard calls noological), one that, because it enables practical superiority and distancing objectivity, is "higher."

In addition to gradual changes and differentiations (races), paleontology as genuine history recognizes leaps that can be attributed to macromutations, key mutations leading into new directions. But it does not recognize levels, even if we may equate evolution with a transition into conditions of greater diversity—in phylogenetic terms, and on the whole, we may make this claim without qualification; ontogenetically, on the contrary, it is complicated by exceptions, particularly in the area of parasitism (as was pointed out to me by Rudolf Reinboth). Level increases do not simply run parallel to the "line" of development and cannot be deduced from the approaching emergence of the human. These increases rather correspond to living matter's few specific modes of organization that we encounter in plants, animals, and humans, as fuzzy as the intermediate forms may be. The fact that they represent level increases, however, can only be made comprehensible by conceptualizing their positionality. That is the purpose of this book.

It is for this reason that I omit any discussion of phylogeny, which in the last decades has made tremendous progress, thanks to radiocarbon dating and exact genetics. The more we find out about the emergence of life on earth, the clearer we will be able to judge whether earthly evolution is the only possible "natural creation story," at least in its main line, or whether it realized only one of many possibilities. Our rapidly increasing contact with other celestial bodies makes this question all the more relevant. Is it permissible to conceive of entirely different types and modes of organization than those found on earth when considering the hardly dismissible possibility that there may be life on other stars? Or would it always have to be protists, plants, animals, and humans, at least [354] in terms of mode of organization? Philosophy cannot answer this question. To date it only has *one* story of life at its disposal, and it must stick to it if it is to restrict itself to an analysis of the conditions of possibility of this story while carefully eliminating every kind of teleological interpretation, even just of its direction.

This concludes my general comments meant to clarify the intention of this book. Following are a few addenda and corrections following the recommendations of Rudolf Reinboth, *Privatdozent* of Zoology at the University of Mainz. I am extremely grateful to him for his help. I would also like to express my thanks to Professor Adolf Portmann (Basel) for drawing my attention to the Moscow symposium on "The Origin of Life on the Earth" (1957).

[The numbers in the following refer to the original pagination, as given in square brackets throughout the text (translator's note).]

[49 and 59] (Chapter 2):

My description of the so-called specific sensory energies here is incorrect. It is the sensory cells and the central fields that carry specific triggering functions for sensations, not the peripheral nerve, which is a neutral conductor. Johannes Müller's law of specific sensory energies (in connection with the distinction between adequate and inadequate stimuli) does not apply universally. See, for instance, Max Hartmann, *Allgemeine Biologie* (Jena: Gustav Fischer, 1925), 822f.

Appendix 329

[62f.] (Chapter 2):
Uexküll's nonconformity and tendency to think in black and white have not been favorable to his posthumous reputation in ethology, a field largely based on his work. His dual animosity toward animal psychology, which in his time operated with anthropomorphic analogies, and American behaviorism, which was guided by chain reflex models, has lost all currency in modern ethology (Nikolaas Tinbergen, Gerard Baerends, Dykstra,[4] Konrad Lorenz, Erich von Holst, etc.). See the symposium on "environment" at the philosophy conference I organized in Bremen in 1950 (Helmuth Plessner, *Symphilosophein* [Munich: Lehnen, 1952]). Important for the key role always played here by the interpretation of instinct is Adriaan Kortlandt, *Aspects and Prospects of the Concept of Instinct* (Leiden: E. J. Brill, 1955).

[99f.] (Chapter 3):
A noteworthy conference took place in Moscow in 1957, a symposium initiated by the International Union of Biochemistry on the topic of "The Origin of Life on the Earth."[5]

The conference was hosted by the Academy of Sciences of the U.S.S.R. In his opening address, Alexander Oparin observed that even twenty or thirty years prior such a symposium would have been impossible and that throughout most of the first half of the century there had been only a few scattered attempts to broach the topic in question. Researchers had been guided by the idea "that living things (though only the most primitive ones) could arise directly from the inorganic 'materials.'"[6] Supposedly successful experiments demonstrating such "spontaneous generation" always turned out to be faulty, however, leading to the conviction that the problem could not be solved and was not worth expending any serious effort on. Today, Oparin argued, it is clear that this predicament arose from an incorrect approach to the question. "The problem of the origin of life cannot be solved in isolation from the study of the whole course of the development of matter which preceded this origin. Life is not separated from the inorganic world by an impassable gulf—it arose as a new quality during the process of the development of that world."[7] An evolutionary approach must be taken in this matter, in collaboration with astrophysics, geophysics, and—chemistry. Oparin postulates three successive stages: (1) simplest organic compounds, hydrocarbons and their closest derivatives; (2) an increase in complexity of these compounds in the lithosphere, atmosphere, and hydrosphere. The result of this process was the appearance of very complicated substances "of high molecular weight, in particular, substances resembling proteins, nucleic acids, and other compounds characteristic of contemporary protoplasm." Based on these, "one may postulate the emergence of some sort of primary systems, . . . which changed under the influence of the external medium and which could undergo selection."[8] The development of these systems constitutes the third stage, which in turn led to the formation of the simplest primary organisms.

Of course, we may ask whether the history of matter proceeded according to the tried-and-true pedagogical principle of simple to complex, but as a guideline for analysis this schema is certainly useful. It used to be assumed that even the simplest organic compounds, hydrocarbons, could only be formed biogenically; today we know that there are other ways. It used to be thought that the asymmetry of the compounds characteristic of plasm was a monopoly of living matter; today we know that asymmetrical syntheses take place under the influence of circularly polarized ultraviolet light, in catalytic reactions on the surface of quartz crystals, spontaneously in slow crystallizations from solutions, etc. We also have the [356] same experimental confirmation of the possibility of abiogenic formation of amino acids, porphyrins, protein-like polynucleotides, and other high-molecular compounds, whose role as, for instance, genetic building material is well known.

The study of genes and viruses, finally, led to the question of what stage "life" can be said to begin in. Oparin asks, "Can life be attributed to individual molecules even if they are very complex, or only to the multimolecular systems which served as a basis for the emergence of life?"[9] The answer depends on what criteria are considered essential for life, and there is no agreement on this. Gerhard Schramm writes, for instance, "By no means is it permissible to regard a virus as a preliminary stage of life, for viral reproduction is contingent upon the existence of living cells. . . . The significance of virology for the problem of the emergence of life seems to me to lie rather in the insight it can give us into the biochemical foundations of reproduction."[10] Wendell M. Stanley is of a different opinion: "The essence of life is the ability to reproduce. This is accomplished by the utilization of energy to create order out of disorder, to bring together into a specific predetermined pattern from semi-order or even from chaos all of the component parts of that pattern with the perpetuation of that pattern with time. This is life."[11] He does not consider the ability to mutate, to change, or to respond to stimuli to be an essential criterion, although it is crucial for evolution. At least nature has "provided a built-in error so that the replication process is not perfect," and we recognize this change as mutation.[12] Accumulated and stabilized into differences, it is called "genes." Thus viruses not only constitute a preliminary stage of life, but in fact live—Stanley counters the common objection made by Schramm by pointing to parasites: some tapeworms are also only able to reproduce in the organism of the host. The morphological differentiation of some viruses "can hardly be called molecular in nature . . . [but] rather more organismal or cell-like."[13]

Only biochemists can judge the verifiability of these arguments, and definitive answers are still outstanding. But the way in which the arguments are presented is itself instructive: in every controversy, there is an attempt to formulate the necessary and sufficient conditions for the emergence of the quality of being alive in chemically defined compounds or systems of compounds. The formal designation of this quality, as, for instance, a "self-perpetuating pattern of chemical reactions"[14] (J. B. S. [357] Haldane) or Bertalanffy's hierarchical order of an open system that, thanks to its system conditions, maintains itself even as it substitutes components, returns again and again as the idea guiding models (Jordan, Schrödinger, Linus Pauling, Melvin Calvin, etc.). Such structural durability, however, is only comprehensible if it is associated with a specific relationship to the surroundings, regardless of the degree of complication the structure possesses in its stability. In order to preserve itself in relation to the surrounding medium while in exchange with it, the structure has to meet certain specific requirements. Haldane writes, "The critical event which may best be called the origin of life was the enclosure of several different self-reproducing polymers within a semipermeable membrane."[15] Impermeable encasing would not suffice. It is rather that "the materials of the membrane must be synthesized or accumulated from the environment, and must be organized in a stable arrangement between the environment and the inner aqueous medium of the organism."[16] Stanley's observations of particular viral forms support this assertion.

The formation of natural membranes, then, is of special significance. They allow organic materials to acquire the character of organisms, to become living beings in the strict sense of the word. Whether this is the first phenomenon to satisfy the definition of life is an open question. The ambivalent nature of viruses (and genes?) enjoins us to caution. What is certain is that with membrane formation, the outline has become stabilized to the point that we call it the "form" of an organism. It is possible that the coherence of large molecules and the chains they form into systems is great enough to allow for a membrane-free living being. Again, we may think here of viruses (and genes?). In any case, however, membrane formation represents a step beyond such a being in the direction of a "higher" level of boundedness. It marks the living "entity" [*Lebe"wesen*"] as individual and has a dual effect: both enclosing/shielding and opening/mediating in relation to the setting.

Membranes are not merely surfaces that every body, depending on its state of matter, has in relation to the media of other states of matter bordering on it. They are mediating surfaces. A membrane is not simply where the body ends, but where it is brought into relation with its medium. The molecular complex (and perhaps also the molecules) forming it maintains its pattern not only in the state of reproduction, but also in its constant contact with the area of influence [358] bordering on it. The body's own area is in contact—at a distance—with this area of influence, shielded from and open to it. It is under these conditions that metabolism takes place, a process that always consists of a selection from among various possibilities of exchange. Evidently, this filter function of semipermeable membranes is also decisive for living substances' responsiveness to stimuli, a property that requires them to have a relatively stable area of their own that is set off from their surroundings. If a body is in contact with its medium, a contact that is distanced by the inhibiting and channeling of its mediating boundary layer, it increases its possibilities of maintaining its own area. This has nothing to do yet with sensation or even consciousness, even if both of these can be understood as evolved forms of contact at a distance, albeit under special organizational conditions. If we want to speak precisely about the emergence of consciousness, this is where we ought to start. To simply tie consciousness to organic matter, as, for instance, Teilhard de Chardin still proposes doing (and he is here a late link in a long chain), is not, as they say, completely unfounded, but is contrary to the narrow, and only meaningful, definition of consciousness.

Membrane or not, "life could originate only as a result of the evolution of a multimolecular organic system, separated from its environment by a distinct boundary, but constantly interacting with this environment in the manner of 'open' systems. Since . . . present-day protoplasm possesses a coacervate structure, the mentioned systems . . . could have been coacervate drops."[17] The effect of such structural cohesive forces on the surface is the decisive factor: they turn the outline into a boundary, in which two mutually affecting areas come to mediate each other without infringing upon the outlined body's structure. This state of affairs gives the body positionality—that is, sets it off against the outside. It thus acquires a *setting* [*Umgebung*] in the medium, in later developmental stages possibly an *environment* [*Umwelt*]. The so-called character of wholeness is given along with positionality, a character that takes the shape of form (not necessarily of a nonvarying nature). Membranes, of course, facilitate the stabilization of form. We can assume that the reactive effects of the bounding form on the system it encloses by virtue of its inevitably both inhibiting and channeling function (whether by means of a membrane or not) in turn create new [359] qualities that do not quite match the original pattern. Evolutionary research may expect discrepancies of this kind. Indeed, it must if it accepts the notion of an open system with a certain degree of constancy.

A whole has parts or components in relation to which it must preserve a certain self-sufficiency; otherwise it is not a whole, but an aggregate. Its configuration provides enough leeway for components to be replaced or to become inoperative. It will be the special task of molecular biology to identify the cohesive forces necessary for such leeway. Vitalists in the spirit of Driesch will then no longer be able to invoke properties of living systems such as the ability to regenerate or change functions (for instance, a section of plasma that has not yet been fully differentiated fills in for another in the early stages of embryo development, or particular tissue parts fill in for others, such as Constantin von Monakow's vicarious function in the central nervous system). The occurrence of a whole that uses its components as a means of self-preservation no longer justifies resorting to a natural factor such as entelechy, which contradicts the basic rules of methodological analysis in the empirical natural sciences. Notwithstanding the self-sufficiency and uniqueness of living systems, their autonomy is lost as soon as they are alleged to have a special agent. Physical and chemical concepts no longer require an additional factor to treat the phenomenon of a whole—that is, of a purposive configuration.

On the topic of viruses, see also Wilhelm Troll, *Das Virusproblem in ontologischer Sicht* (Wiesbaden: Steiner, 1951), and Wolfhard Weidel, *Virus*, (Berlin and Heidelberg: Springer, 1957). On the problem of the defining characteristics of living being, see Theodosius Dobzhansky, *Die Entwicklung zum Menschen* (Hamburg and Berlin: Parey, 1958); Cyril Dean Darlington, *The Facts of Life* (London: Allen and Unwin, 1953); Bernhard Rensch, *Neuere Probleme der Abstammungslehre* (Stuttgart: Enke, 1954); and Hans Vogel, *Vom Kristall zum Lebewesen* (Nuremberg: Hans Carl, 1952).

[124] (Chapter 4):

See Erwin Bünning, *Entwicklungs- und Bewegungsphysiologie der Pflanze* (Berlin and Heidelberg: Springer, 1953), and Bünning, *Die physiologische Uhr* (Berlin and Heidelberg: Springer, 1963).

[144] (Chapter 4):

See especially Alfred Kühn, *Vorlesungen über Entwicklungsphysiologie* (1955; repr. Berlin and Heidelberg: Springer, 1965), particularly the thirtieth lecture.

[146] (Chapter 4):

Not enough consideration was given here to Rudolf Ehrenberg, *Theoretische Biologie: Vom Standpunkt der Irreversibilität des elementaren Lebensvorganges* (Berlin and Heidelberg: Springer, 1923).

[360]; [198] (Chapter 5):

Erwin Schrödinger, *What Is Life?* (1944; repr. New York: Cambridge University Press, 1967), and Max Hartmann, "Prozeß und Gesetz in Physik und Biologie," *Philosophia Naturalis* 2, no. 3 (1953) cover similar ground.

[217] (Chapter 5):

I was cautious enough to refer here to the conditional apriority of sexual differentiation. Rudolf Reinboth comments, "Sexuality is not always a prerequisite for reproduction. Asexual reproduction is very widespread among invertebrates and in some cases is the rule under normal conditions. I am skeptical, at least from a biological perspective, of the notion of an 'absolute qualitative difference' between male and female being. Even just the catchphrases 'bisexual potency' and 'relative sexuality' point to the problem. Finally, the many examples of *natural* sexual inversion in the animal kingdom ought not to be overlooked. I myself have long been concerned with sex reversal in fish, where this phenomenon is widespread and, furthermore, appears in the most varied forms." See also Max Hartmann, *Die Sexualität* (Jena: Fischer, 1956).

[222] (Chapter 5):
According to more recent measurements, it is incorrect to speak in general of a predominance of assimilation in plants and a predominance of dissimilation in animals. It is only the type of assimilation that differs between plants and animals. Furthermore, there is a distribution of the metabolic stages in higher plants at least, and a division of labor between roots and leaves.

[223] (Chapter 5):
Conrad-Martius's characterization of plant movement is only valid in light of the notion of the open form. At issue here is the physiological comparability of stimulus processes, given that plants do not have nervous tissue—that is, no reflexes (although they do have certain tissues that conduct stimuli), and no centers. Plants as open forms do not have centrally mediated control of their movements and lack central representation of their organism. This does not mean, however, that stimulus, stimulus conduction, and movement in plants and animals cannot be physiologically and chemically compared and contrasted. On the other hand, the decentralized type of the closed form—lower animals, in other words—regardless of how "primitive" they are (nerve nets, nerve rings), [361] possess central coordination of their movements and a representation of their organism. They "have" their bodies. This is why Uexküll's statement

[248] (Chapter 6)
"When a sea urchin runs, the legs move the animal" is a humorous exaggeration. Of course, sea urchins have control and coordination. What other purpose could their nerve rings serve? See Libbie Henrietta Hyman, *The Invertebrates*, vol. 4 (New York: McGraw-Hill, 1955). This is also why

[251] (Chapter 6)
there is no need to further examine Uexküll's claim that the receptor apparatus knows nothing of the activities of the motor apparatuses. Buytendijk's experimental findings on octopuses and F. K. Sanders and J. Z. Young's research on cuttlefish provide evidence to the contrary. The notion that all invertebrates are of the decentralized organization type and all vertebrates of the centralized type is too good to be true. The boundaries are likely blurry and very complicated—consider, for instance, Karl von Frisch's bee

studies. Pure types do not occur; nor are they meant to. It is rather ideal types of trends that are expressed to a greater or lesser degree in the species in question.

[257] (Chapter 6):

Without going into recent discoveries of the astonishing orientation abilities of insects and birds, I would like to call into question Volkelt's hypothesis of the complex quality of the environmental structure: it is quite certainly not generally valid and does not constitute an alternative to the human mode of perception. Our way of perceiving, as we know from Gelb and Goldstein's aphasia studies, is sustained and guided by language. Words as carriers of meaning and means of objectification help structure and stabilize the contents of perception. Where language is absent—not communication as between animals—, other "methods" of structuring and stabilizing must be in place, depending on the demands of the biological environment, of course: the method of complex qualities, however, is surely but one of many.

[288f.] (Chapter 7):

I here refer the reader to "Conditio Humana," my introduction to *Propyläen Weltgeschichte*, vol. 1, *Vorgeschichte—Frühe Hochkulturen*, ed. Golo Mann and Alfred Heuß (Berlin, Frankfurt, and Vienna: Propyläen, 1961), 33–86, which also appeared in *Opuscula aus Wissenschaft und Dichtung*, vol. 14 (Pfullingen: Neske, 1964).

GLOSSARY

Translator's note: The glossary is compiled in view of important German terms. The English terms listed occasionally render nontechnical German words other than those given here. It should generally be clear from the context whether usage of a term is technical or merely colloquial.

ENGLISH TO GERMAN

abut: *anstoßen*
actual: *wirklich*, occasionally *aktuell*
actuality: *Wirklichkeit*, occasionally *Aktualität*
adaptation: *Anpassung*
adaptedness: *Angepaßtheit*
ahead, being ahead: *vorweg, vorweg sein*
and-connection: *Undverbindung*
anticipate: *vorgreifen, vorwegnehmen*
anticipation: *Vorweg*
appear: *erscheinen*
appearance: *Erscheinung*
awareness: *Bewußtheit*

beginning-something: *Ausgangsetwas*
behavior: *Verhalten, Gebaren, Gehabe*
being: *Sein*
body (on its own): *Körper*
   bodily: *körperlich*
   bodily thing, physical thing: *Körperding*

bodily object: *Körpergegenstand*
lived body: *Leib*
lived physical body: *Leibkörper*
physical body and lived body (contrast between the two): *Körper* and *Leib*
physical lived body: *Körperleib*
border (n.): *Begrenzung*
   bordered: *grenzhaft*
   border on (v.): *angrenzen*
boundary: *Grenze*
   bounded: *begrenzt*
   boundedness: *Begrenzung*
   boundlessness: *Unbegrenztheit*
   bounds: *Begrenzungen*

center, central point: *Mitte, Mittelpunkt*
central position: *Mittelstellung*
conditioned, conditional: *bedingt*
conform/conformity: *Eingepasstheit, eingepasst*
confront: *entgegentreten, gegenübertreten*
consolidation: *Zusammenfassung*
contraposition: *Entgegensetzung, Gegenüberstellung*
cooperation and conflict: *Miteinander und Gegeneinander*
counterworld: *Gegenwelt*
crossing over: *Hinüber*
custom: *Sitte*

337

detach: *ablösen*
development: *Entwicklung*
diversity: *Mannigfaltigkeit*
dividedness: *Hälftenhaftigkeit, Halbheit*
dual aspect: *Doppelaspekt*

edge: *Rand*
embodied: *leibhaft, leiblich*
empirical, empirically: *empirisch,* occasionally *erfahrungsmäßig*
end-something: *Endetwas*
environment: *Umwelt*
essential: *wesentlich, wesenhaft*
  essential characteristics: *Wesensmerkmale*
  essential trait: *Wesenszug*
ethical: *sittlich*
ethical life: *Sittlichkeit*
evolution: *Evolution,* occasionally *Entwicklung*
excentric: *exzentrisch*
execution: *Vollzug*
existence: *Existenz, Dasein,* occasionally *Sein,* occasionally *Bestand*
experience: *Erfahrung,* occasionally *Erlebnis*
  by experience: *erfahrungsmäßig*
  experiential position: *Erfahrungsstellung*
  lived experience: *Erlebnis, Erleben*
  mode of experience: *Erfahrungsrichtung*

factual: *faktisch*
fellow human being: *Mitmensch*
field conditions: *Feldverhalte*
final cause: *Zweckursache*
fit into, fitting in: *Eingepasstheit, eingepasst*
form of order: *Ordnungsform*
functional unity: *Wirkeinheit*
function-circle: *Funktionskreis*
fundamental essentiality, fundamental characteristic: *Grundwesenheit*

gestalt: *Gestalt*
  beyond gestalt: *übergestalthaft*
given beforehand, pregiven, already given: *vorgegeben*
  prior givenness: *Vorgegebenheit*
grasp: *fassen, erfassen*
grip: *greifen*

historical basis of reacting: *historische Reaktionsbasis*
human, human being: *Mensch, menschliches Wesen*
humanities: *Geisteswissenschaften*

I: *Ich*
Illusion: *Schein*
in-between: *das Zwischen*
into it: *in es hinein*
integration: *Eingliederung*
interior aspect: *Innenaspekt*
interiority: *Innerlichkeit*
interplay: *Widerspiel*
interval: *Pause*
intuition: *Anschauung, Intuition*
  ordinary intuition: *gewöhnliche Anschauung*
  sensory intuition: *sinnliche Anschauung*

level: *Stufe*
life: *Leben*
  life circle, circle of life: *Lebenskreis*
  living being: *Lebendiges, lebendiges Sein, Lebewesen*
  living space: *Lebensraum*
  living unit: *Lebenseinheit*
  phenomenon of life: *Lebenserscheinung*
location: *Lage, Lagerung*
locomotion: *Ortsbewegung*

manifold: *Mannigfaltigkeit*
manifold unity: *Mannigfaltigkeitseinheit*
matter: *Stoff, Materie, Substanz*
means: *Mittel*

measurable: *meßbar, feststellbar, faßbar*
mediacy: *Mittelbarkeit*
   mediate, mediately: *mittelbar*
   mediation: *Vermittlung*
moral code, more: *Sitte*
motor skills/functions: *Motorik*
mutually inconvertible, not mutually convertible: *ineinander nicht überführbar*

notice: *merken*

objectivity: *Gegenständlichkeit*
observable: *feststellbar, beobachtbar*
one: das Eine, Ein
opposition: *Gegeneinander*
   opposing field: *Gegenfeld*
   opposing position: *Gegenstellung*
   opposing sphere: *Gegensphäre*
   oppositional, opposing: *gegensinnig*
   relation of opposition: *Gegenüberbeziehung*
out beyond it: *über es hinaus*
outer field: *Außenfeld*
outer world: *Außenwelt*
outline, outlining: *Umrandung*
outward appearance: *Aussehen*

passing over, passage: *Übergehen*
passing through: *Hindurch*
perceptual field: *Wahrnehmungsfeld*
perceptual thing: *Wahrnehmungsding*
peripheral value: *Randwert*
periphery: *Rand*
persist: *beharren*
personal, person-bound, of persons: *personal*
philosophy of nature: *Naturphilosophie*
physical: *körperlich, physisch*
place (v.): *setzen*
   placed: *gestellt sein*
   without place: *ortlos*
posit: *setzen*
   posited, set: *gesetzt*
   positedness: *Gesetztsein*

posit out beyond: *hinaussetzen über*
position: *Stellung, Lage*
   position of the gaze: *Blickstellung*
   positional field: *Positionsfeld*
   positionality: *Positionalität*
   positioned: *gestellt sein*
   positioning, being positioned: *Gestelltheit*
   positioning/being positioned over against: *Entgegengestelltheit*
potency, potential: *Potenz*
proper, properly: *eigentlich*
psyche: *Seele, Psyche*
punctiformity: *Punktualität*

radius: *Umkreis*
real, really: *real, wirklich*, occasionally *reell* (where indicated)
reality: *Realität, Wirklichkeit*
   the real, the real thing: *das Reale*
realization: *Realisierung, Verwirklichung*
reflexive: *rückbezüglich*
relation-against: *Gegenverhältnis*
relation-with: *Mitverhältnis*
representation: *Vorstellung, Darstellung*, occasionally *Vertretung*
return: *Rückwendung*

see, view: *erschauen, sehen*
   seeable: *erschaubar*
self: *Selbst*
   being itself/oneself, being a self: *Selbstsein*
   self-givenness: *Selbstgegebenheit*
   self-modification: *Selbstveränderung*
   self-position, self-positioned: *Selbststellung*
   self-sufficient, self-sufficiency: *selbständig, Selbständigkeit*
sensation: *Empfindung*
senses: *Sinne*
   sense data: *Sinnesdaten*
   sense perception: *sinnliche Wahrnehmung*
   sense quality: *Sinnesqualität*

340 *Glossary*

sensory energies: *Sinnesenergien*
sensory quality: *sinnliche Qualität*
set apart from: *abheben*
  set-apartness: *Abgehobenheit*
  setting-apart: *Abhebung*
set off against: *gegen etwas absetzen*
set out of: *heraussetzen*
shared surroundings: *Mitfeld*
shared world: *Mitwelt*
situation: *Lage*
space-claiming: *raumbehauptend*
spacelike, spacelikeness: *raumhaft, Raumhaftigkeit*
spatial thing: *Raumding*
sphere of effecting: *Wirkungssphäre*
sphere of noticing: *Merksphäre*
spirit: *Geist*
  world of the spirit: *geistige Welt*
spiritual: *geistig*
spirituality, spiritual nature, mind: *Geistigkeit*
standing on its own, stand-alone: *selbständig, Selbständigkeit*
stand over against: *gegenüberstehen*
stock: *Bestand*
substance: *Substanz, Stoff*
substantive character: *Sachcharakter*
surrounding field: *Umfeld*
surroundings: *Umgebung*

thing: *Ding*
  thing-body: *Dingkörper*
  thinghood: *Dinglichkeit*
  thingly: *dinglich*
  thingness: *Dinghaftigkeit*
  world of things: *Dingwelt*
through-point: *Durchgangspunkt*
timelike, timelikeness: *zeithaft, Zeithaftigkeit*
total (n.): *Insgesamt*
totality: *Gesamtheit*, occasionally *Ganzheit*
transgredience: *Transgredienz*
transition: *Übergang*

unitary form: *Einheitsform*

variety: *Mannigfaltigkeit*
vitality: *Lebendigkeit*

whole, wholeness: *Ganzheit*
  the whole/a whole: *das Ganze/ein Ganzes*
whole body: *Gesamtkörper*

GERMAN TO ENGLISH

abheben, Abhebung, Abgehobenheit: set apart from, setting-apart, set-apartness
ablösen: detach
absetzen (gegen etwas): set off against
Aktualität: actuality [*Aktualität*], currency
aktuell: actual [*aktuell*], current
Angepaßtheit: adaptedness
angrenzen: border on
Anpassung: adaptation
Anschauung: intuition
  gewöhnliche Anschauung: ordinary intuition
  sinnliche Anschauung: sensory intuition
anstoßen: abut
Ausgangsetwas: beginning-something
Aussehen: outward appearance
Außenfeld: outer field
Außenwelt: outer world

bedingt: conditioned, conditional
begrenzt: (generally) bounded
Begrenzung: boundedness; occasionally border
Begrenzungen: bounds
beharren: persist
Bestand: existence, stock
Bewußtheit: awareness
Blickstellung: position of the gaze

Darstellung: representation, presentation, display
Dasein: existence, being

Dinghaftigkeit: thingness
Dingkörper: thing-body
dinglich: thingly
Dinglichkeit: thinghood
Dingwelt: world of things
Doppelaspekt: dual aspect
Durchgangspunkt: through-point

eigentlich: proper, properly
das Eine, Ein: *one*
Eingepasstheit, eingepasst: conform/conformity, fit into, fitting in
Eingliederung: integration
Einheitsform: unitary form
Empfindung: sensation
Endetwas: end-something
Entgegengestelltheit: positioning/being-positioned over against
Entgegensetzung: contraposition
entgegentreten: confront
Entwicklung: development, occasionally evolution
Erfahrung: experience (see also *Erlebnis*)
erfahrungsmäßig: by experience, occasionally empirical, empirically
Erfahrungsrichtung: mode of experience
Erfahrungsstellung: experiential position
erfassen: (generally) grasp
Erlebnis, Erleben: lived experience, occasionally experience (see also *Erfahrung*)
erschaubar: seeable
erschauen: see, view
erscheinen: appear
Erscheinung: appearance, occasionally phenomenon
exzentrisch: excentric

faktisch: factual
faßbar: able to be grasped, measurable
fassen: comprehend, grasp

Feldverhalte: field conditions
feststellbar: observable, measurable
Funktionskreis: function-circle

das Ganze/ein Ganzes: the whole/a whole
Ganzheit: wholeness, whole, occasionally totality
Gegebenheitsweise: manner in which something is given, mode of givenness
Gegeneinander: opposition
Gegenfeld: opposing field
gegensinnig: oppositional, opposing
Gegensphäre: opposing sphere
Gegenständlichkeit: objectivity
Gegenstellung: opposing position
Gegenüberbeziehung: relation of opposition
gegenüberstehen: stand over against
Gegenüberstellung: contraposition
gegenübertreten: confront
Gegenverhältnis: relation-against
Gegenwelt: counterworld
Geist: spirit
Geisteswissenschaften: humanities
geistig: spiritual
geistige Welt: world of the spirit
Geistigkeit: spirituality, spiritual nature, mind
Gesamtheit: totality
Gesamtkörper: whole body
gesetzt: posited, set
Gesetztsein: positedness
Gestalt: gestalt
gestellt sein: positioned, occasionally placed
Gestelltheit: positioning, being positioned
greifen: grip
Grenze: boundary
grenzhaft: bordered
Grundwesenheit: fundamental essentiality, fundamental characteristic

Hälftenhaftigkeit, Halbheit: dividedness
heraussetzen: set out of
hinaussetzen über: posit out beyond
Hindurch: passing through
Hinüber: crossing over
historische Reaktionsbasis: historical basis of reacting

Ich: I
ineinander nicht überführbar: not mutually convertible, mutually inconvertible
in es hinein: into it
Innenaspekt: interior aspect
Innerlichkeit: interiority
Insgesamt: total

Körper und Leib (contrast between the two): physical body and lived body
Körper (on its own): body
Körperding: bodily thing, physical thing
Körpergegenstand: bodily object
Körperleib: physical lived body
körperlich: bodily, physical

Lage: location, situation, position
Lebendiges: living being
lebendiges Sein: living being
Lebendigkeit: vitality, being alive
Lebenseinheit: living unit
Lebenserscheinung: phenomenon of life
Lebenskreis: circle of life, life circle
Lebensraum: living space
Lebewesen: living being
Leib: lived body
leibhaft: embodied
leiblich: embodied
Leibkörper: lived physical body

Mannigfaltigkeit: manifold, variety, diversity
Mannigfaltigkeitseinheit: manifold unity
Mensch: human, human being
menschliches Wesen: human being

merken: notice
Merksphäre: sphere of noticing
Miteinander und Gegeneinander: cooperation and conflict
Mitfeld: shared surroundings
Mitgehen: accompany, accompaniment
Mitmensch: fellow human being
Mitte: center
Mittel: means
mittelbar: mediate, mediately
Mittelbarkeit: mediacy
Mittelpunkt: center, central point
Mittelstellung: central position
Mitverhältnis: relation-with
Mitwelt: shared world
Motorik: motor skills/functions

Naturphilosophie: philosophy of nature

Ordnungsform: form of order
ortlos: without place
Ortsbewegung: locomotion

Pause: interval
personal: personal, person-bound, of persons
Positionalität: positionality
Positionsfeld: positional field
Potenz: potency, potential
Punktualität: punctiformity

Rand: edge, periphery
Randwert: peripheral value
raumbehauptend: space-claiming
Raumding: spatial thing
raumhaft, Raumhaftigkeit: spacelike, spacelikeness
das Reale: the real, the real thing
real: real, really (see also *wirklich*)
Realisierung: realization
Realität: reality (see also *Wirklichkeit*)
reell: real, really [*reell*] (see also *wirklich*)
rückbezüglich: reflexive
Rückwendung: return

Sachcharakter: substantive character
Schein: illusion
Seele: psyche, occasionally soul or spirit
Sein: being, occasionally existence
selbständig, Selbständigkeit: self-sufficient, self-sufficiency, standing on its own, stand-alone
Selbstgegebenheit: self-givenness
Selbstsein: being itself/oneself, being a self
Selbststellung: self-position, self-positioned
Selbstveränderung: self-modification
setzen: posit, place
Sinnesdaten: sense data
Sinnesenergien: sensory energies
Sinnesqualität: sense quality
sinnliche Qualität: sensory quality
sinnliche Wahrnehmung: sense perception
Sitte: more, custom, moral code
sittlich: ethical
Sittlichkeit: ethical life
Stellung: position
Stoff: substance, matter
Stufe: level

Transgredienz: transgredience

über es hinaus: out beyond it
Übergang: transition
Übergehen: passing over, passage
übergestalthaft: beyond gestalt
Umfeld: surrounding field
Umgebung: surroundings
Umkreis: radius

Umrandung: outline, outlining
Umwelt: environment
Unbegrenztheit: boundlessness
Undverbindung: and-connection

Verhalten: behavior
Vermittlung: mediation
Verwirklichung: realization
Vollzug: (generally) execution
vorgegeben: given beforehand, pregiven, already given
Vorgegebenheit: prior givenness
vorgreifen: anticipate
Vorstellung: representation, notion, idea
Vorweg: anticipation
vorwegnehmen: anticipate
vorweg sein: being ahead

Wahrnehmungsding: perceptual thing
Wahrnehmungsfeld: perceptual field
wesenhaft: essential
Wesensmerkmale: essential characteristics
Wesenszug: essential trait
wesentlich: essential
Widerspiel: interplay
Wirkeinheit: functional unity
wirklich: actual, real, really
Wirklichkeit: actuality, reality [*Wirklichkeit*] (see also *Realität*)
Wirkungssphäre: sphere of effecting

zeithaft, Zeithaftigkeit: timelike, timelikeness
Zusammenfassung: consolidation
Zweckursache: final cause
das Zwischen: in-between

NOTES

PREFACE TO THE FIRST EDITION (1928)

1. Helmuth Plessner, *Die Einheit der Sinne: Grundlinien einer Ästhesiologie des Geistes* (Bonn: F. Cohen, 1923).
2. All page references to the *Levels* are to the second German edition on which this translation is based. Throughout the body of the translation, the reader will find page references to the German edition in square brackets.
3. Georg Misch, "Die Idee der Lebensphilosophie in der Theorie der Geisteswissenschaften," *Kant-Studien* 31, nos. 1–3 (1926).
4. Martin Heidegger, *Being and Time*, trans. Joan Stambaugh (Albany: State University of New York Press, 2010). First published as *Sein und Zeit* in 1927.
5. Josef König, *Der Begriff der Intuition* (Halle: Niemeyer, 1926).

PREFACE TO THE SECOND EDITION (1965)

1. Plessner is referring here to Max Scheler, *Formalism in Ethics and Non-Formal Ethics of Values: A New Attempt toward the Foundation of an Ethical Personalism*, trans. Manfred S. Frings and Roger L. Funk (Evanston, Ill.: Northwestern University Press, 1973); originally published in 1913 (Part 1) and 1916 (Part 2) (translator's note).
2. The quote is from Goethe's *Faust*, Part 2 (translated by A. S. Kline), line 4727 (translator's note).
3. Plessner is referring here to Max Scheler, "On the Idea of Man," trans. Clyde Nabe, *Journal of the British Society for Phenomenology* 9, no. 3 (October 1978), although he is mistaken about the original publication date, which was 1915 (translator's note).
4. The judgment of an anthropologist as experienced as Egon Freiherr von Eickstedt on this matter must carry weight precisely because he does not belong to any philosophical party but in distinction to some of his peers

understands the need for a philosophical anthropology. He describes the situation in the following words: "The sociologist and philosopher Pleßner, coming from Driesch and Windelband both, in 1928 published the first closed system of a quite original bio-philosophy, in which the human appears as the central figure. The human's restless versatility (plasticity) in a 'surrounding field' (positionality) full of relationships and tensions leads him beyond himself, thus away from himself, and thereby into an organically unique dynamic of being. According to Pleßner, this dynamic can only be understood if facts and interpretations—that is, anthropology and philosophy—proceed together. Unfortunately, this intrepid venture was immediately overshadowed by the revolutionary effects of an essay by the older Scheler, who had long since achieved success elsewhere and whose famous Darmstadt outline sketch, also from 1928, for its part exhibits the realism/idealism antagonism of the confused philosophy of its time. Scheler's lifelong battle for the is and the ought of human beings and to get beyond his (and, among others, my) teacher Husserl's phenomenology in the guise of an applied, closer-to-life phenomenology caused him to return again and again to the problem of the human. The result is a philosophy which, along with its at times surprising turns, reflects with rare vividness the self of its creator: it fluctuates between idealism and realism"; Egon Freiherr von Eickstedt, "Anthropologie mit und ohne Anthropos," *Homo* 14, no. 1 (1963).

5. Karl Löwith, "Natur und Humanität des Menschen," in *Wesen und Wirklichkeit des Menschen: Festschrift für Helmuth Plessner* (Göttingen: Vandenhoeck & Ruprecht, 1957), 75.

6. Löwith, "Phaenomenologische Ontologie und protestantische Theologie," *Zeitschrift für Theologie und Kirche*, n.s. (1930).

7. Arnold Gehlen, *Anthropologische Forschung* (Reinbek: Rowohlt, 1961), 17.

8. Paul Alsberg, *In Quest of Man: A Biological Approach to the Problem of Man's Place in Nature* (Oxford: Pergamon, 1970), 35; first published as *Das Menschheitsrätsel* in 1922.

9. Gehlen, *Anthropologische Forschung*, 112.

10. Gehlen, 109ff.

11. Frederik J. J. Buytendijk, *Algemene theorie der menselijke houding en beweging* (Utrecht: Het Spectrum, 1948). Plessner references the German translation published in 1956 (translator's note).

12. The page number here refers to the original pagination (translator's note).

13. Adolf Portmann, "Die Erscheinung der lebendigen Gestalten im Lichtfeld," in *Wesen und Wirklichkeit des Menschen*, ed. Klaus Ziegler (Göttingen: Vandenhoeck & Ruprecht, 1957), 40.

14. Portmann, "Die Erscheinung der lebendigen Gestalten im Lichtfeld," 39.

INTRODUCTION

1. Even those who cannot themselves participate in a norm-governed life—infants and very young children, the comatose, the incapacitated elderly—biologically live through the mediations of norm-saturated social practices.

2. The two competing texts are Max Scheler, *The Human Place in the Cosmos*, trans. Manfred S. Frings (1928; repr. Evanston, Ill.: Northwestern University Press, 2009), and Arnold Gehlen, *Man: His Nature and Place in the World* (1940; repr. New York: Columbia University Press, 1988). For a useful survey of philosophical anthropology in its narrow and wide sense, see Axel Honneth and Hans Joas, *Social Action and Human Nature*, trans. Raymond Meyer (Cambridge: Cambridge University Press, 1988).

3. First published in 1924, Helmuth Plessner's *The Limits of Community: A Critique of Social Radicalism*, trans. Andrew Wallace (Amherst, N.Y.: Humanity, 1999) accurately and presciently critiqued the repudiation of modernity and the longing for community prevalent in Weimar Germany, a longing that would, in just a few years, become the ethnic nationalism of German fascism. That we are once more in a time of longing for community and ethnic nationalism makes Plessner's subtle study urgent once more. The idea of longing for community, under the heading of "the law of the utopian standpoint," became an integral element of Plessner's philosophical anthropology. The translator's Introduction to *Limits of Community* provides a helpful overview of this aspect of Plessner's social philosophy.

4. See Thomas Nagel, "What Is It Like to Be a Bat?," *Philosophical Review* 83, no. 4 (Oct. 1974): 435–50.

5. For this and much more, one can now consult Richard O. Prum, *The Evolution of Beauty: How Darwin's Forgotten Theory of Mate Choice Shapes the Animal World—and Us* (New York: Doubleday, 2017).

6. I owe this way of making this claim to Karen Ng.

7. Immanuel Kant, *Critique of Judgment*, trans. Werner S. Pluhar (Indianapolis: Hackett, 1987), 400.

8. See here the instructive essay by Gesa Lindemann, "From the Critique of Judgment to the Principle of the Open Question," trans. Millay Hyatt, *Ethical Theory and Moral Practice* 18, no. 5 (November 2015): 891–907. Plessner's *Habilitationsschrift* "Untersuchungen zu einer Kritik der philosophischen Urteilskraft" addressed the Third *Critique*, primarily in order to elaborate the relation between the First and the Second *Critique*, which is, in effect, the very dialectic I am here canvassing.

9. Kant, *Critique of Judgment*, 370.

10. For this and the next quote, see Kant, 371.

11. For reasons that will become evident, I am focusing on Kant's account of the characteristic properties of living organisms rather than his defense of

the claim that the form of reflecting judging attending to these properties employs a form of teleological judgment that assesses organisms as acting in *purposive* ways. For a defense of the thesis that the "concept of a purpose is closely associated with the concept of normativity; to regard something as a purpose is to regard it as subject to normative laws, standards, and constraints," see Hannah Ginsborg, *The Normativity of Nature: Essays on Kant's Critique of Judgment*, part 3 (Oxford: Oxford University Press, 2015). The quoted phrase appears on p. 275.

12. Tibor Gánti, *The Principles of Life* (Oxford: Oxford University Press, 2003), Chaps. 1–2. For a philosophical account that uses Gánti's principles as its starting place, see Mark A. Bedau, "What Is Life?," in *A Companion to the Philosophy of Biology*, ed. Sahotra Sarkar and Anya Plutynski (Malden, Mass.: Wiley-Blackwell, 2011), 455–71.

13. This is not the only way to pick out evolutionary units. Peter Godfrey-Smith, *Philosophy of Biology* (Princeton: Princeton University Press, 2014), Chapter 3, "Evolution and Natural Selection," follows Richard Lewontin's account of natural selection in which genes are optional (31). Godfrey-Smith goes on to argue against gene-centric views that "it is not possible to *explain* what is happening by staying at the genetic level. Changes to gene frequencies are usually a result of the lives and deaths of whole organisms, and are sometimes affected by competition between larger units such as families and tribes" (44). Part of the pushback against reductionism in biology involves making organisms the primary unit of accounting, a thesis that Plessner depends on and defends. For a wide-ranging consideration of the relation between units of life and units of evolution, see John Dupré, *Processes of Life: Essays in the Philosophy of Biology* (Oxford: Oxford University Press, 2014).

14. I am of course borrowing the phrase and idea from Hans Jonas, *The Phenomenon of Life: Toward a Philosophical Biology* (New York: Harper & Row, 1966).

15. This is the strategy employed, subtly, by Thomas Nagel in *Mind and Cosmos: Why the Materialist Neo-Darwinian Conception of Nature Is Almost Certainly False* (Oxford: Oxford University Press, 2012) in arguing against materialist reductionism and for the necessity of teleological explanation in the physical and biological sciences.

16. For skeptical views of physics, see Nancy Cartwright, *How the Laws of Physics Lie* (Oxford: Clarendon, 1983), and Ronald N. Giere, *Science without Laws* (Chicago: University of Chicago Press, 1999).

17. All the quotes in this sentence are from Godfrey-Smith's *Philosophy of Biology*, Chapter 2. Although Godfrey-Smith has distinctive views of his own, I take this work to be a fair state-of-the-art statement of the philosophy of biology today and hence to represent if not quite the consensus view, one close enough to that. For an early and powerful critique of the reductionist

program—the grand idea of a unified science in which biology would be reduced wholly to physics and chemistry—see John Dupré, *The Disorder of Things: Metaphysical Foundations of the Disunity of Science* (Cambridge, Mass.: Harvard University Press, 1993).

18. Dupré, *Philosophy of Biology*, 3. Just to make this thought literal, in, for a prime example, *The Cambridge Companion to the Philosophy of Biology*, ed. David Hull and Michael Ruse (Cambridge: Cambridge University Press, 2007), the chapter entries include: Adaptation; Population Genetics; Units and Levels of Selection; Gene; Information in Biology; Evolutionary Developmental Biology; and Molecular and Systems Biology and Bioethics.

19. Evan Thompson, *Mind in Life: Biology, Phenomenology, and the Sciences of Mind* (Cambridge, Mass.: Harvard University Press, 2007), ix.

20. Godfrey-Smith, *Other Minds: The Octopus, the Sea, and the Deep Origins of Consciousness* (New York: Farrar, Strauss and Giroux, 2016), 79.

21. Although it is not inaccurate to say that Plessner is engineering a novel combination of the methods of transcendental reconstruction, phenomenology, and hermeneutics—life as always the interpretation of life—, the most natural reading of the text involves its withdrawing from methodological prescriptivism. On my reading, which is not everyone's, philosophical anthropology is a content-based philosophical paradigm rather than a method-based philosophical practice. If pressed, I would be tempted to argue that the early chapters of the *Levels* involve a complex effort to sublate philosophical method within the orientational structure of life itself, thus making life processes the structural premise of rationality.

22. In this respect, the idea of positionality refers back to and corrects Fichte's thesis that, transcendentally, we should regard the self as a product of its own self-positing. The primordial act of self-consciousness for Fichte is the self positing itself, being both subject and object of its primitive act of self-positing. Plessner means for us to come to recognize how the self of idealism derives its fundamental structure from the living thing.

23. This contrast between the thing with properties and the organism as a whole with parts involves a running dialogue with Kant and Hegel's account of the object of perception. If I am understanding Plessner correctly, he is claiming that the presumptive transcendental force of the substance-property/subject-predicate construction is a deflated version of the logic of organisms *having* property-parts that first occurs with the entry of the excentric positionality of the human. This is at least part of what I meant earlier in saying that the structure of life is a premise for the operation of reason.

24. For a thoughtful account of the evolutionary theory of senescence, as part of his effort to understand how an organism that is as breathtakingly complex as the octopus could be subject to such a remarkably short life, in nearly every case under two years, see Godfrey-Smith's *Other Minds*, Chapter 7.

That aging and death can be explained does entail that these are not modal features of living organisms. Again, Plessner assumes the possibility of explanation, which does nothing to remove, disappear, the phenomena in question.

25. With the rare exceptions: Venus fly traps are the most famous of carnivorous plants; among the animals, the sea slug *Elysia chlorotica* and the spotted salamander partially manage photosynthesis through a symbiotic relationship with algae cells, while the Oriental hornet has its own complex chemical means. These exceptions should also form a powerful reminder that there really are almost no strict laws in the organic world; almost always, at the level of the organism, biology works with rough generalizations and differences in degree rather than kind.

26. Although Plessner does not press the thought, his mention of a plant being a segment of a corresponding life cycle entails that there are other fundamental units of biological life beyond the individual organism. A more comprehensive analysis would have to show how his categorial scheme might handle, and be expanded in order to handle families, populations, habitats, ecological niches, and, generally, ecosystems of different dimensions.

27. In *Laughing and Crying: A Study of the Limits of Human Behavior*, trans. James Spencer Churchill and Marjorie Grene (Evanston, Ill.: Northwestern University Press, 1970), 37, Plessner states that in one fundamental respect "man is inferior to the animal since the animal does not experience itself as shut off from its physical existence."

28. There is a slowly emerging, much criticized, and disputed effort to consider plant life as manifesting a distinctive form of life rather than just a stage on the evolutionary ascent to the animal. For an overview of the argument in biology, see Michael Pollan, "The Intelligent Plant: Scientists Debate a New Way of Understanding Flora," *New Yorker*, December 23 and 30, 2013. For an inaugural philosophical effort to dignify plant life, see Michael Marder, *Plant-Thinking: A Philosophy of Vegetal Life* (New York: Columbia University Press, 2013). On the question of constraint, Godfrey-Smith's *Other Minds* is also implicitly concerned with this issue, since he takes the cephalopods to be the expansive life form that appears after the life-tree "fork" that separates the bilaterians into vertebrates, on the one hand, and mollusks and arthropods on the other. That is the perhaps too precious but also heuristically powerful thought driving his interrogation: the octopus is us if we had been formed along a different evolutionary pathway. That both pathways generated something like expanded nervous systems enabling, well, consciousness or its like, is working through the idea of evolutionary development in relation to the problem of constraint.

29. I presume that the animal experience of having a body is bound in the first instance to feelings of pain: are not animals that protect an injured limb

or nurse a wounded body part implicitly aware of having a body with which they are not fully at one?

30. See in this regard, Matthew Boyle, "Additive Theories of Rationality: A Critique," *European Journal of Philosophy* 24 (2016): 527–55.

31. In J. M. Bernstein, *Torture and Dignity: An Essay on Moral Injury* (Chicago: University of Chicago Press, 2016), I argue that the exemplary or paradigm cases of the utter dislocation of the dual aspects governing the human bodily scheme are torture and rape. A similar argument is provided by Janna van Grunsven, "The Body Exploited: Torture and the Destruction of Selfhood," in *Plessner's Philosophical Anthropology: Perspectives and Prospects*, ed. Jos de Mul (Amsterdam: Amsterdam University Press, 2014). This immense collection of twenty-six essays provides a consistently useful series of considerations of Plessner's thought in English. Although the German literature on Plessner is large, a good place to start is with the collective commentary *Helmuth Plessner: Die Stufen des Organischen und der Mensch*, ed. Hans-Peter Krüger (Berlin: De Gruyter, 2017).

32. Plessner, *Laughing and Crying: A Study of the Limits of Human Behavior*, trans. James Spencer Churchill and Marjorie Grene (Evanston, Ill.: Northwestern University Press, 1970). This work will be republished by Northwestern University Press in 2019 with a foreword by me.

33. I owe this clause to Janna Van Grunsven.

34. For this phrasing, see Hans-Peter Krüger, "The Nascence of Modern Man," in de Mul, *Plessner's Philosophical Anthropology*, 51.

35. Originally published under the title *Macht und Menschliche Natur* [Power and Human Nature], this is now available under Plessner's intended title, *Political Anthropology*, trans. Nils F. Schott (Evanston, Ill.: Northwestern University Press, 2018); the translation includes an extended introduction by Heike Delitz and Robert Seyfert and an epilogue by Joachim Fischer.

36. I assume working through the relation between biology and consciousness will be a consuming and central enterprise in the near future. On competition and altruism in evolution, see the fine work by Elliott Sober and David Sloan Wilson, *Unto Others: The Evolution and Psychology of Unselfish Behavior* (Cambridge, Mass.: Harvard University Press, 1998). On the significance of microbiological life and symbiosis, see Dupré, *Processes of Life*, Part III.

37. Pluralism and promiscuous realism are ideas that John Dupré has been pressing and defending for thirty years. If bio-philosophy ever escapes from the philosophical shadows within which it still lives, either Dupré's view will become philosophical orthodoxy or philosophy will have sunk into an even deeper crisis than its present one.

38. For invaluable comments and suggestions on the first draft of this Introduction, I must thank Adam Gies, Karen Ng, Janna Van Grunsven, Gesa Lindemann, and Hans-Peter Krüger. It was a great good fortune that

Hans-Peter and I met at a moment when we both had been pursuing, without result, the project of getting Plessner's masterwork translated into English. Together we have managed it. For his immense contributions to Plessner studies and for this translation, it is a pleasure to here acknowledge Hans-Peter's tireless efforts.

EPIGRAPH

1. Julius Löwenberg, Robert Avé-Lallemant, and Alfred Dove, *Life of Alexander von Humboldt*, ed. Karl Bruhns, trans. Jane Lassell and Caroline Lassell (New York: Cosimo Classics, 2009), 1:204.

1. AIM AND SCOPE OF THE STUDY

1. For this entire section, see Helmuth Plessner, *Die Einheit der Sinne: Grundlinien einer Ästhesiologie des Geistes* (Bonn: F. Cohen, 1923), 118–37 and 258–67.
2. Georg Misch, "Die Idee der Lebensphilosophie in der Theorie der Geisteswissenschaften," *Österreichische Rundschau* 20, no. 5 (1924); reprinted in *Kantstudien* 31, no. 1–3 (1926).
3. Misch, "Die Idee der Lebensphilosophie."
4. Misch.
5. Misch.
6. Misch.
7. See Plessner, *Die Einheit der Sinne*, xiii–xiv.
8. Cf. Plessner, *Die Einheit der Sinne*, part V, in particular 268–71, 276–81, 285–88. See also Frederik J. J. Buytendijk and Helmuth Plessner, "Die Deutung des mimischen Ausdrucks," *Philosophischer Anzeiger* 1, no. 1 (1926).

2. THE CARTESIAN OBJECTION AND THE NATURE OF THE PROBLEM

1. Fichte's concept of the absolute I includes other elements not addressed here.
2. See on this point Frederik J. J. Buytendijk and Helmuth Plessner, "Die Deutung des mimischen Ausdrucks," *Philosophischer Anzeiger* 1, no. 1 (1926).
3. Paul Claudel, *Poetic Art*, trans. Renée Spodheim (Madison: University of Wisconsin, 1948); first published in French in 1913.
4. The phrase derives from Aristotle and refers to an unwarranted leap from one genus to another (translator's note).

3. THE THESIS

1. Wolfgang Köhler, "Some Gestalt Problems," in *A Source Book of Gestalt Psychology*, ed. Willis D. Ellis (London: Routledge & Kegan Paul, 1999),

55–70; originally published in *Jahresbericht über die gesamte Physiologie* 3 (1922): 512–39. This essay, among other things, applies the ideas Köhler developed on the problem of the organic in his book *Die physischen Gestalten in Ruhe und im stationären Zustand* (Braunschweig: Friedrich Vieweg & Sohn, 1920).

2. Hans Driesch, "'Physische Gestalten' und Organismen," *Annalen der Philosophie* 5 (1925/26).

3. Driesch, "'Physische Gestalten,'" 42.

4. Driesch, 3–4.

5. Driesch, 161.

6. Driesch, 5.

7. Driesch, 7.

8. Driesch, 8.

9. Adolf Meyer, *Logik der Morphologie im Rahmen einer Logik der gesamten Biologie* (Berlin: Springer, 1926), 30.

10. Meyer, *Logik der Morphologie*, 30.

11. From here through the end of the section "Definitions of Life," unless otherwise indicated, reality and variants render *Wirklichkeit/wirklich* (translator's note).

12. In order to understand this comment in relation to Chapter 1, I ask the reader to bear in mind that in terms of a theory of modals (not to be confused with modality!), the current study with its justification of modals of living being is the counterpart to Helmuth Plessner, *Die Einheit der Sinne: Grundlinien einer Ästhesiologie des Geistes* (Bonn: F. Cohen, 1923), which uses the aesthesiological method to attempt a theory of the modals of nonliving being.

13. Meyer, *Logik der Morphologie*, 41.

14. See in particular Wilhelm Roux, *Terminologie der Entwicklungsmechanik* (1912) and Roux, "Das Wesen des Lebens," in *Kultur der Gegenwart* (Leipzig and Berlin: Abteilung Allgemeine Biologie, 1915).

15. Immanuel Kant, *Critique of Pure Reason*, trans. Norman Kemp Smith (New York: St. Martin's, 2007), 128; first published as *Kritik der reinen Vernunft* in 1787. Emphases are Plessner's (translator's note).

16. See Plessner, *Die Einheit der Sinne*, 63ff. The concept of representation is conceived more broadly here.

17. A contrast between *representable* contents that are attainable in intuiting encounters and *distinct* contents rendered comprehensible by being fully viewed can already be found in *Die Einheit der Sinne*. *Specifiable* contents of intuition are separate from both of these. Since the distinction between perception and viewing is crucial for the current study, I use the concept of representation more broadly than in *Einheit*, extending beyond the "possibility of reproduction." Thus a law of arithmetic can be geometrically *represented*, although this second mode of givenness is precisely *not* a reproduction

of the first. This does not, however, affect the validity of the classification given in *Die Einheit der Sinne*.

4. THE MODES OF BEING OF VITALITY

1. On the following, see in particular Johannes von Kries, "Über Merkmale des Lebens," *Freiburger Wissenschaftliche Gesellschaft* 6 (1919), and Frederik J. J. Buytendijk, "Anschauungskriterien des Organischen," *Philosophischer Anzeiger* 3 (1928).
2. See Chapter 4, in Driesch's classical work *Der Begriff der organischen Form*, Abhandlungen zur theoretischen Biologie 3 (Berlin: Gebrüder Borntraeger, 1919), 35ff. Driesch is concerned here with the penetration of the dual category of thing-property by the dual category of whole-parts.
3. See Driesch, "Der Begriff der organischen Form," 39ff.
4. The setting of the hiatus shows it to be a relationship of *"interconnection,"* which Josef König was the first to describe (in its highest form, as the essence of intuition [*Intuition*]); König, *Der Begriff der Intuition* (Halle: Max Niemeyer, 1926).
5. On this point, see my criticism of Driesch's concept of entelechy in Helmuth Plessner, "Vitalismus und ärztliches Denken," *Klinische Wochenschrift* 1, no. 89 (1922).
6. Here is a key instance where Plessner's pointed use of prepositions cannot be captured in English. *Zu*, which in this context has been rendered as "for," also means "to" and "toward" and is, as it were, inclined forward into the future, whereas *nach*, rendered here as "according to," also means "after," "past," or "following" and is thus inclined toward the past (translator's note).

5. THE ORGANIZATIONAL MODES OF LIVING BEING: PLANTS AND ANIMALS

1. Wilhelm Roux, *Der Kampf der Teile im Organismus* (Saarbrücken: Verlag Dr. Müller, 2007); first published in 1881.
2. Armin von Tschermak, *Allgemeine Physiologie* (Berlin: Springer, 1916), 1:3.
3. Hans Driesch, "Studien über Anpassung und Rhythmus," *Biologisches Zentralblatt* 39, no. 10 (1919).
4. From here to the end of this section, "reality" and variants render *Wirklichkeit/wirklich* (translator's note).
5. The aesthesiological investigation of sensory modalities I undertake in Helmuth Plessner, *Die Einheit der Sinne: Grundlinien einer Ästhesiologie des Geistes* (Bonn: F. Cohen, 1923), allowed me to prove this for the relationship between subject and object in perception.
6. Driesch, *The Science and Philosophy of the Organism* (Aberdeen: Printed for the University, 1908), 1:48–49.

7. Plessner is citing Goethe's poem "True Enough: To the Physicist" here (translator's note).
8. Armin von Tschermak provides an overview in his *Allgemeine Physiologie*, vol. 1 (Berlin: Julius Springer, 1916), in particular 423ff. Von Tschermak rejects the equation of plant and animal modes of reaction to external stimuli (see 1:51 also for the literature on psychovitalism and tropism theory).
9. See Heinrich Andres, "Versuch einer charakterologischen Analyse der Lebensfunktionen der Pflanze," *Abhandlungen zur theoretischen Biologie*, no. 26 (1927).

6. THE SPHERE OF THE ANIMAL

1. Jakob von Uexküll, *Umwelt und Innenwelt der Tiere*, 2nd ed. (Berlin: Springer, 1921), 46.
2. Uexküll, "Environment (Umwelt) and Inner World of Animals," trans. Chauncey J. Mellor and Doris Grove, in *Foundations of Comparative Ethology*, ed. Gordon M. Burghardt (New York: Van Nostrand Reinhold, 1985), 231.
3. Uexküll, *Umwelt und Innenwelt der Tiere*, 177–78.
4. Uexküll, 177.
5. See Helmuth Plessner, *Die Einheit der Sinne: Grundlinien einer Ästhesiologie des Geistes* (Bonn: F. Cohen, 1923), 90ff.
6. Julius Pikler, *Theorie der Empfindungsqualität als Abbildes des Reizes: Schriften zur Anpassungstheorie des Empfindungsvorgangs* (Leipzig: Barth, 1922), 4:50: "Accordingly the basic qualities, in other words the toned basic sensations or undertones of all the senses that possess qualities of any kind, in other words, toned sensations or tones, are four in number, and they are, in the order of their relatedness, mild, harsh, sharp, and coarse. The following table emerges: sweet-sour-salty-bitter, caressing-tickling-itching-pain, flowery-vinegary-burnt-putrid, blue-green-yellow-red, in which the elements of every horizontal row (or in the corresponding position in the list) are one and the same for different senses." See also 4:71.
7. Here and in the following two paragraphs, Plessner's term *Lage* (see glossary) is translated as "position" (translator's note).
8. Uexküll, *Umwelt und Innenwelt der Tiere*, 171ff.
9. Uexküll, 179.
10. Uexküll, *Environment (Umwelt) and Inner World of Animals*, 234–35.
11. Uexküll, *Umwelt und Innenwelt der Tiere*, 182.
12. Reality renders *Wirklichkeit* from here to the end of the section (translator's note).
13. Hans Volkelt, *Über die Vorstellungen der Tiere* (Leipzig: W. Engelmann, 1914).

14. Plessner is quoting from *Hamlet* here, Act 3, scene 1 (translator's note).
15. Volkelt, *Über die Vorstellungen der Tiere*, 85–86.
16. Volkelt, 87–88.
17. Volkelt, 89–90.
18. Wolfgang Köhler, *The Mentality of Apes*, trans. Ella Winter (Oxon: Routledge, 2001); first published in German in 1917.
19. Hans Driesch, *The Science and Philosophy of the Organism* (London: Adam and Charles Black, 1908), 2:54.
20. Driesch, *Science and Philosophy of the Organism*, 2:60.
21. Driesch, 2:71ff.

7. THE SPHERE OF THE HUMAN

1. In this chapter, "center" renders *Mitte* unless otherwise noted (translator's note).
2. *Sich ausleben* means to live it up or enjoy life to the fullest; literally it is "to live oneself out" (translator's note).
3. Sigmund Freud, *On the History of the Psycho-Analytic Movement*, trans. Joan Riviere (New York: Norton, 1966), 67; first published in German in 1914.
4. Alfred Seidel, *Bewußtsein als Verhängnis*, ed. Hans Prinzhorn (Bonn: Schulte-Bulmke, 1927).
5. This synthetic proof of the mediatedness of the subject-object relation in comprehending knowledge or cognition can be seen as a confirmation of the analytical proof given by Nicolai Hartmann in his *Grundzüge einer Metaphysik der Erkenntnis* (1921; Berlin: De Gruyter, 1965). Attacking his theory of the image with phenomenological arguments as Paul Linke does ("Bild und Erkenntnis," *Philosophischer Anzeiger* 1, no. 2 [1926]) is bound to fail, no matter how astutely reasoned. The appeal to the intentional structure and its self-evidence is of no use, as it sees only one side of cognition.
6. Helmuth Plessner, *Die Einheit der Sinne: Grundlinien einer Ästhesiologie des Geistes* (Bonn: F. Cohen, 1923), 276ff.
7. Plessner is quoting here from Friedrich Schiller's *Walleinstein: A Dramatic Poem*, Act 3, scene 2 (translator's note).
8. The socio-philosophical consequences of this degradation of the physical world can be found in anti-society, pro-community radicalism; see Plessner, *The Limits of Community: A Critique of Social Radicalism*, trans. Andrew Wallace (Amherst, N.Y.: Humanity, 1999); first published as *Grenzen der Gemeinschaft* in 1924.
9. From here to the end of this section, "reality" renders *Wirklichkeit* (translator's note).
10. Cf. Plessner, *Die Einheit der Sinne*, 137ff. and 146–60. The fact that my theory of types of meaning is not tied to a particular standpoint can be

seen by comparing it to Hans Freyer's *Theory of Objective Mind: An Introduction to the Philosophy of Culture*, trans. Steven Grosby (Athens: Ohio University Press, 1991), first published in German in 1923), and Julius Stenzel's *Sinn, Bedeutung, Begriff, Definition* (Darmstadt: H. Gentner, 1958), first published in 1925.

11. The phrase is attributed to Archimedes and is commonly translated as "Give me a place to stand, and I shall move the world" (translator's note).

12. From here until the end, "reality" and its variants render *Wirklichkeit* unless otherwise noted (translator's note).

13. Plessner, *Limits of Community*. See in particular Chapter 4, pp. 108ff. The reception of this book, which was first published in 1924, has been strangely volatile, and not only because of its subtitle and outward form. Many today are still not able to distinguish the subtle boundaries between sociology and social philosophy. The clarification of the domain of philosophical anthropology will surely sharpen awareness of the purely theoretical nature of this work's criticism of social radicalism and its aim to establish the *"public sphere" as a mode of realization* of the human.

14. The reference here is to Adolf von Harnack, *Marcion: The Gospel of the Alien God*, trans. John E. Steely and Lyle D. Bierma (Durham, N.C.: Labyrinth, 1990), 139, first published in German in 1924 (translator's note).

APPENDIX

1. See Helmuth Plessner, "Vitalismus und ärztliches Denken," *Klinische Wochenschrift* 1, no. 39 (1922).

2. Cf. Frederik J. J. Buytendijk and Paul Christian, "Kybernetik und Gestaltkreis als Erklärungsprinzipien des Verhaltens," *Der Nervenarzt* 34 (March 1963).

3. Gerhard Frey, *Gesetz und Entwicklung in der Natur* (Hamburg: Meiner, 1958), 180–81.

4. Plessner does not provide a first initial for this individual, whom we were not able to identify (translator's note).

5. Alexander I. Oparin, A. E. Braunshteĭn, A. G. Pasynskiĭ, and T. E. Pavlovskaya, eds., *Proceedings of the First International Symposium on "The Origin of Life on the Earth"* (New York and London: Pergamon, 1959).

6. Oparin et al., *Origin of Life on the Earth*, 1.

7. Oparin et al., *Origin of Life on the Earth*, 1–2.

8. Oparin et al., *Origin of Life on the Earth*, 2.

9. Oparin et al., *Origin of Life on the Earth*, xi.

10. Oparin et al., *Origin of Life on the Earth*, 311.

11. Oparin et al., 311.

12. Oparin et al., 317.

13. Oparin et al., 316.
14. Oparin et al., 437.
15. Oparin et al., 440.
16. Peter Mitchell, in Oparin et al., 440.
17. Oparin et al., 428.

INDEX

a priori, xxxi, xlvii, lxiv, 4, 30, 59, 60f., 72f., 100, 102, 104, 106f., 109–111, 113, 123, 136, 139f., 155, 157, 193f., 200f., 203, 253, 255, 261, 265, 298, 300, 317
absolute, idea of the, 317, 320f., 352
abstraction, xliv, 85, 218, 236, 253–256
act, 42–44, 51f., 117, 120f., 140, 143, 184f., 192f., 198, 223, 276, 280, 299, 313, 315, 349
action, xi, xxi, xxvii–xxx, xxxviii, xlix, lvii, lxif., 2, 6, 15f., 55, 60, 90, 105, 182, 190, 208, 223, 228–236, 240f., 243f., 248f., 251f., 270, 288f., 293f., 296–298, 309f., 326. *See also* auturgy; self-action
actuality, xi, lvi, 5, 82, 89, 92, 134, 148, 156, 158, 160–166, 168, 174, 176, 179, 196, 200, 215, 219, 241, 256, 261, 274, 276, 299, 305, 308, 312
adaptation, xxviii, xlii, l, 4f., 50, 110, 117, 153f., 179, 186–197, 200, 240, 246, 257f., 299, 312
adaptedness, 60, 179, 186–196, 223, 240, 265
Adler, Alfred, 292f.
adumbration (*Abschattung*), 78, 83, 274
aesthesiology of spirit, 22, 29
aesthetics, 15, 28, 100, 187. *See also* art
affect, 58, 271, 275. *See also* emotion; feeling
aging, lii, 108f., 137–140, 157, 170, 197–199, 350

ahead, 117, 132–135, 164–168, 171, 178, 193, 196–198, 221, 223, 236, 259, 261f., 264f., 278, 296
alien(ness), lvii, 19, 40, 54, 137, 140, 182, 184, 189, 221–223, 226, 237, 241f., 270, 280, 302, 314, 316
Alsberg, Paul, xxviii, 346
ambiguity, xxixf., lviii, 43, 45, 83, 95, 121, 180, 220, 274, 319
amino acids, 330
analogy, 23, 90, 92, 136, 184, 237, 278f., 303, 329
anatomy, xxiii, 21, 64, 69, 203, 233, 237, 247, 274
and-connection (*Undverbindung*), 85, 155, 165
anesthesiology, 22, 29–32, 72, 299f., 308, 353f.
angel, xxvi, 143
animacy (*Belebtheit*), xxiv, xxx–xxxiii, 23, 65, 84, 123, 195, 262, 279
animal, lii–lviii, 3, 50, 57–59, 61–65, 184, 188, 202–209, 212, 215–218, 219–272, 277f., 284–293, 296–298, 301–305, 327–329, 334–336, 350, 355
animal psychology, 57–64, 242f., 254, 329
animism, 279, 282
antagonism, 211f.
anthropocentrism, 51, 74
anthropoid, 64, 247f., 250
anthropology, vii, xvff., 20–22, 24, 27, 32, 69–71

359

Index

anthropomorphism, xxvi, xlv, 57f., 63, 243, 279, 320, 329
anticipation, lv, 79, 139, 164, 167, 192–196, 199, 232, 259f., 264–266, 296, 312f.
antipathy, 276. *See also* hate; sympathy
antiquity, xx, 22
appearance (*Erscheinung*), xxix, xxxiv, xlviii, 26, 34–46, 51f., 56, 76–86, 93, 96, 98, 100, 110, 112f., 118, 120, 149f., 152f., 170, 235–237, 245f., 274–278, 304f., 312, 325f.
apperception, 8, 56, 265, 278
approachability, 237, 286
apriorism, 3f., 11, 19, 140f., 194
apriority, 30, 70, 82, 140, 334
Aristotle, xxi, 171, 325, 353
art, 8, 15, 17, 28, 68, 291, 293, 300. *See also* aesthetics; culture
artificiality, lxi, 287–300, 314, 319
aspect, 4, 11, 27f., 32, 77–79, 224, 247, 255; dual aspect, xlvii, l, livf., lviii, lx, 4, 28, 43, 53, 55, 65f., 72, 75ff., 84ff., 93–99, 120–122, 149f., 219–221, 224, 271f., 273–275; interior/internal aspect, 61, 73, 226
aspect boundary, 96f.
aspect divergence, 83f., 96
aspectivity, 78f., 224f.
assimilation, xliii, 105, 108, 179, 182–184, 189, 206, 210f., 335
association, 56, 82, 208, 230, 243, 254, 257f., 261f., 265. *See also* memory
astrophysics, 329
atheism, 320. *See also* religion; theism
atmosphere, 286, 329
atomism, xli, 56, 243
Auerbach, Felix, 105, 183f.
Augustinus, Aurelius, xxv
autarky, 179
autonomy, xxxiiif., xlv, 14, 65, 86, 90, 92, 99, 102f., 118, 137, 152f., 180, 186, 190, 196, 245, 259, 333
auturgy, 90, 105. *See also* action; self-action

awareness, xxv, lvii, 15, 31, 50, 62, 225, 243, 250, 256, 270; of others 50

Baden School, 16
Baerends, Gerard, 329
balance, 184f., 190f., 202f., 210–212, 240, 248, 278, 289, 291–293, 297f., 315, 321
Baltzer, Fritz, 247, 324
barrier (vs. boundary), lxixf., 145
Bauer, Erwin, 105
Becher, Erich, 60, 208, 257
becoming, 1, 87, 117, 124–132, 136, 144f., 147, 160, 171, 197, 259f. *See also* development; persistence; process
Beer, Theodor, 58
beginning-something, 126, 131f.
behavior, xxvi–xxx, 31, 58, 62–64, 190, 202, 242–245, 247, 249f., 254, 256, 258, 266, 285, 308, 312, 327
being-at-hand (*Vorhandensein*), xxiv
Bergson, Henri, x, 4–7, 9f., 19, 134, 208, 240, 262, 327
Bernard, Claude, 105
Bertalanffy, Ludwig von, 324, 331
Bethe, Albrecht, 58
Bichat, Xavier, 105
Bierens de Haan, Johan, 254
biochemistry, xx, xxxi, xxxiii, 323f.
biology, xvf., xxxif., xlf., xliiif., xlvi, lviii, lxv, 33, 38, 50, 58f., 61, 63, 71, 87, 102f., 106, 108–110, 112, 128, 135, 151, 185, 192, 198, 201, 242, 265, 324, 327, 333, 348–351
bio-philosophy, xvi, xlvi, l, lviii, lxv, 346, 351
birth, xxiv, lvi, 24
bodily object, 224, 272
bodily thing, xxiv, 120f., 123, 126f., 146, 149, 152, 268, 273f. *See also* thing
body, xi, xiii, xvf., xxxiif., l–lx, lxiiif., 40–44, 47–51, 53f., 57, 59, 67, 84, 93, 97f., 103, 112f., 115, 119–121, 129–131, 135–138, 142–152, 155–281 (passim), 332f.; animal body, 222, 244, 268, 270; inorganic body, 115, 121, 186; lived

body (*Leib*), xi, xvf., xxvi, lv–lx, 18, 20, 23, 28, 30f., 62, 111, 159, 180, 185, 205f., 213–215, 219–223, 226f., 233f., 236, 249, 252, 260, 267f., 270–275, 279, 281, 304, 308f., 312, 315; living body, 86f., 93, 111, 121, 128, 131, 144, 147–149, 157–191, 202, 211, 213f., 217, 221, 223, 234, 236f., 268; organic body, liii, 115, 121, 158, 165, 167, 170f., 178, 186, 193, 210, 301; organized body, 178, 181, 210, 225 (*see also* organ); physical body, xi, lvif., lixf., 31, 37, 112, 121, 135, 156, 159, 169, 171, 202, 220, 226, 273–275; physical lived body (*Körperleib*), lx, 159, 180, 221f., 267, 272, 308; thing-body, 84, 93–95, 98, 224, 274
body/mind, xxxviii, xl, xlv, 4, 35
body/soul, xxvif., 47, 69, 72f., 281f., 304
Bolk, Louis, xxvif.
border, xlix, 97f., 178. *See also* barrier; boundary; contour
boundary, x, xii, xxxii, xxxivf., xlviii–lvii, lxiv, 48, 53–55, 90, 92–99, 103. 111–121, 124–129. 135, 137, 143–148, 150, 160, 171, 178, 182f., 185f., 189, 202, 210, 218, 227, 234, 269, 274, 301, 332f.
bounds (*Begrenzungen*), 124, 127, 176. *See also* boundary
brain, xxix, 37, 62, 226, 233, 237–239, 241f., 290, 296. *See also* central nervous system; central organ
breeding, 5, 200, 326
Breysig, Kurt, 68
brokenness, 287, 309, 311
Brown, John, 105
Brünning, Erwin, 334
Bütschli, Otto, xv, 92
Buttel-Reepen, Hugo von, 245
Buytendijk, Frederik J. J., xxx, xxxviii, 64, 116, 188, 254, 257, 335, 346, 352, 354, 357

caducity, 105
Caesar, Julius, 14
calculability, 20
Calvin, Melvin, 331
capacity, xliii, lv, 37, 49, 151f., 160–163, 194, 197, 213, 230, 258
capitalism, 2
categorical subjunctive, 200
category, xlvii–lv, lxivf., 4–9, 20, 22f., 59–62, 68, 71, 87, 105–109, 114, 134, 142, 156, 160, 265, 281, 299f.
causal determination, 4, 134, 200
causa efficiens, 134
causa finalis/final cause, 134f., 142, 153, 199
causal explanation, xliii, xlvi, 17, 242
causality, lx, 5, 7, 80, 86f., 92, 100, 102, 134f., 137, 142, 153, 166, 182, 186, 190, 200f. *See also* entelechy
cell, xliii, 89, 92, 136, 152, 154f., 194f., 198, 200–203, 205–207, 224, 324, 328, 330f.
censorship, 291
center, liii–lvii, 27, 40, 42–44, 48, 53, 77–80, 148f., 151, 155, 173–175, 189f., 192, 203, 207f., 211–215, 217, 219–227, 230f., 233–235, 242–244, 260f., 264, 267–274, 277–281, 287, 294, 302, 304f., 309f., 312, 315, 318, 335. *See also* core; periphery
central nervous system, liii, 155, 224, 228, 231, 237, 251, 333
central organ, 204, 211, 213f., 217, 219, 234, 237–239, 242
central point, excentric, 188
central position, 221f., 242
centralization/decentralization, 216, 227, 301f.
cephalization, 326
cerebral parasitism, 291
cerebralization, xxiii, 326
cerebrum, 290f.
chemistry, xx, xxxii, 107, 110, 201, 323, 329, 349
chemistry, colloidal, 91
child, 31, 122, 248, 257, 279, 347
chimpanzee, xxiii, 249f., 253

choice, free, 15, 223. *See also* freedom; volition; will
circle of life, xxxiv, 172ff., 178f., 182, 204, 210, 213, 215, 266
circulation (of matter and energy), 182, 196f.
civilization, 2, 291. *See also* culture
Claudel, Paul, 63
cogitatio/cogitation, 36, 42–44, 51. See also *res cogitans*
cognitio rei/circa rem, 17
cognition, xi, 9, 19, 61, 72, 106, 148, 188, 309, 356
Cohen, Hermann, 15
colors, x, xxxiv, 25f., 40, 76f., 85–87, 93, 95, 101, 103, 206, 217, 241, 254
communication, xxviii, lvi, 23, 57f., 336
community, 198, 282, 287, 319f., 347, 356
compensation, xxvii, 133, 278, 295. *See also* overcompensation
complex quality (Ehrenfels), 85, 242ff., 245–247, 251, 254f., 336
compromise, 311f.
Comte, Auguste 14
conceptualization, 26, 101, 109f., 253
conflict, lxi, 284, 294
conformity, xl, 8f., 187f. *See also* adaptedness
Conrad-Martius, Hedwig, xvi, 207, 335
conscience (*Gewissen*), 3, 5, 294
consciousness, xf., xxiif., xxvf., xxix, xlf., xliii–xlv, 1, 3–6, 11, 19f., 24, 27f., 35–37, 40, 42, 44f, 47, 51–54, 56, 61–67, 74, 77–80, 82f., 100, 106, 148, 167, 185, 208f., 215, 223–228, 230f., 233, 240–247, 249, 251f., 254–258, 264–266, 269f., 276, 279, 282, 287, 289, 291, 295, 304–310, 312, 316f., 320, 325, 332, 349–351
consolidation, 211, 216, 223
construction, xxx, lix, lxiv, 3, 8, 53
construction plan, 158, 165, 183, 189, 229, 231, 244
contemplation, 309
contingency, 8, 59, 317, 319f.

contour, xxxii, xlix, 61, 72, 79, 87, 95–98, 112f., 116, 120, 124, 182, 238
contouring, xxxii, 127
contraposition, xii, 41, 43, 144f., 161, 173, 176, 184f., 211, 215, 226, 287
convergence, 49f., 53, 80, 270f., 273, 317
cooperation/conflict, 61, 181, 284, 286
Copernican revolution, xxv
core, xii, liv, 39f., 76–83, 94, 96, 109, 112, 120–122, 126, 148–151, 155, 160, 173, 175, 206, 213f., 219f., 224–226, 231, 235, 245f., 251, 260, 269, 274, 276, 312
corporeality, 38, 127, 130, 137, 139, 203, 210. *See also* body
correctability (of reactions), 257, 261
counterworld (*Gegenwelt*), 61f., 240
craniometry, 69
creativity, 298f., 311, 314
critique: of reason, science, knowledge, 12, 15f., 27, 29f., 325; of values, 16; of the senses, 29f., 72
Critique of Judgment, xlii
Critique of Pure Reason, ix, 87
crying, xxx, lxiiif., 277
culture, xlf., lxii, 3, 7–10, 14–16, 21f., 24, 28f., 67–69, 71, 73, 279, 289–292, 294–300, 311, 317
custom, lxii, 291–293. *See also* morality; mores
cybernetics, xxi

Dacqué, Edgar, 272
Darlington, Cyril Dean, 334
Darwin, Charles, 35, 187, 191, 290
darwinism, xxi, 23, 325
Dasein, xxii–xxv
death, xxii, xxiv, xlii, lii, lxiii, 24, 105, 131, 137–143, 183f., 192, 197f., 202, 238, 275, 295f., 318, 350
decentralization. *See* centralization/decentralization
decision, xxii, lvii, 232, 277
deduction/deductive, xxxiif., xxxix, xlvif., 3–5, 23, 39, 87, 102, 106, 108,

113f., 116, 118, 135, 141, 198, 209, 218, 290, 293, 299, 327
deficient being (*Mängelwesen*), xxvif.
depth (*Tiefenhaftigkeit*), 49, 79f., 95, 309
description, phenomenological, 20, 27
determinacy, 126f. *See also* indeterminacy
determination (causal), xxxviii, 4, 15, 80, 97–99, 134, 166, 293
development, xxvii, l, lii, 6, 87, 103–105, 108f., 113, 116, 129–144, 152, 157, 184, 187f., 197–205, 311, 325, 327, 329, 350; embryonic xxxiii, 333
developmental physiology, 87, 323
developmental psychology, 81
Dewey, John, xxvii
dialectic, 107, 140–142, 183, 278
dichotomy (Cartesian), 36, 40, 42, 48, 50f., 54, 56, 58, 67f., 70f., 74
differentiation, xxvii, xxix, liiif., 31, 91, 136, 154f., 157f., 185, 202, 205f., 216, 238, 324, 327, 331
Dilthey, Wilhelm, x, xvii, xxii, 16–19, 21f., 24f., 33, 68, 308
directness, indirect, 241, 288, 301–303, 305–310
disenchantment (*Entzauberung*), 279
disintegration (into spheres, zones, worlds), 46, 51, 73, 75, 181, 183, 185, 211f.
dissimilation, xliii, 179, 182, 184, 189, 206, 210f., 335. *See also* assimilation
divergence (aspect divergence), 83, 96, 130
diversity, 8, 105, 129, 136, 283, 312, 317, 327
dividedness, 289, 294, 297
Dobzhansky, Theodosius, 334
double-sidedness, 224–226
doubt, 36f., 305, 320
Driesch, Hans, xv, xx, xxxiii, 84, 87–93, 96, 98–100, 103, 105, 134, 136, 151–153, 186, 190, 203, 258, 323f., 333, 346, 353f., 356

drive (*Trieb*), xxiiif., xxvii, xxixf., lxv, 58, 83, 194, 207, 215f., 222, 228, 231, 234, 251f., 254f., 263, 265, 271, 289, 291–298, 307, 309f.
dual aspect (*Doppelaspekt*). *See* aspect
dualism, xxxvii–xli, xlv, liv, lxivf., 26, 53f., 69, 225, 242, 281, 311
Dubois, Eugène, 325
duty, lxi, 294

economy, 8, 25, 68, 73, 293
ectropism/ectropy, 117, 184
edge, 94. *See also* border; boundary; contour
Edison, Thomas, xxiii
effecting, 227–235, 252. *See also* noticing; sphere: of effecting
Ehrenberg, Rudolf, 334
Ehrenfels, Christian von, 246
Eickstedt, Egon Freiherr von, 345f.
Einheit der Sinne, xv, 22, 29f., 32f., 71, 299f., 308, 345, 352–356
Einstein, Albert, 35
ekstasis, 306
elimination (*Ausscheidung*), 105, 185
embodiment, xxi, liv, 240, 274, 280. *See also* body
embryo, 162, 198, 205, 333
emergence, 5–9, 62, 191, 213, 289, 330–332
emotion, xvi, xxi, 37, 57, 307–309. *See also* affect; feeling; mood
empathy, 14, 19, 23, 27, 63f., 208, 279
empiricism (vs. apriorism), 3f., 11, 19, 47, 71, 99, 104, 140, 161, 293
emptiness, 252, 256, 272f., 283. *See also* negative, sense of; space; time
enchantment (*Verzauberung*), 2. *See also* disenchantment
end-something, 126, 131f.
Engels, Friedrich, 8
entelechy, xxxiii, 87f., 103, 105, 133, 135, 137, 152f., 323, 325, 333, 354
entropy, 109, 184
envelop, 48, 226

environment, xii, xvi, xlvii–l, 28, 30, 32, 50, 58–63, 154, 187f., 192, 229, 231, 240, 245–247, 266, 270, 273–275, 279, 284, 286, 289, 296, 317f., 326, 329, 331–333, 336
epigenesis, 135
epistemology, 11, 26, 35–37, 66, 73, 81, 273, 307f.
equilibrium, xlii, lxif., 94, 179, 188, 192, 196, 238f., 266, 288f. See also balance
equipotentiality, 149, 152, 154, 156f., 174, 177, 205
Erlebnisphilosophie, 277
essence (passim, esp.), xxiif., 10, 26, 40–42, 73, 78, 111, 131–133
ethics, xxi, 4, 10, 15, 21, 28, 223, 294. See also morality; mores
ethnicity, ethnic, 14. See also people; race
ethnology, xx, 21
ethology, xxvi, 327, 329. See also behavior
evidence, 30, 37, 47, 52, 168, 277, 306f., 310, 315
evolution, xxi, xli, 3, 9, 134f., 247, 290, 326–329, 331–333, 348–351; creative evolution (Bergson), 7, 134, 327
evolutionary history, xxvi, 7, 21, 69, 326f.
evolutionary psychology, xxxvii, xl
evolutionism, xxvi, 2, 23
excentricity, xii, xxx, lii, lv–lxi, lxiii, 188, 192, 267–289, 293–308, 312, 314–321, 349
excessiveness, 297
excitation, xxix, 190, 217, 227, 230, 233, 241f., 264. See also nervous system; stimulus
execution (*Vollzug*), 42–44, 97, 221, 223f., 233, 260, 267, 269–271, 275, 277f., 287, 305. See also action
existential analysis (*Existentialanalytik*), xvii, xxiif.
experience, xf., xl, xlviif., lv–lix, 3, 5, 10f., 19–32, 34f., 37, 40, 45–47, 52, 54, 60, 63, 65, 68–73, 81f., 111, 169, 232, 258, 263–280

experiential position (*Erfahrungsstellung*), 11, 19, 21, 23f., 46f., 55
experiment, x, 12, 38, 58–61, 63f., 86, 118, 135, 242, 244
explanation, causal, 17, 242
expression, xxviii, 8, 19f., 23, 29, 64, 71, 94, 105, 116, 159, 209, 271, 292–294, 299f., 309, 313, 315; facial, xxix, 23, 85, 300, 313
expressivity, 298ff., 300f., 309, 314f.
extension, 34–42, 46, 49f., 54, 272. See also physicality; *res extensa*
extrauterine spring, xxvii
eye, 24f., 47, 102, 148, 193, 195, 224, 226, 239, 269, 306, 309. See also gaze

familiarity, 286
family, 287
fate, 8f., 15, 68, 138, 142, 144, 197, 200, 317
Faustian, 9f.
fear, xxvf., 290, 295f.
Fechner, Gustav, 76
feeling, xxiv, xxix, 26, 49, 58, 79, 271, 275–278, 280, 300, 307–310, 319, 350. See also affect; emotion; mood
fellow human being, 284
Fichte, Johann Gottlieb, 18, 52, 66, 106, 120, 140, 224, 280, 302, 349
field: opposing (*Gegenfeld*), 185f., 188, 215; positional, 178f., 185f., 188–193, 195–197, 204, 215, 220, 223, 231, 236, 239, 242, 274; shared, 239, 285; surrounding (*Umfeld*), lv, lx, 147, 186, 188, 190, 192f., 213f., 222f., 226, 228f., 231–236, 238, 240–242, 244, 248, 251–253, 255–257, 260, 267, 270–274, 283–286, 301–305, 346
field condition (*Feldverhalt*), 253, 256f.
field structure, 55, 249f., 253, 255
final cause. See *causa finalis*/final cause
finitude, xxii–xxv, 13, 223

force, 17, 54, 80f., 86f., 89, 133, 137, 139, 142, 209, 211, 300, 333, 349. *See also* entelechy; substance
form, 93f., 115–118, 127f., 188, 192, 199, 202f., 218; closed/open, 203–226; symbolic, 29. *See also* life form; space; time
form-idea, 127f., 130, 132f., 139, 141–143, 201
forward displacement (*Vorgelagertheit*), 41f., 45f.
freedom, xlii, lxf., 1, 14, 117f., 222, 232, 271, 278, 288, 294
Freud, Sigmund, xxiv, xxvi, xxix, 35, 291f., 297, 356
Frey, Gerhard, 324, 357
Frisch, Karl von, 335
frontality, lv, 219ff., 223, 226–228, 248, 253, 255, 257, 261, 270f., 283f.
functional system, xxviii
functional unity, 116, 149–151, 155–157, 172, 174, 178, 181, 211, 217
function-circle, 178, 205, 212, 229f., 240, 251, 255, 265
future, 9, 15, 130f., 160–168, 193, 197, 259–262, 284, 296, 315, 321, 354. *See also* past; time

Galilei, Galileo, 12
Gánti, Tibor, xliif.
gap, xxix, lvii, 241, 256, 270, 276, 283. *See also* hiatus
gaze, 36, 42–44, 46, 57, 72, 75–78, 83, 225f., 233, 270, 276, 286; position of the gaze, 72, 75f., 83, 225f. *See also* perception; sense perception
Gehlen, Arnold, xxvi–xxx, 346f.
Gelb, Adhémar, 262, 336
genes, 324, 326f., 330–332, 348
genetics, xx, 3, 134f., 326–328, 330, 348f. *See also* germ cells; heredity; phylogeny
geometry, Euclidian, 168, 300
geophysics, 329
germ cells, 198, 200f., 205

germ plasm, 134, 198
gestalt, xxxiiif., xlviii, 18, 33, 55, 64, 76f., 83, 85f., 88–99, 101, 112–115, 117f., 120, 127–129, 134–136, 139, 141, 143, 149f., 154f., 157, 165, 172, 174, 189, 196, 199, 202, 217f., 235, 245f., 249, 253f., 256, 269, 272f., 324, 352f.
gestalt-idea, 127–129, 134f., 141
gestalt quality, 246, 254
gignomen, 26, 55
given(ness), 41–44, 51, 64, 84, 101, 111f., 192, 223, 226, 231, 241, 246, 253, 268–270, 304, 353
God, xxiii, 2, 158, 194, 240, 289, 317, 320; *See also* religion
Goethe, Johann Wolfgang von, 21, 30, 193, 195, 345, 355
Golden Age, 287. *See also* history; philosophy: of history
Goldstein, Kurt, 262, 336
goods, 16, 69
Greek antiquity, xx, 22
growth, xliii, l, lii, 86, 103, 105, 110, 116, 136, 183, 205, 207f.
Grünbaum, Abraham Anton, 262

Haberlandt, Gottlieb, 207
habitat, lxiv, 186, 350. *See also* living space
habitus, 64, 75, 257
habitus picture, 64
Haering, Theodor, xix
Haldane, J. B. S., 331
handiness (*Griffigkeit*), 235
Hartmann, Eduard von, 324f.
Hartmann, Max, 328, 334f.
Hartmann, Nicolai, xix, xxii, 201, 356
hate xxi. *See also* antipathy
Hauptmann, Carl, 105
having, livf., lviii, 148–151, 156, 161, 164, 174–176, 214, 220f., 225f., 236, 268, 350f.
Hegel, Georg Wilhelm Friedrich, xxv, 30f., 66f., 78, 80, 83, 106, 140f., 283, 325–327, 349

Heidegger, Martin, xvii, xx–xxvi, lxv, 345
Helmholtz, Hermann von, xxxi, 100, 109
Herbst, Curt, xv
Herder, Johann Gottfried, xxvif.
here/now, lv, lvii, 221–223, 226, 231, 239, 259, 267–271, 273, 275, 277, 317. *See also* position; space; time
heredity, xliii, 5, 87, 103–105, 108–110, 196, 201, 232
Hering, Ewald, 168, 257
hermeneutic, 17, 19f., 24, 27, 32, 349
heteronomy, 196. *See also* autonomy
hiatus, xxix, 117, 140f., 144, 189, 198f., 217, 223, 227, 231, 240, 271, 354. *See also* gap
hiddenness, 246
historicity, xxii, 14, 314
history, 2–21, 24–28, 34, 38, 42, 51, 66–69, 71, 73, 87, 293, 311, 314–317, 327, 330. *See also* evolutionary history
historical base of reacting, 208, 258, 262–265, 279. *See also* memory
Holst, Erich von, 329
homelessness, lxi, 287f.
homo faber, xxiii
Homo sapiens, 326f.
horizon of possibilities, 247, 318
horizontal/vertical, 28f., 32, 355
human, the (*der Mensch*), 22–33, 267–321
humanities, xx, xxxi, lxiv, 10–12, 14–18, 21–24, 27, 67–69, 71–73, 293
humanity (*Menschheit*), 318
hunger, 111, 234
Husserl, Edmund, xi, xxif., xxv, xxxviii, 25, 68, 78f., 153, 253, 346
Huxley, Aldous, 326
Hyman, Libbie Henrietta, 335
hypertrophy (of drives), 292f., 295. *See also* drive

I (*Ich*), xxi, lvii, lxiii, 4, 36f., 40–54, 62, 220–222, 225, 267–272, 274, 279–281, 286, 302, 304f., 318

idealism, xxif., xxiv, xxxviii, xlvii, lx, lxii, lxiv, 3, 5, 11, 27, 36, 38, 44–46, 51, 53, 66, 79, 120, 148, 240, 276, 282, 306–308, 311, 325, 346, 349
ideation, 253
identity, 19, 39, 43, 69, 112, 127f., 302f.
ideology, xli, 1f.
illness, 105, 291
illusion, 45f., 50, 82, 118, 150, 291, 314
image, mental (*Vorstellung*), 52, 54, 77f.
imagination (*Phantasie*), 14, 19, 142
immanence, lxi, 2, 8, 41, 45–48, 50–54, 65, 188, 298, 306–310, 315
immediacy, ix, x, lviii, lxi–lxiii, 23, 25, 158, 173, 177, 210, 240f., 288, 298, 301–309, 312, 314f.
impulse, xv, 49, 53f., 207, 209, 222, 226, 229, 232, 235, 268, 278, 291f., 308
inadequacy, 309f.
in-between, xxxii, 94, 97f., 119, 124, 126, 145, 182, 214, 261, 307
independence (*Selbständigkeit*), xxxii, xliii, liv, 77, 97, 155, 169, 174, 189, 204, 216, 298
indeterminacy, 103
indirectness, xxviii, 50, 302f., 306–309, 319. *See also* artificiality; mediacy
individual(ity), xxii, xlii, xliv, lvi, 13f., 18, 32, 65, 68, 123ff., 128f., 141, 172, 184, 197–201, 203–205, 232, 254–256, 266f., 272, 279, 317–320, 332
inference, 23, 307
infinity, 13, 321
inhibition, 137, 173, 227, 242. *See also* stimulus
inside (*das Innen*), xxix, xlix, l, 49–55, 76, 81, 94, 148f., 220f.
instinct, xxvii, xxix, lvif., 194, 196, 207, 216, 222, 228, 232f., 243–245, 247, 256, 265f., 285, 288, 292, 308, 327, 329; instinct reduction, xxix
institution, xxviif., 326
integration, 124, 147, 153, 179, 185, 189, 205f., 210, 213, 215, 266, 273, 317

See also assimilation; central organ; wholeness

intellect, 3, 5–7, 9f., 19, 60, 282

intelligence, xxvi, lv, 63f., 244, 247–250, 252f., 256–258, 289–291, 293, 295–297

intention, 27, 43, 52, 83, 167, 208, 310–316

intentionality, xxv, xlii, 284, 306f.

interest, 263, 309. *See also* drive; volition

interiority, 34–46, 48–50, 54, 61, 65, 68, 208, 222, 270, 274, 311

internalization, xxviii

intuition (passim, esp.), ixf., xvii, 5f., 8, 10f., 19f., 23, 25f., 29, 32, 45, 66, 73, 81–84, 92–96, 99–101, 104–113, 122f., 188, 224f., 232, 235f., 254f.. *See also* sensation; sense; space; time

invention, xxviii, 296, 298f.

irrationalism, 10, 141

James, William, xxvii, 260

Jaspers, Karl, xx, 33

Jennings, Herbert Spencer, 257

Jordan, Pascual, 324, 331

judgment (logic), xli, 17, 51, 82, 106, 109, 185, 194, 196, 325

Kant, Immanuel, xxi, xxv, xliif., xlv–xlviii, 11–16, 20, 27, 29f., 38, 45, 52, 55, 60, 63, 66f., 70, 76, 78, 87, 105f., 109, 142, 188, 276, 306, 309, 349

Katz, David, 64

Klaatsch, Hermann, xxvii, 290

knowledge, xvi, xxix, lxiv, 4, 8–11, 15–19, 21f., 27, 29, 37f., 45, 47, 51f., 60, 70, 106, 112, 148, 166, 188, 276, 283f., 288, 291, 304–307, 309, 356

Koenigswald, Gustav Heinrich Ralph von, 325

Köhler, Wolfgang, xxxiii, 64, 84–92, 96, 99f., 247–250, 252f., 256–258, 296

König, Josef, xvii, 354

Körper, xi, lvi, lviii, 23, 31, 213, 273. *See also* body

Kraus, Alfred, 33

Kries, Johannes von, 105, 116, 354

Krüger, Felix, 85

Kühn, Alfred, 203, 334

Lamarck, Jean-Baptiste de, 191, 194

Lamprecht, Karl, 68

language, xxvii–xxx, xl, xliii, xlvii, lvi, lxii, 19, 22, 24, 28, 63, 95, 233, 300, 315f., 326, 336

Lask, Emil, xv

laughing, xxx, lxiii, lxiv

Laughing and Crying, lvii, lxiii, 350f.

law(s), xvi, xxxviiif., xli–xliv, lxi.lxiii, 2f., 8, 12, 14f., 20, 24, 28–31, 45f., 52, 54, 56, 60–63, 68, 71–73; fundamental laws of anthropology, 287–321; laws of thermodynamics, 109, 184; moral law, lxi, 294

layers of being, 28, 103, 107, 110, 152f., 201, 281

learning, lvf., lix, 230, 327

Lebensphilosophie, xvif., 3, 5, 8–11, 352

Lehmann, Emil, 92

Leib, xi, xv, lvi, 23, 31, 213, 273. *See also* body: lived body

Leibniz, Gottfried Wilhelm, 29, 67, 141, 320, 325

level, xvi, xix, xxi, xxiv, liii, 28, 102, 107, 128, 133, 136, 142, 155, 210, 213, 217f., 225, 265f., 268–270, 301, 325–328

life, xvi–xviii, xxiv–xxvi, xxxi–xxxv, xlii–lii, lxv, 1–10, 104–110

life circle, liii, 180, 182, 202f., 209f., 294

life form (*Lebensform*), 7f., 183, 195, 247, 318

life horizon, 20, 28, 32

life plan, 59–63, 192

life plan research (*Lebensplanforschung*), 59f., 63, 242. *See also* Uexküll, Jakob Johann von

light, 25, 193–196, 207f., 239, 325. *See also* eye

Limits of Community, xv, 319, 347, 356f., 359

Lindworsky, Johannes, 256

Lipps, Theodor, xxi

368   Index

Litt, Theodor, xix
lived body (*Leib*). *See* body: lived body
living space, 186
living unit, personal, 28f., 32f., 193, 284. *See also* unity
location. lvi, 48, 123, 150, 168–170, 188, 207, 238, 245. *See also* frontality; orientation; space
Locke, John, x, 39, 46
locomotion, 54, 59, 206f., 213
logic, 15–17, 19f., 22, 28, 47, 108f., 113, 301
Lorenz, Konrad, xxvi, xxix, 329
Lotze, Hermann, 15
love, xxi, 63, 276. *See also* hate; sexuality; sympathy
Löwith, Karl, xxiv–xxvi, 346

Mach, Ernst, 39
machine, xxxiii, 87–91, 134f., 152, 243, 258, 324. *See also* mechanism
manifestation, 10, 51, 78, 102, 107, 127f., 141f., 152f., 202, 215, 232, 266, 272, 292, 299–301, 305, 309. *See also* appearance
manifold, lvii, 25, 48–50, 172–175, 211, 213. *See also* unity
Marcion of Sinope, 321, 357
Marx, Karl, 8, 14, 35, 293
material, xliii; vs. form, 80; inorganic, xlii, liii, 329; organic, liii, 62, 331; sensory, lvii, 53–56, 68; structural, of a gestalt, 89, 91
materialism, xxxviif., xli, lxv, 3, 10, 14, 293
materiality, xv, 87, 315
mathematics, xxxix, xl–xliii, xliv, xlvi, lxi, lxvii, 11–13, 16, 25, 29, 35f., 56, 101–103, 136, 153, 169, 273
matter: circulation, of 182, 196f.; vs. form/gestalt, 115, 135; inorganic, xxxiii; organic/living, 168, 182–184, 189, 196f., 201, 206, 213, 257, 327, 330, 332; sensible, 29f.
Mayer, Robert, 109

meaning (language) xxviii, 16f., 19–32, 309, 314f., 356
meaning of being (*Sinn von Sein*), xxiif.
means, lii, 120, 154, 156, 158f., 172–182, 214, 226, 233, 289, 297, 316. *See also* tool
measurability, 20, 36, 38f., 169
measurement, 35f., 38, 45f., 48, 65, 86f., 109, 112, 123, 153, 168–170, 238f.
mechanics, classical, 16. *See also* physics
mechanism, xxxiii, xli, xliv, 5–7, 37, 56, 58f., 86, 90, 102f., 152, 200, 255, 261, 324, 326. *See also* vitalism
mediacy, 175. *See also* immediacy
medicine, xvi, xx, 21, 33, 38
medium, xxxii, xxxiv, xlixf., lii–liv, 95–98, 119, 141, 178–180, 182f., 185–200, 205–207, 209f., 212–216, 221, 267, 301, 310, 330–333
membrane, xxxii, 154, 331–333
memory, xxi, lv, 5, 168, 208, 243, 246, 257ff., 262ff., 275, 278f.
mental image (*Vorstellung*), 52–54, 77f.
Merleau-Ponty, Maurice, xxxv, lxv
metabolism, xliii, 86, 105, 108–110, 176, 179, 184, 189, 196, 206, 211f., 332, 335
metaphysics, xvii, xxvii, xxxi, xxxviii, 9, 11, 19, 53, 67, 70, 81, 106, 113, 223, 241, 273, 295, 319f.
method, philologico-historical, 13
methodology, xv–xvii, xx–xxii, xxivf., xxxiii, 15, 17, 22, 36f., 41, 50f., 55–57, 64f., 68f., 74, 82, 100, 109, 225, 278, 285, 323, 333
Meyer, Adolf, 100, 105, 353
milieu, xxxiv, 192, 286, 327. *See also* surrounding field
mimesis, xxix, 142
mind, xxxviif., xlf., xlv, liv, lxiv, 4, 26, 35, 56, 290, 349
Misch, Georg, xvii, 17–19, 21, 345, 352
modal (*Modal*), xxiv, lif., 100–105, 107, 110f., 113f., 202, 350, 353
modality, sensory, xv, 30, 72, 111, 308, 353f.
monadology, 188, 192, 307, 320
Monakow, Constantin von, 333

monism, 11, 22, 70, 142
mood, xxvf., 26, 63, 111, 275. *See also* emotion; feeling
morality, 292, 294. *See also* ethics
mores (*Sitte*), 68, 294f., 297f.
Morgan, Thomas Hunt, 245
morphogenesis, 99, 161, 195, 203, 205
morphology, xxi, 64, 105, 136, 155, 157, 173, 179, 191, 203, 210, 212f., 331, 353
motive, xxvii, 23, 289, 294. *See also* action; purpose
motor skills (*Motorik*), xxviiif., lii–lx, 62, 212, 215, 226–237, 240–242, 252, 254–256, 265, 309, 335. *See also* locomotion
movement (bodily), l, 23, 49, 54, 59, 105, 107, 116f., 122f., 129, 167, 170, 185, 197, 207f., 213, 227–229, 231–239, 242, 248, 255–259, 261, 309, 335. *See also* action; locomotion
Müller, Johannes, 195, 328
multicellular organism, xliv, 154f., 198, 202f., 227
Münsterberg, Hugo, 55
mutation, 326f., 331

nakedness, 40, 288f., 294, 297
narrowing (*Einengung*), 227f.
natural history, 3f.
natural science, x, xii, xv, xx, xxiv, xxxi, xl, lix, lxiv, 10–13, 16f., 21–25, 27–29, 32, 36–38, 45, 64f., 67, 69, 71–74, 81f., 87, 99, 101f., 107, 134, 200f., 293, 324f., 333. *See also* biology; physics
naturalism, xli, xliv, lxv, 3, 71, 289–292, 295. *See also* spiritualism
negative, sense of the, 250, 252f. *See also* emptiness; intelligence
neo-Kantianism, x, xxii, 15f., 38, 55, 109, 276
nerve, 26, 31, 37, 45, 54, 101
nervous system, xxxiv, liii, 37, 45, 155, 185, 207, 212, 224, 226, 228, 230f., 233, 237, 241f., 251, 261f., 264, 328, 333, 335, 350

neurosis, 291
Newton, Isaac, xxxix, xlif., 16, 325
Nietzsche, Friedrich, xxiiif., lxv, 2, 16, 292f., 296
norm, xxii, xxxviii, xlvf., lvi, lxii, lxiv, 5, 28, 294, 347f. *See also* ethics; morality; mores
nothingness lvii, lxi, 139f., 271f., 288, 294, 316
noticing, 226–231, 234–238. *See also* effecting; sphere of noticing
not-yet (mode), 160–164, 168, 193, 197
Novalis, 166
now (mode), 124–126, 161–169, 221, 260f. *See also* here/now
nucleic acids, 330
nullity (of the world), lxi, lxiii, 316–321
nutrition, 103–105, 185, 197, 200, 203, 205f., 213, 216, 292f., 324

object, xxii, xlvii, xlix, 17f., 27f., 39, 41, 43–45, 55–57, 60, 62, 66f., 72f., 84, 156, 174–176, 220, 228f., 251, 270f., 283f., 305, 311f., 349, 356
objectification, 20, 208, 253, 284, 310f., 313, 316, 318, 336
objectivity, xvi, xlviii, lx, 7, 15, 19, 41, 43, 45, 63, 72, 167, 223, 229, 251, 275, 285f., 298, 305f., 308, 327
obstruction (*Hemmung*), 287, 294
occasionalism, 48, 53
olfactory (organs, data), 195, 235, 254
ontology, xxi–xxiii, 20, 27, 325
Oparin, Alexander, 329f., 357
openness, xxiii, xxvii, 178f., 202–210, 215, 301
opposing field (*Gegenfeld*), 185f., 188, 215
opposing position (*Gegenstellung*), 40–44, 78
opposing sphere (*Gegensphäre*), 36, 41–44, 188, 271
optics (physical), 25f., 45, 102, 111, 242. *See also* gaze
ordering type (*Ordnungstyp*), 110, 112

organ, xxviii, xxxiv, 25, 31, 49, 54, 59, 136, 152, 154–159, 172–182, 191, 194f., 201, 203f., 206, 210–213, 216f., 226–228, 230, 238f. *See also* central organ; sense organs
organic, 99ff., 110ff.
organism, xxviii, xxxiiif., li–lv, 28, 32, 152–160, 173–182
Orwell, George, 326
Ostwald, Wolfgang, 105
other, the, liv, 14, 23, 37, 47, 50, 55–58, 279f.
ought (*Sollen*), lxi, 294
outer field (*Außenfeld*), 185f., 221f., 226, 237–240, 251, 270. *See also* surrounding field
outline/outlining, xxxii, 96, 116, 134, 141, 332f. *See also* contour
outside (*das Außen*), xlix, 37, 52, 84, 135, 145, 148, 309, 333
overcompensation, 292f., 295. *See also* compensation; sublimation

pain, 49, 141, 275, 311, 350
paleontology, xxvi, 5, 7, 166, 325, 327
panlogism, 114
panpsychism, xxxviii, xlv, 63
paradise, 287
parallelism, theory of, 241f.
parasitism, 217, 291, 327
Parmenides, 19
Pascal, Blaise, xxi
passage, 124–126, 130f., 144. *See also* boundary
passing through (*Hindurch*), 159, 175, 180, 221, 260, 267, 269, 271, 305
past, 7, 9, 15, 19, 130f., 160, 163f., 166, 168, 257–265, 278, 284. *See also* memory; time
Pauling, Linus, 331
Pavlov, Ivan Petrovich, 262
people, 28, 279
perception, xxix, xxxi, xlviif., 3, 13, 25, 36f., 43, 50, 52–54, 56–58, 61, 70, 72, 76f., 79f., 83, 85, 96, 101, 112, 160, 185, 194, 207, 225, 236, 240, 245f., 249f., 252, 254, 256, 272, 278, 280, 285, 304, 306, 308, 336, 349, 353
perceptual field, 254, 261
perceptual thing, 76, 78, 81, 84
periphery, 48, 53f., 79f., 120, 189, 211; peripheral value, 90f., 115, 135f. *See also* body; boundary
persistence, xliii, 124–126, 165, 259. *See also* becoming
person, xvi, 4, 10, 16, 20f., 24, 28, 30–33, 35, 44, 68f., 72f., 267, 272, 276, 279–282, 284, 309–313, 315
Petersen, Hans, 105
Pflüger, Eduard, 105
phenology, 58
phenomena of life, xxxiiif., xliii–l, lxivf., 86, 100, 107, 113, 143, 208, 225, 325
phenomenology, xvif., lxv, 20, 25–27, 107f.
philosophy, xv–xviii, 2f., 10–22, 24f., 27, 36–38, 50–53, 55, 65–68, 71, 73f.; of biology, xliv, xlvi, lviii, lxv; cultural, 7; of history, 7f., 24; of language, 22 of life, xxxviii, 2, 27, 32, 114; of nature, xlvi, 21f., 38, 51, 71, 74, 100; social, xvi, 347, 357
phototropism, 207
phylogeny, xx, xxi, xxix, 4f., 10, 328
physicality (*Körperhaftigkeit*), 6, 11, 35f., 39f., 46, 65, 119, 179, 197. *See also* extension; *res extensa*; vitality
physics, xx, xxxix, xl–xlii, xlvi, lxi, 12f., 17, 25, 37, 64, 70, 87, 102f., 107, 110, 154, 168f., 183, 186, 250, 273, 281, 323, 329
physiology, 3, 21, 29, 37, 56, 59, 64, 69, 87, 135, 192, 237, 242, 261, 274, 281, 323; developmental, 87, 135, 323
Pick, Arnold, 262
Pikler, Julius, 236, 355
placelessness, 271, 279, 319
plan, unitary, 158–160, 165, 234
Planck, Max, 35
plant, liii, 202–209, 215–217, 257, 266, 301, 335, 350

plasma, 92, 134, 154, 198, 201, 203, 207, 330, 333
plasticity, xxviii, xlii, 107, 116, 346
pleasure, 49
Plotinus, 193
population genetics, 326. *See also* genetics
Portmann, Adolf, xxvif., xxxiv, 324, 328, 346
positedness, 122, 147, 149, 158f., 166, 260, 269
position, xii, xlix, 49, 123, 189f., 220–223, 239, 261, 270; *See also* frontality; positionality; self-position
position of the gaze, 72, 75f., 83, 225f.
positional field, 178f., 185f., 188–193, 195–197, 204, 215, 220, 223, 231, 236, 239, 242, 274
positionality, xli, xxx, xxxiv, xlix, l, lii, lxiv, 118, 121–123, 145–147, 165, 171, 174, 180, 185–196, 213f., 219–226, 237, 242f., 252, 261, 267–269, 272, 275, 280–283, 286–288, 293–304, 307f., 320f., 328, 333, 349; excentric, xxx, lii, lv–lxiii, 296f., 302, 304, 307, 349
positivism, 14, 23, 54f., 109
potency, 151, 153, 156, 160–165, 180. *See also* capacity
power, xl, lxv, 56, 292f., 296f., 311, 326
pragmatism, xxvii, lxv, 292, 296, 309
preformationism, 134
prehistory, xx, xxvi, 290
presence (*Gegenwart*), 40f., 43–45, 130f., 167f., 174f., 186, 193, 259–261
process, xlv, xlix, 31, 58f., 86, 88, 91, 103, 117, 123–138, 144f., 323f., 349
progress, 1f., 9, 130f., 137f., 311, 314
property (*Eigenschaft*), 40, 77, 79–87. *See also* core
proteins, 206, 216f., 330
protoplasm, 203, 330, 333
psyche, xv, 25f., 28, 35, 54–57, 61, 76, 247, 256, 271f., 274–279, 281f., 284, 289, 308, 310f., 321. *See also* consciousness; soul

psychoanalysis, xvi, lxv, 275, 291, 295
psycho-Lamarckism, 191, 194
psychologism, 29, 45, 293
psychology, xx, xxiv, xxx, 15, 17, 21, 25f., 28–30, 33, 37f., 50, 56–64, 81, 85, 118, 242f., 246f., 254, 276, 278, 306, 329; animal, 57–64, 242f., 254, 329; developmental, 81; evolutionary, xxvii, xl
psychopathology, xx, 21, 62
psychophysically neutral/indifferent, 24, 28, 32, 55, 226, 271
psychovitalism, 194, 355
purpose, xlvf., 59, 176, 202, 298. *See also* intention
Pütter, August, 105

quale, 101, 136, 235
quality, xxxiv, 26, 37, 40, 101–103, 134, 160f., 166, 168, 246, 254, 324
quantum physics, xx

race, racial, 14, 28, 326f.
radius, xxi, 24, 197, 232, 242, 279
rationalism, 1, 11, 25
realism, lx, lxii, lxv, 44, 46f., 55, 306, 308, 346, 351. *See also* idealism
reality, x, xi, 50, 52, 102, 218, 225, 240f., 277, 279–282, 299, 304–321
reason, xl, xliii, xlvii, lxii, 1, 8, 11f., 15f., 27, 66
reciprocity (*Wechselwirkung*), 5, 187, 241f.
reduction (methodology), xl–xliv, 21
reflex, xxx, 207, 228, 230, 234, 237, 243f., 266, 329, 335. *See also* instinct; nervous system
reflexivity, lv, 268–270
regeneration (*Verjüngung*), xlii, 152, 197f., 324
regulation, xxi, xlviii, 104f., 108–110, 149–154, 187, 197, 199, 216, 239, 327. *See also* self-regulation
Reinboth, Rudolf, 327f., 334
relation, reflexive, xii, lv, 260
relativism, xxii, 8

relativity, theory, of 35, 166, 168, 170
relief (*Entlastung*), xxviiif.
religion, 8, 14f., 17, 68, 291, 317
Rensch, Bernhard, 334
representation, 42, 51–54, 62, 65, 111, 149f., 156f., 169, 212f., 216, 219, 226, 232f., 237f., 252, 260, 306, 309, 335, 353
repression of drives, 291, 295. *See also* culture
reproduction (*Fortpflanzung*), xliii, 86f., 105, 113, 176, 196–198, 201f., 205f., 216, 330, 332, 334
res extensa/res cogitans, xl, 35–37, 39–47, 67
residue, 131, 262, 265. *See also* anticipation
resonance, 13f.
respect (for the other), 27, 320
response, liv, lvii, 207f., 227, 230f., 233, 241f., 257f., 301f. *See also* stimulus
responsibility, lxiv, 4, 10, 24
Reuter, Fritz, 295
Révész, Géza, 257
revolution, political, 320
Rhumbler, Ludwig, 92
rhythm, l, 55, 107, 116f.
Rickert, Heinrich, x, 15, 69
Romanes, George, 246
romantic(ism), 63, 70, 208f., 266
rootedness, 294
Roux, Wilhelm, 90, 105, 353f.

sameness, 43f., 126f., 280. *See also* identity
Sanders, F. K., 335
Sartre, Jean-Paul, xxv, lxv
Scheler, Max, xvif., xix, xxi–xxiv, xxvif., xxxviii, lxv, 33, 69, 320, 345–347
Schelling, Friedrich Wilhelm Joseph, 325–327
Schiller, Ferdinand Canning Scott, xxvii
Schopenhauer, Arthur, xxiii, 292, 325
Schramm, Gerhard, 330f.

Schrödinger, Erwin, 324, 331, 334
science xvi, xx, xxxviii–xl, xliv, xlvi, 7, 12, 14f., 17, 20–26. *See also* biology; humanities; natural science; physics; physiology
seeing, 62, 71, 224f. *See also* eye; gaze
Seidel, Alfred, 292, 356
selection, liii, 5, 35, 187, 191, 196, 199f., 295, 330, 348
self, xxv, xlviii, lif., lvii, 4, 18, 36f., 39–48, 51–54, 65, 148f., 156, 159f., 185, 214f., 220, 222, 224f., 242, 267, 275, 277f., 291, 293
self-abolition, 140, 184
self-action, 90, 105. *See also* auturgy
self-awareness, 270
self-choice, xxii
self-consciousness, lvi–lviii, lxiif., 52f., 62, 106, 307, 349
self-differentiation, 91, 136
self-groundedness, 140
self-knowledge, 19, 283
self-mediation, lii, 165, 172, 260
self-modification, 105, 183
self-negation, 139
self-observation, 58, 237
self-position, 36f., 42–44, 48f., 52, 61f., 276f.
self-potency, 180
self-preservation, xxxiv, 105, 184, 333
self-realization, 264
self-referentiality, 121
self-reflection, xii, 270, 276
self-regulation, 105, 149
self-representation, 156
self-sufficiency (*Selbständigkeit*), xxxi, xxxiv, xxxviii, liii, 98, 149f., 154f., 176f., 179f., 202, 209, 211, 215, 252, 259, 266, 327, 333
self-transcendence, 19
sensation (*Empfindung*), 25f., 30–32, 37, 39, 49, 51, 54–56, 71, 111, 208, 213, 228f., 231f., 235–237, 240f., 251, 254f., 306, 311, 328, 332, 355
sense data, 77, 81, 231, 235, 246f.

sense organs, 13, 45, 49f., 53f., 59, 62, 101, 207, 216, 228, 231, 237, 239, 241f. *See also* organ; sensory modality
sense perception, 58, 61, 76, 83, 96
sensory energies, 37, 45, 54, 195, 328
sensory modality (*Sinnesmodalität*), xv, 111, 308, 354. *See also* quality
sensualism, 50, 55–57, 65, 81f.
serum studies, 69
set-apartness, 117, 123, 143, 177, 214, 220, 222f., 266, 273, 312
setting (*Umgebung*), 95, 147, 179, 189, 193, 232, 284f., 332f. *See also* field: surrounding
sex, xxix, 201f., 234, 292, 334
sexuality, 201, 205, 292f., 298, 334
shame, 288, 319
shared world (*Mitwelt*), 47, 272, 279–286, 294, 300, 317f., 320f.
shell, 245, 274
sidedness, 78–80. *See also* depth; spatiality
signal, 228–231, 234, 245, 252. *See also* stimulus
Simmel, Georg, 293
skin, xlix, 31, 49, 115f., 146f.
socialism, 2
sociality, 300, 319f.
sociology, x, xvi, xxvii, 8, 14, 21, 25, 33, 68f., 289, 319, 357
soul, xxvi, 8–10, 17, 23, 47, 60, 63, 65, 69, 72f., 122, 209, 240, 270, 279, 281f., 304; supra-individual, 60; world soul, 240. *See also* body; consciousness; psyche; self
soul form (*Seelenform*), 9
space, xif., xlixf., 48–50, 76, 79–81, 87, 98f., 117f., 123, 143, 147, 149–152, 165–171, 182, 186–188, 202, 213f., 219–221, 239, 252, 256, 270, 272–274
spacelikeness, 49, 80f., 99, 118ff., 171, 214, 239, 273
space-time continuum, lvi, 182, 273
spatiality, l, 49, 79f., 123, 147, 151, 171, 186, 239

species, xxvii, xxxiv, xlii, xliv, 57, 65, 187f., 192, 197–200, 205, 218, 290, 295, 326, 336
Spemann, Hans, xxxiii, 324
Spencer, Herbert, 4–6, 9, 105, 194
Spengler, Oswald, x, 8–10, 19
sphere, xvi, xxxiv, 3, 11f., 17, 19, 21, 24, 30, 33, 35f., 39, 41–46, 53, 60, 62, 76, 96, 109, 118, 140, 153, 186, 222, 228–237, 271–273, 278, 280f., 283; of action, 60, 228, 232, 235; of culture, 289, 298; of effecting, 229–234; of life, xvi, 186, 192, 208; noological, 327; of noticing, 228–237; opposing, 36, 41–44, 188, 271; of spirit, 283
Spinoza, Baruch de, 76
spirit, xxiii, xxiv, xxvi, lxi, 2–4, 6, 8–14, 16–19, 21f., 24f., 29f., 32, 65–69, 127, 139–141, 281–284, 289–291, 293–295, 308–310, 312f., 316f., 321
spiritualism, 11, 289f., 295
spirituality, xxiii, 29, 295
spontaneity, 58, 89–91, 99, 117, 213, 222f., 227, 233, 244, 278, 329f.
Stanley, Wendell M., 330f.
state (*Staat*), 2, 15, 68, 73, 198, 286
Stern, William, 33
stimulability, 105, 113
stimulus, lvii, 49, 58f., 63, 110, 190, 194f., 203, 205, 207, 216, 227, 231, 233, 241f., 255, 257f., 301f., 335
Storch, Otto, xxix
strangeness, 321
stream of consciousness, 28
struggle, 181, 187, 191f., 196, 200, 240, 271, 290, 292, 295
subject, xxviii, xlvii, xlix, 4, 10, 19, 27f., 31, 33, 47f., 51f., 55f., 61f., 67, 69, 148–151, 173–176, 188, 194, 220, 225, 229, 252, 269, 282–284, 305–308, 310–313; living, xi, lxii, 60–63, 213, 133, 262–264, 276, 301–303. *See also* I, self
subjectivism, xvii, 66, 277–279

374 Index

subjectivity, xlviii, 6, 31, 78f., 140, 226, 282, 308
subject-object, xlix, 19, 28, 43, 45, 283
subject-objectivity, 19, 43, 45
sublimation, lxv, 35, 45, 71, 216, 290f., 293, 295, 297. See also drive
substance, 5, 40, 67, 77, 79–82, 86f., 94, 116. See also form
superstructure (*Überbau*), 3, 8, 35
supra-individual, 60, 198
surrounding field (*Umfeld*), lv, lx, 147, 186, 188, 190, 192f., 213f., 222f., 226, 228f., 231–236, 238, 240–242, 244, 251–253, 255–257, 260, 267, 270–274, 283–286, 301–305, 346
sympathy, xxi, 73, 276, 279
system, xxviii, xlii, lvi, 4, 88, 90f., 144–149, 172, 176, 191f., 324, 330–333; of conformity (*Konformitätssystem*), 7–9, 29; harmoniously equipotential (Driesch), 151f., 157, 174, 205, 324; living, xlii, xlvi, lif., lxv, 98, 137, 269, 271, 333. See also nervous system

tactility, 31
technization, 2
Teilhard de Chardin, Pierre, 327, 332
teleologism, xlvi, liii, 106, 134, 158, 165, 176f., 198, 325, 328. See also entelechy; purpose
temperament, xxv, 278
temporality, l, 149, 171, 239. See also time
theism, xxii, xxxviii
theomorphism, xxiii, 320
theory of knowledge, xvi, 15f., 18, 52, 188. See also epistemology; knowledge
theory of science, 16f., 74
thing: living, xxxiii, xlvii, xlix, li, liv, 85, 93, 96, 98, 107, 120f., 123, 125–129, 142–144, 147, 155, 167, 174, 185, 187, 189, 196, 202f., 210, 213–215, 217–220, 224, 268, 270–272, 349; perceptual, 78, 81; physical, xxxiv, lviii, 43, 65, 83, 86, 90, 96, 112–114, 119f., 124–126, 128, 130, 142, 149, 151, 155f., 158, 161f., 180f., 185, 187, 196, 274
thing-body, 84, 93–95, 98, 224, 274
thinghood, 80, 84, 94, 96, 98, 114, 127f., 137, 179, 229, 236, 245f., 268
thingness (*Dinghaftigkeit*), 124, 181, 196, 235
thought, 10, 29, 36, 40, 45, 60, 79, 109, 277, 294, 310. See also intelligence
thrownness, xxiv
time, 9, 150, 160f., 163–168, 239. See also space-time continuum; temporality
timelessness, 271, 279
timelikeness, xii, lvii, 99, 165, 167, 171, 221, 224–227, 239, 266
Tinbergen, Nikolaas, 329
tissue, 81, 136, 152, 154, 173, 203, 206, 208, 216, 333, 335
Tönnies, Ferdinand, 319f.
tool, xxviii, xlv, lii, lxii, 175f., 244, 249, 289f., 295–298, 326. See also artificiality; culture; means; organ
topography, 89–91, 94
totality, 15, 40, 90, 94, 129, 156, 159, 180, 284, 318. See also gestalt; organism
transcendence, lxi, lxiii, 2, 19, 130, 139f., 189, 306, 316
transcendental idealism, xxif., xlvii, 5, 20, 45f., 106, 349
transgredience, 78f., 83
trauma, 275
Troeltsch, Ernst, xv
Troll, Wilhelm, 334
truth, 8, 10f., 45, 82f., 109, 327
Tschermak, Armin von, 105, 354f.
type, 123, 127f., 130, 134, 141–143, 157, 336. See also form; form-idea; gestalt-idea
typicality, 128

Uexküll, Jakob Johann von, xv, xxvi, 50, 58f., 62f., 135, 158, 192, 212, 229–231, 238, 240, 242, 329, 335, 355
unconsciousness, xxix, 244, 278, 295

understanding (*Verstehen*), x, xv, xx, xxiii, xli, xlvi, 5, 14–20, 30, 86, 304
unfolding (*Entfaltung*), 135, 151, 157
Ungerer, Emil, 105
unicellular organism, 154f., 198, 202f., 227
uniqueness, 14f., 90, 104, 201, 317f., 324, 333
unitary plan, 158–160, 165, 234
unity, xlix, lif., 30f., 43f., 59, 62, 72, 75f., 78, 90–92, 116, 148–151, 154–165, 172–176, 180–182, 211–213, 262, 271. See also self; thing; wholeness
universe, xli, xliv, 184, 318
upright gait, xxiii, 3, 290
urge (*Drang*), 289, 300, 310, 319. See also drive, volition
utopian standpoint, lxi, 316, 320, 326

value, xxif., xliii, liii, 9, 13, 15f., 29, 31, 52f., 55, 68f., 72, 327
variability, 20, 116, 139, 192, 218
variety (*Mannigfaltigkeit*), 49f., 53, 105, 116, 136, 149–152, 154, 157f., 172–177, 211
viability (*Lebensfähigkeit*), 57, 187
virology, xxxiii, 330
virus, xx, xxxiv, xliii, 324, 330–332
vitalism, xx, xxxiii, xxxviii, 58, 86–88, 90, 93, 100, 102f., 134f., 152f., 194, 292, 323–325, 333
vitality (*Lebendigkeit*), xxv, xxxi, xxxiiif., l, 63, 68, 86, 93, 107, 115f., 123, 144, 153, 177, 179, 181, 183, 197, 269, 278, 287f.
Vogel, Hans, 334

volition, 207, 263, 275, 277, 289, 308. See also will
Volkelt, Hans, 245–247, 250–252, 336, 355f.

weapon, 290, 296, 298
Weber, Max, xxvii
Weidel, Wolfhard, 334
Weismann, August, 134, 198
Wettstein, Fritz von, 201
wholeness (*Ganzheit*), xxxiii, xlviii, xlix, lii, 87–93, 97, 99, 110, 112f., 156f., 172–174, 176, 210, 215, 218, 333
will, free, 4
will to power, lxv, 292f., 296f.,
Windelband, Wilhelm, xv, 15, 69, 346
world: inner, lxiii, 46–48, 50–55, 57f., 67, 230f., 272, 274f., 277–281, 306, 318f.; outer, lxf., lxiii, 46, 48–54, 56, 154, 185, 188, 191, 216, 233, 272–275, 277, 279, 281, 284, 302, 306, 317–319; physical, 3, 13, 22f., 42, 46–48, 51f.; shared, 47, 272, 279–286, 294, 300, 317f., 320f.; spiritual, 7, 13f., 17f., 68
world history, 27, 311, 316
world-openness, xxiii, xxvii
world system (*Weltsystem*), 7, 9
worldview (*Weltanschauung, Weltbild*), 9, 20, 36, 51, 55, 66f., 273, 279

Young, J. Z., 335
youth, 137, 139, 141, 144. See also aging

zeitgeist, 8
zoology, xx, xxxviif., 5, 7, 240, 284, 326
zoon politikon, 300

HELMUTH PLESSNER (1892–1985) was a German philosopher and sociologist. From 1953–59, he was president of the German Sociological Association. Three of his many books have appeared in English: *Political Anthropology* (Northwestern, 2018), *The Limits of Community* (Humanity Books, 1999) and *Laughing and Crying* (Northwestern, 1970).

J. M. BERNSTEIN is University Distinguished Professor in Philosophy at The New School for Social Research in New York City.

MILLAY HYATT is a writer and translator based in Berlin. Her dissertation, "No-Where and Now-Here: Utopia and Politics from Hegel to Deleuze," received the University of Southern California's doctoral research prize.

forms of living

Stefanos Geroulanos and Todd Meyers, series editors

Georges Canguilhem, *Knowledge of Life*. Translated by Stefanos Geroulanos and Daniela Ginsburg, Introduction by Paola Marrati and Todd Meyers.

Henri Atlan, *Selected Writings: On Self-Organization, Philosophy, Bioethics, and Judaism*. Edited and with an Introduction by Stefanos Geroulanos and Todd Meyers.

Catherine Malabou, *The New Wounded: From Neurosis to Brain Damage*. Translated by Steven Miller.

François Delaporte, *Chagas Disease: History of a Continent's Scourge*. Translated by Arthur Goldhammer, Foreword by Todd Meyers.

Jonathan Strauss, *Human Remains: Medicine, Death, and Desire in Nineteenth-Century Paris*.

Georges Canguilhem, *Writings on Medicine*. Translated and with an Introduction by Stefanos Geroulanos and Todd Meyers.

François Delaporte, *Figures of Medicine: Blood, Face Transplants, Parasites*. Translated by Nils F. Schott, Foreword by Christopher Lawrence.

Juan Manuel Garrido, *On Time, Being, and Hunger: Challenging the Traditional Way of Thinking Life*.

Pamela Reynolds, *War in Worcester: Youth and the Apartheid State*.

Vanessa Lemm and Miguel Vatter, eds., *The Government of Life: Foucault, Biopolitics, and Neoliberalism*.

Henning Schmidgen, *The Helmholtz Curves: Tracing Lost Time*. Translated by Nils F. Schott.

Henning Schmidgen, *Bruno Latour in Pieces: An Intellectual Biography*. Translated by Gloria Custance.

Veena Das, *Affliction: Health, Disease, Poverty.*

Kathleen Frederickson, *The Ploy of Instinct: Victorian Sciences of Nature and Sexuality in Liberal Governance.*

Roma Chatterji, ed., *Wording the World: Veena Das and Scenes of Inheritance.*

Jean-Luc Nancy and Aurélien Barrau, *What's These Worlds Coming To?* Translated by Travis Holloway and Flor Méchain. Foreword by David Pettigrew.

Anthony Stavrianakis, Gaymon Bennett, and Lyle Fearnley, eds., *Science, Reason, Modernity: Readings for an Anthropology of the Contemporary.*

Richard Baxstrom and Todd Meyers, *Realizing the Witch: Science, Cinema, and the Mastery of the Invisible.*

Hervé Guibert, *Cytomegalovirus: A Hospitalization Diary.* Introduction by David Caron, Afterword by Todd Meyers, Translated by Clara Orban.

Leif Weatherby, *Transplanting the Metaphysical Organ: German Romanticism between Leibniz and Marx.*

Fernando Vidal and Francisco Ortega, *Being Brains: Making the Cerebral Subject.*

Mirko D. Grmek, *Pathological Realities: Essays on Disease, Experiments, and History.* Edited, translated, and with an Introduction by Pierre-Olivier Méthot, Foreword by Hans-Jörg Rheinberger.

Helmuth Plessner, *Levels of Organic Life and the Human: An Introduction to Philosophical Anthropology.* Introduction by J. M. Bernstein, Translated by Millay Hyatt.

CPSIA information can be obtained
at www.ICGtesting.com
Printed in the USA
FSHW021508240619
59368FS